MW00998334

Student's Solutions Manual

UNIVERSITY PHYSICS

ninth edition

Volume Two

Student's Solutions Manual

YOUNG • FREEDMAN

UNIVERSITY PHYSICS

ninth edition
Volume Two

A. Lewis Ford
Texas A&M University

Addison-Wesley Publishing Company
Reading, Massachusetts • Menlo Park, California • New York • Don Mills, Ontario
Harlow, United Kingdom • Amsterdam • Bonn • Sydney • Singapore • Tokyo
Madrid • San Juan • Milan • Paris

ISBN 0-201-44168-3

4 5 6 7 8 9 10 - VG - 00 99

CONTENTS

Chapter			Page
36	Exercises	3, 5, 9, 11, 13, 17, 19, 23, 27	420
	Problems	31, 33, 37, 39, 41	
37	Exercises	1, 7, 9, 11, 13, 15, 17, 21, 27, 33	426
	Problems	37, 39, 41, 45, 47, 49, 51	
38	Exercises	3, 5, 9, 11, 13, 15, 17, 19, 23, 27, 31, 33	433
	Problems	37, 39, 41, 47, 51, 53	
39	Exercises	1, 3, 7, 11, 13, 19, 21, 23, 25, 27, 31, 33, 39, 41	439
	Problems	43, 45, 47, 53, 55, 57, 63	
40	Exercises	1, 5, 7, 9, 11, 13, 17, 21, 23, 25, 27, 33, 39	447
	Problems	43, 45, 49, 51, 55, 57, 59, 61, 65	
41	Exercises	3, 5, 7, 9, 15, 17, 21	455
	Problems	27, 29, 31, 35, 37, 39, 41	
42	Exercises	3, 7, 9, 17, 19, 21, 23, 25	460
	Problems	27, 31, 33, 37, 41, 43	
43	Exercises	3, 5, 11, 15, 17, 19, 21, 25, 27, 29	467
	Problems	33, 35, 37, 39, 43, 47	
44	Exercises	5, 7, 9, 11, 13, 19, 23, 25, 27	474
	Problems	29, 31, 35, 37, 39	
45	Exercises	3, 5, 9, 11, 15, 17, 21, 25, 27, 31, 33, 35	482
	Problems	41, 45, 47, 49, 51, 53, 55, 57	
46	Exercises	3, 7, 9, 11, 15, 19, 23, 25, 27, 29, 31	490
	Problems	37, 39, 41, 45, 49	

CHAPTER 22

Exercises 1, 3, 9, 11, 13, 15, 19, 25, 27, 29, 31, 35, 37, 47, 49, 51

Problems 55, 57, 61, 63, 65, 67, 75, 77, 79, 81, 83, 85

Exercises

22-1

$N = nN_A = (3.00\text{ mol})(6.022\times10^{23}\text{ atoms/mol}) = 18.06\times10^{23}$ atoms

1 proton per hydrogen atom and each proton has charge $+e = 1.602\times10^{-19}$ C

$\Rightarrow Q = (1.602\times10^{-19}\text{ C/proton})(18.06\times10^{23}\text{ protons}) = \underline{2.89\times10^{5}\text{ C}}$

(This is a huge amount of charge.)

22-3

a) The charge of one electron is $-e = -1.602\times10^{-19}$ C. The number of excess electrons needed to produce net charge q is

$$\frac{q}{-e} = \frac{-2.50\times10^{-9}\text{ C}}{-1.602\times10^{-19}\text{ C/electron}} = \underline{1.56\times10^{10}\text{ electrons}}$$

b) Find the number of lead atoms in 10.0×10^{-3} kg of lead. The atomic mass of lead is 207×10^{-3} kg/mol, so the number of moles in 10.0×10^{-3} kg is

$$n = \frac{m_{tot}}{M} = \frac{10.0\times10^{-3}\text{ kg}}{207\times10^{-3}\text{ kg/mol}} = 0.04831\text{ mol}$$

N_A (Avogadro's number) is the number of atoms in 1 mole, so the number of lead atoms is $N = nN_A = (0.04831\text{ mol})(6.022\times10^{23}\text{ atoms/mol}) = 2.909\times10^{22}$ atoms

The number of excess electrons per lead atom is $\dfrac{1.56\times10^{10}\text{ electrons}}{2.909\times10^{22}\text{ atoms}} = \underline{5.37\times10^{-13}}$.

22-9

a) $q_1 = q_2 = q$

$$F = \frac{1}{4\pi\epsilon_0}\frac{|q_1q_2|}{r^2} = \frac{q^2}{4\pi\epsilon_0 r^2} \Rightarrow q = r\sqrt{\frac{F}{(1/4\pi\epsilon_0)}} = (0.300\text{ m})\sqrt{\frac{0.150\text{ N}}{8.988\times10^{9}\text{ N}\cdot\text{m}^2/\text{C}^2}} = \underline{1.23\times10^{-6}\text{ C}}$$

(on each)

b) $q_2 = 3q_1$

$$F = \frac{1}{4\pi\epsilon_0}\frac{|q_1q_2|}{r^2} = \frac{3q_1^2}{4\pi\epsilon_0 r^2} \Rightarrow q_1 = r\sqrt{\frac{F}{3(1/4\pi\epsilon_0)}} = \frac{1}{\sqrt{3}}(1.23\times10^{-6}\text{ C}) = \underline{7.10\times10^{-7}\text{ C}}$$

and $q_2 = 3q_1 = \underline{2.13\times10^{-6}\text{ C}}$

22-11

The weight of an electron is $m_e g$. The nucleus of a hydrogen atom is a single proton with charge $+e$. The Coulomb force between the electron and the

22-11 (cont)

nucleus is $\frac{1}{4\pi\epsilon_0}\frac{e^2}{r^2}$.

Equating these two forces gives $m_e g = \frac{1}{4\pi\epsilon_0}\frac{e^2}{r^2}$

$$r = e\sqrt{\frac{1/4\pi\epsilon_0}{m_e g}} = 1.602\times10^{-19}\,C\sqrt{\frac{8.988\times10^{9}\,N\cdot m^2/C^2}{(9.109\times10^{-31}\,kg)(9.80\,m/s^2)}} = \underline{5.08\,m}$$

22-13

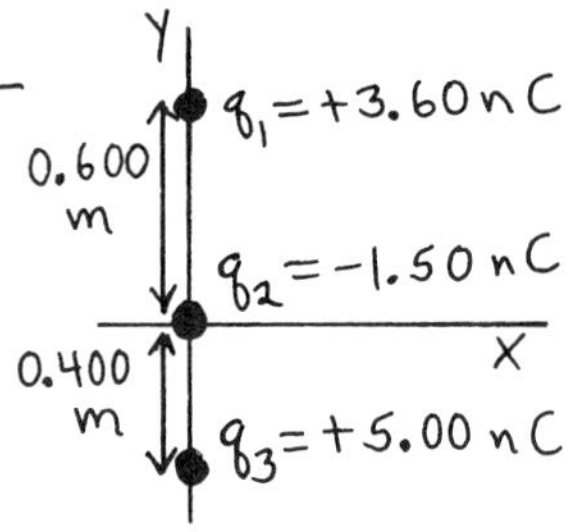

Like charges repel and unlike attract, so the free-body diagram for q_3 is

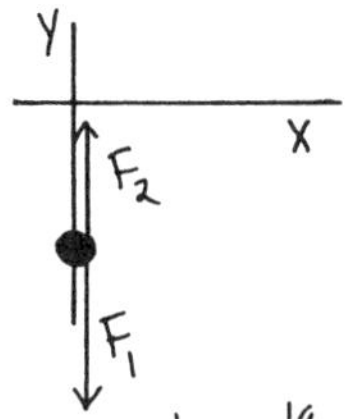

$$F_1 = \frac{1}{4\pi\epsilon_0}\frac{|q_1 q_3|}{r_{13}^2}$$

$$F_1 = (8.988\times10^{9}\,N\cdot m^2/C^2)\frac{(3.60\times10^{-9}C)(5.00\times10^{-9}C)}{(1.00\,m)^2}$$

$$F_1 = 1.618\times10^{-7}\,N$$

$$F_2 = \frac{1}{4\pi\epsilon_0}\frac{|q_2 q_3|}{r_{23}^2} = (8.988\times10^{9}\,N\cdot m^2/C^2)\frac{(1.50\times10^{-9}C)(5.00\times10^{-9}C)}{(0.400\,m)^2}$$

$$F_2 = 4.213\times10^{-7}\,N$$

The resultant force is $\vec{R} = \vec{F}_1 + \vec{F}_2$.

$R_x = 0$

$R_y = F_2 - F_1 = 4.213\times10^{-7}\,N - 1.618\times10^{-7}\,N = 2.60\times10^{-7}\,N$

The resultant force has magnitude $\underline{2.60\times10^{-7}\,N}$ and is in the $\underline{+y\text{-direction}}$.

22-15

a)

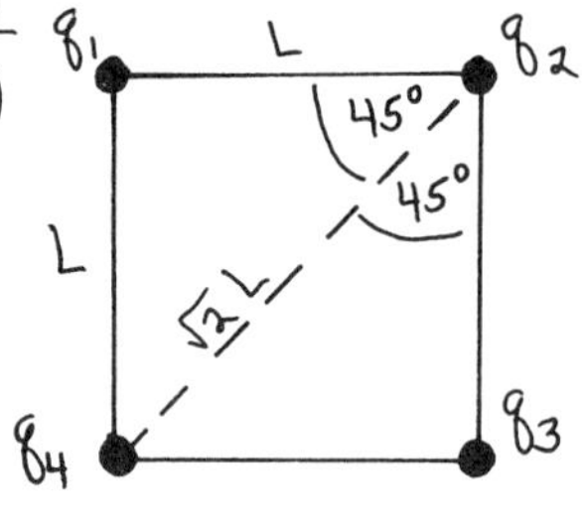

$q_1 = q_2 = q_3 = q_4 = q$

b) Consider forces on q_2. Take the y-axis to be parallel to the diagonal between q_2 and q_4 and let +y be away from q_2. Then F_2 is in the +y-direction.

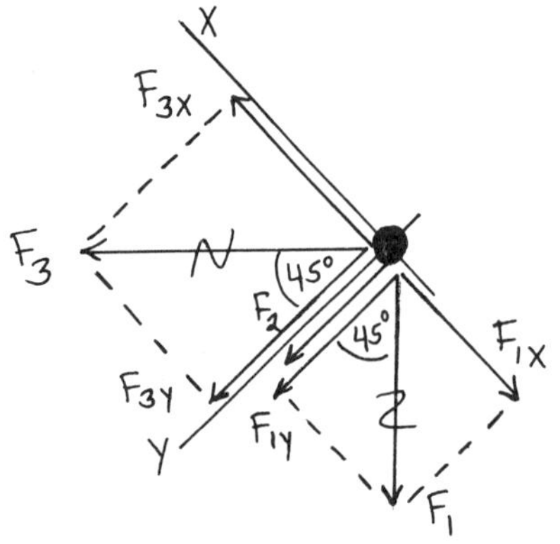

$$F_3 = F_1 = \frac{1}{4\pi\epsilon_0}\frac{q^2}{L^2}$$

$$F_2 = \frac{1}{4\pi\epsilon_0}\frac{q^2}{2L^2}$$

$$F_{1x} = -F_1\sin45^\circ = -\frac{F_1}{\sqrt{2}}$$

$$F_{1y} = +F_1\cos45^\circ = +\frac{F_1}{\sqrt{2}}$$

$$F_{3x} = +F_3\sin45^\circ = +\frac{F_3}{\sqrt{2}}$$

$$F_{3y} = +F_3\cos45^\circ = +\frac{F_3}{\sqrt{2}}$$

$$F_{2x} = 0,\quad F_{2y} = F_2$$

$R_x = F_{1x} + F_{3x} = 0$

$$R_y = F_{1y} + F_{3y} + F_{2y} = \frac{2}{\sqrt{2}}\frac{1}{4\pi\epsilon_0}\frac{q^2}{L^2} + \frac{1}{4\pi\epsilon_0}\frac{q^2}{2L^2} = \frac{q^2}{8\pi\epsilon_0 L^2}(1+2\sqrt{2})$$

$R = \frac{q^2}{8\pi\epsilon_0 L^2}(1+2\sqrt{2})$. Same for all four charges. In general the resultant force on one of the charges is directed away from the opposite corner.

22-19

a) Gravitational force exerted by the earth is $w_e = m_e g$.

$$w_e = (9.109 \times 10^{-31}\,\text{kg})(9.80\,\text{m/s}^2) = \underline{8.93 \times 10^{-30}\,\text{N}}$$

In Examples 22-7 and 22-8, $E = 1.00 \times 10^4$ N/C, so the electric force on the electron has magnitude

$$F_E = |q|E = eE = (1.602 \times 10^{-19}\,\text{C})(1.00 \times 10^4\,\text{N/C}) = 1.602 \times 10^{-15}\,\text{N}.$$

$$\frac{w_e}{F_E} = \frac{8.93 \times 10^{-30}\,\text{N}}{1.602 \times 10^{-15}\,\text{N}} = 5.57 \times 10^{-15}$$

The gravitational force is <u>much</u> smaller than the electric force and can be neglected.

b) From part (a) we know that this object must have mass much greater than an electron's mass.

$$mg = |q|E$$

$$m = \frac{|q|E}{g} = \frac{(1.602 \times 10^{-19}\,\text{C})(1.00 \times 10^4\,\text{N/C})}{9.80\,\text{m/s}^2} = 1.64 \times 10^{-16}\,\text{kg}$$

$$\frac{m}{m_e} = \frac{1.64 \times 10^{-16}\,\text{kg}}{9.109 \times 10^{-31}\,\text{kg}} = 1.80 \times 10^{14}\,; \quad m = 1.80 \times 10^{14}\,m_e.$$

So m is much larger than m_e.

c) $\vec{E}$ is uniform between the plates (for points not too close to the edges). So no, the answer to (b) does not depend on where between the plates the object is placed.

22-25

a) For a negative charge the force $\vec{F}$ and field $\vec{E}$ are in opposite directions. $\vec{F}$ is downward so $\vec{E}$ is upward.

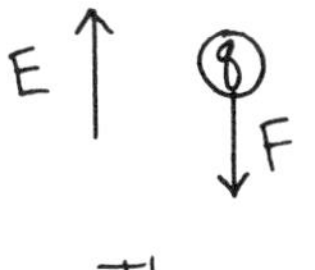

$$E = \frac{F}{|q|} = \frac{5.00 \times 10^{-8}\,\text{N}}{4.00 \times 10^{-9}\,\text{C}} = \underline{12.5\,\text{N/C}}$$

b) The proton has charge $+e$. For a positive charge the field and force are in the same direction so $\vec{F}$ is <u>upward</u> when $\vec{E}$ is upward.

$$F = |q|E = (1.602 \times 10^{-19}\,\text{C})(12.5\,\text{N/C}) = \underline{2.00 \times 10^{-18}\,\text{N}}$$

22-27

The electrical force has magnitude $F_E = |q|E = eE$.
The weight of a proton is $w = mg$.

$$F_E = w \Rightarrow eE = mg$$

$$E = \frac{mg}{e} = \frac{(1.673 \times 10^{-27}\,\text{kg})(9.80\,\text{m/s}^2)}{1.602 \times 10^{-19}\,\text{C}} = \underline{1.02 \times 10^{-7}\,\text{N/C}}$$

(This is a very small electric field.)

22-29

a)

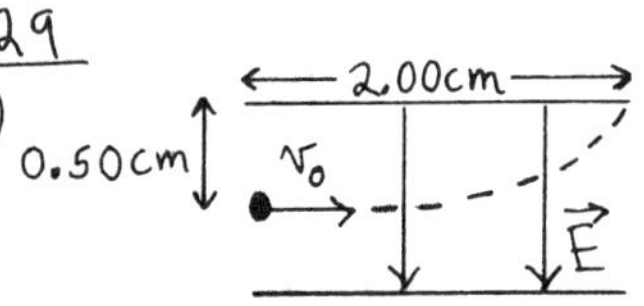

electron $\Rightarrow q = -e$

$\vec{F} = q\vec{E}$ and q negative $\Rightarrow \vec{F}$ and $\vec{E}$ are in opposite directions

$\Rightarrow \vec{F}$ is upward

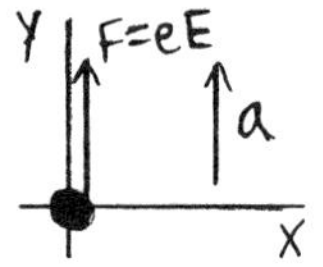

$\sum F_y = ma_y$

$eE = ma$

Solve the kinematics to find the acceleration of the electron:

Just misses upper plate $\Rightarrow x - x_0 = 2.00$ cm when $y - y_0 = +0.500$ cm.

<u>x-component</u>

$v_{0x} = v_0 = 4.00 \times 10^6$ m/s

$a_x = 0$

$x - x_0 = 0.0200$ m

$t = ?$

$x - x_0 = v_{0x} t + \frac{1}{2}a_x t^2$ (the $\frac{1}{2}a_x t^2$ term crossed out)

$t = \frac{x - x_0}{v_{0x}} = \frac{0.0200 \text{ m}}{4.00 \times 10^6 \text{ m/s}} = 5.00 \times 10^{-9}$ s

In this same time t the electron travels 0.0050 m vertically:

<u>y-component</u>

$t = 5.00 \times 10^{-9}$ s

$v_{0y} = 0$

$y - y_0 = +0.0050$ m

$y - y_0 = v_{0y}t + \frac{1}{2}a_y t^2$ (the $v_{0y}t$ term crossed out)

$a_y = \frac{2(y - y_0)}{t^2} = \frac{2(0.0050 \text{ m})}{(5.00 \times 10^{-9} \text{ s})^2} = 4.00 \times 10^{14}$ m/s²

(This analysis is very similar to that used in Chapter 3 for projectile motion, except that here the acceleration is upward rather than downward.)

This acceleration must be produced by the electric-field force:

$eE = ma$

$E = \frac{ma}{e} = \frac{(9.109 \times 10^{-31} \text{ kg})(4.00 \times 10^{14} \text{ m/s}^2)}{1.602 \times 10^{-19} \text{ C}} = \underline{2.27 \times 10^3 \text{ N/C}}$

Note that the acceleration produced by the electric field is <u>much</u> larger than g, the acceleration produced by gravity, so it is perfectly ok to neglect the gravity force on the electron in this problem.

b) $a = \frac{eE}{m_p} = \frac{(1.602 \times 10^{-19} \text{ C})(2.27 \times 10^3 \text{ N/C})}{1.673 \times 10^{-27} \text{ kg}} = 2.174 \times 10^{11}$ m/s²

This is much less than the acceleration of the electron in part (a) so the vertical deflection is less and the proton won't hit the plates.

The proton has the same initial speed, so the proton takes same time $t = 5.00 \times 10^{-9}$ s to travel horizontally the length of the plates. The force on the proton is downward (in the same direction as $\vec{E}$, since q is positive), so the acceleration is downward and $a_y = -2.174 \times 10^{11}$ m/s².

$y - y_0 = v_{0y}t + \frac{1}{2}a_y t^2 = \frac{1}{2}(-2.174 \times 10^{11} \text{ m/s}^2)(5.00 \times 10^{-9} \text{ s})^2 = -2.72 \times 10^{-6}$ m (the $v_{0y}t$ term crossed out)

The displacement is 2.72×10^{-6} m, downward.

c) The displacements are in opposite directions because the electron has negative charge and the proton has positive charge. The electron and proton have the same magnitude of charge, so the force the electric field exerts has

22-29 (cont)

the same magnitude for each charge. But the proton has a mass larger by a factor of 1836 so its acceleration and its vertical displacement is smaller by this factor.

22-31

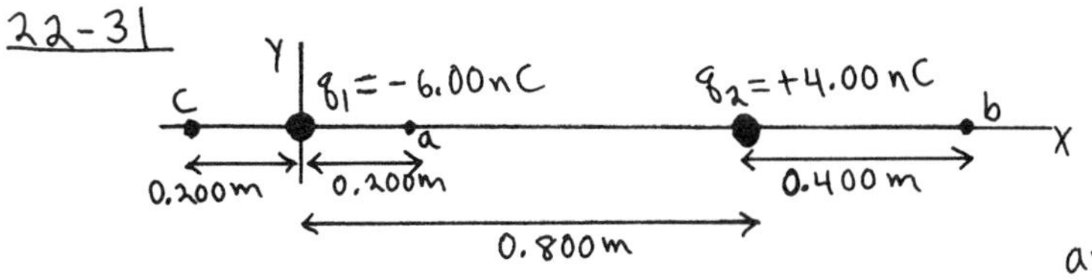

The electric field of a point charge is directed away from the point charge if the charge is positive and toward the point charge if its charge is negative. The magnitude of the electric field is $E = \frac{1}{4\pi\epsilon_0}\frac{|q|}{r^2}$, where r is the distance of the point where the field is calculated from the point charge.

a) At point a the fields $\vec{E}_1$ of q_1 and $\vec{E}_2$ of q_2 are:

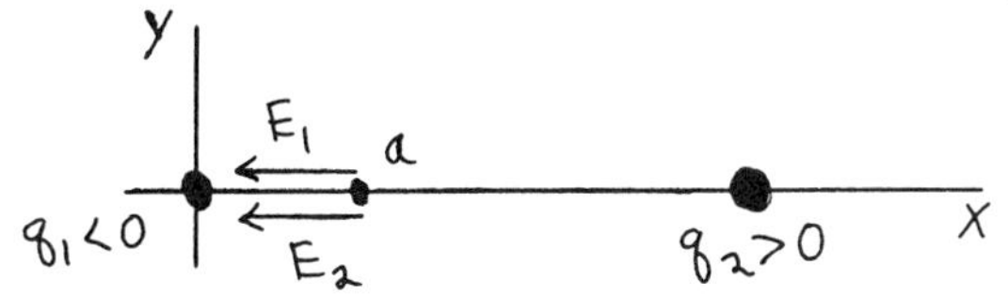

$E_1 = \frac{1}{4\pi\epsilon_0}\frac{|q_1|}{r_1^2} = (8.988\times10^9\,N\cdot m^2/C^2)\frac{6.00\times10^{-9}C}{(0.200\,m)^2}$

$E_1 = 1348\ N/C$

$E_2 = \frac{1}{4\pi\epsilon_0}\frac{|q_2|}{r_2^2} = (8.988\times10^9\,N\cdot m^2/C^2)\frac{4.00\times10^{-9}C}{(0.600m)^2}$

$E_2 = 100\ N/C$

$E_{1x} = -1348\ N/C\ ,\ E_{1y} = 0$

$E_{2x} = -100\,N/C\ ,\ E_{2y} = 0$

$E_x = E_{1x} + E_{2x} = -1348\,N/C - 100\,N/C = -1448\,N/C$

$E_y = E_{1y} + E_{2y} = 0$

The resultant field at point a has magnitude <u>1450 N/C</u> and is in the <u>-X-direction</u>.

b) At point b the fields $\vec{E}_1$ and $\vec{E}_2$ are:

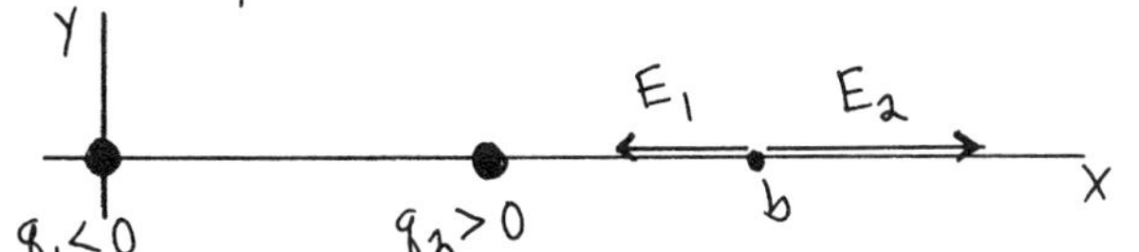

$E_1 = \frac{1}{4\pi\epsilon_0}\frac{|q_1|}{r_1^2} = (8.988\times10^9\,N\cdot m^2/C^2)\frac{6.00\times10^{-9}C}{(1.20m)^2}$

$E_1 = 37.4\ N/C$

$E_2 = \frac{1}{4\pi\epsilon_0}\frac{|q_2|}{r_2^2} = (8.988\times10^9\,N\cdot m^2/C^2)\frac{4.00\times10^{-9}C}{(0.40m)^2}$

$E_2 = 224.7\ N/C$

$E_{1x} = -37.4\ N/C,\ E_{1y} = 0$

$E_{2x} = +224.7\ N/C,\ E_{2y} = 0$

$E_x = E_{1x} + E_{2x} = -37.4\ N/C + 224.7\ N/C = +187.3\ N/C$

$E_y = E_{1y} + E_{2y} = 0$

The resultant field at point b has magnitude <u>187 N/C</u> and is in the +X-direction.

22-31 (cont)

c) At point c the fields $\vec{E}_1$ and $\vec{E}_2$ are:

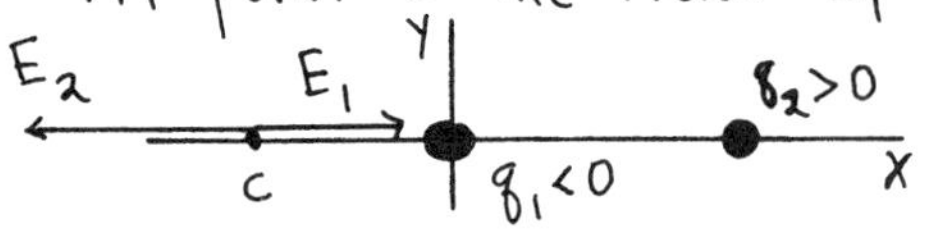

$$E_1 = \frac{1}{4\pi\epsilon_0}\frac{|q_1|}{r_1^2} = (8.988\times10^9\,N\cdot m^2/C^2)\frac{6.00\times10^{-9}C}{(0.200m)^2}$$
$$E_1 = 1348\ N/C$$
$$E_2 = \frac{1}{4\pi\epsilon_0}\frac{|q_2|}{r_2^2} = (8.988\times10^9\,N\cdot m^2/C^2)\frac{4.00\times10^{-9}C}{(1.00m)^2}$$
$$E_2 = 36.0\ N/C$$

$E_{1x} = +1348\ N/C,\ E_{1y} = 0$ $\qquad$ $E_{2x} = -36.0\,N/C,\ E_{2y} = 0$

$$E_x = E_{1x} + E_{2x} = 1348\,N/C - 36.0\,N/C = +1312\,N/C$$
$$E_y = E_{1y} + E_{2y} = 0$$

The resultant field at point c has magnitude <u>1310 N/C</u> and is in the <u>+x-direction.</u>

22-35

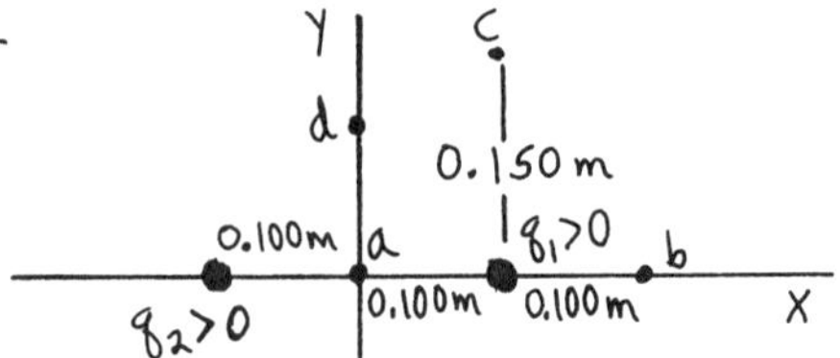

The electric field of a positive charge is directed radially outward from the charge and has magnitude $E = \frac{1}{4\pi\epsilon_0}\frac{|q|}{r^2}$. The resultant electric field is the vector sum of the fields of the individual charges.

a)

$$E_1 = E_2 = \frac{1}{4\pi\epsilon_0}\frac{|q|}{r^2} \text{ with } r = 0.100\ m$$
$$E = E_2 - E_1 = 0$$

b)

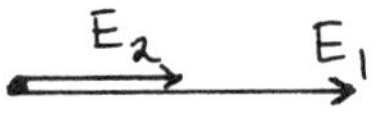

$E = E_1 + E_2$, in the <u>+x-direction</u>

$$E_1 = \frac{1}{4\pi\epsilon_0}\frac{|q_1|}{r_1^2} = (8.988\times10^9\,N\cdot m^2/C^2)\frac{2.00\times10^{-8}C}{(0.100\,m)^2} = 1.80\times10^4 N/C$$
$$E_2 = \frac{1}{4\pi\epsilon_0}\frac{|q_2|}{r_2^2} = (8.988\times10^9\,N\cdot m^2/C^2)\frac{2.00\times10^{-8}C}{(0.300\,m)^2} = 2.00\times10^3 N/C$$
$$E = E_1 + E_2 = 1.80\times10^4\,N/C + 2.00\times10^3\,N/C = \underline{2.00\times10^4\,N/C}$$

c)

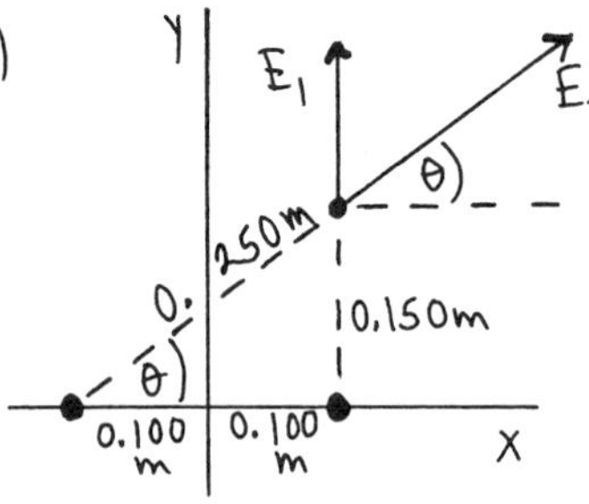

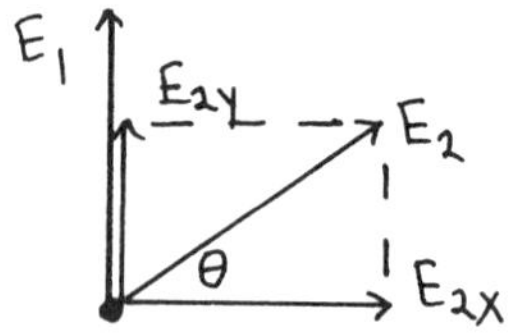

$$E_1 = \frac{1}{4\pi\epsilon_0}\frac{|q_1|}{r_1^2} = (8.988\times10^9\,N\cdot m^2/C^2)\frac{(2.00\times10^{-8}C)}{(0.150m)^2}$$
$$E_1 = 7.998\times10^3 N/C$$

$$\sin\theta = \frac{0.150m}{0.250m} = 0.600$$
$$\cos\theta = \frac{0.200m}{0.250\,m} = 0.800$$

$$E_2 = \frac{1}{4\pi\epsilon_0}\frac{|q_2|}{r_2^2} = (8.988\times10^9\,N\cdot m^2/C^2)\frac{(2.00\times10^{-8}C)}{(0.250\,m)^2}$$
$$E_2 = 2.876\times10^3\,N/C$$

$$E_{1x} = 0,\quad E_{1y} = E_1 = 7.998\times10^3\,N/C$$
$$E_{2x} = E_2\cos\theta = (2.876\times10^3 N/C)(0.800) = 2.301\times10^3 N/C$$
$$E_{2y} = E_2\sin\theta = (2.876\times10^3 N/C)(0.600) = 1.726\times10^3 N/C$$

22-35 (cont)

$$E_x = E_{1x} + E_{2x} = 2.301\times10^3\ N/C$$
$$E_y = E_{1y} + E_{2y} = 7.998\times10^3\ N/C + 1.726\times10^3\ N/C = 9.724\times10^3\ N/C$$

$$E = \sqrt{E_x^2 + E_y^2} = \sqrt{(2.301\times10^3\ N/C)^2 + (9.724\times10^3\ N/C)^2} = \underline{9.93\times10^3\ N/C}$$

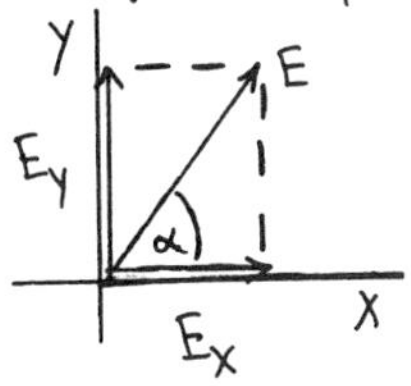

$$\tan\alpha = \frac{E_y}{E_x} = \frac{9.724\times10^3\ N/C}{2.301\times10^3\ N/C} = 4.226$$

$\alpha = \underline{76.7^\circ}$, counterclockwise from +x-axis

d)

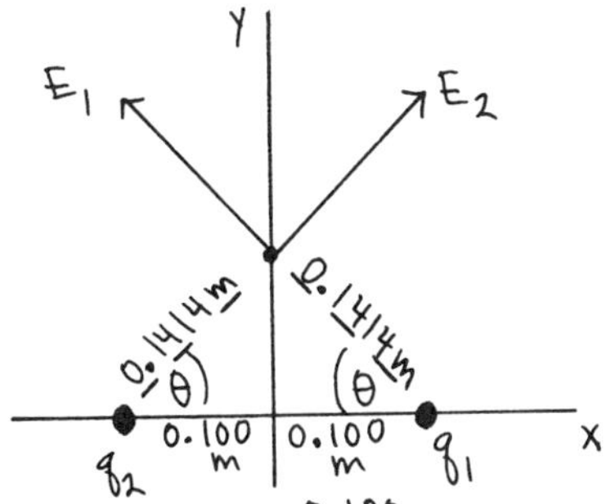

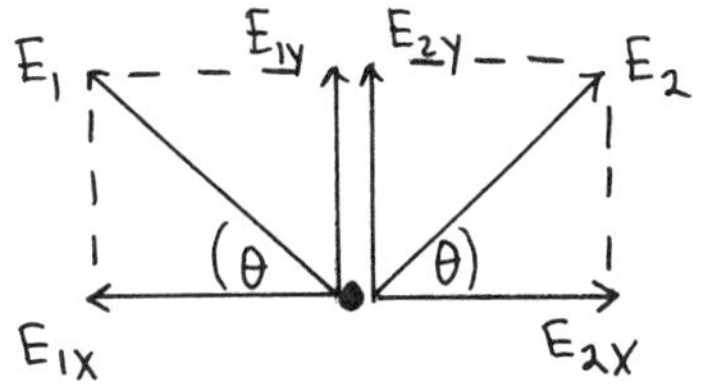

$$E_1 = E_2 = \frac{1}{4\pi\epsilon_0}\frac{|q|}{r^2} = (8.988\times10^9\ N\cdot m^2/C^2)\frac{(2.00\times10^{-8}\ C)}{(0.1414\ m)^2}$$
$$E_1 = E_2 = 8.991\times10^3\ N/C$$

$$\sin\theta = \cos\theta = \frac{0.100}{0.1414} = 0.7072$$

$$E_{1x} = -E_1\cos\theta,\quad E_{2x} = +E_2\cos\theta$$
$$E_x = E_{1x} + E_{2x} = 0$$

$$E_{1y} = +E_1\sin\theta,\quad E_{2y} = +E_2\sin\theta$$
$$E_y = E_{1y} + E_{2y} = 2E_{1y} = 2E_1\sin\theta = 2(8.991\times10^3\ N/C)(0.7072) = 1.27\times10^4\ N/C$$

$E = \underline{1.27\times10^4\ N/C}$, in the $\underline{+y\text{-direction}}$

22-37

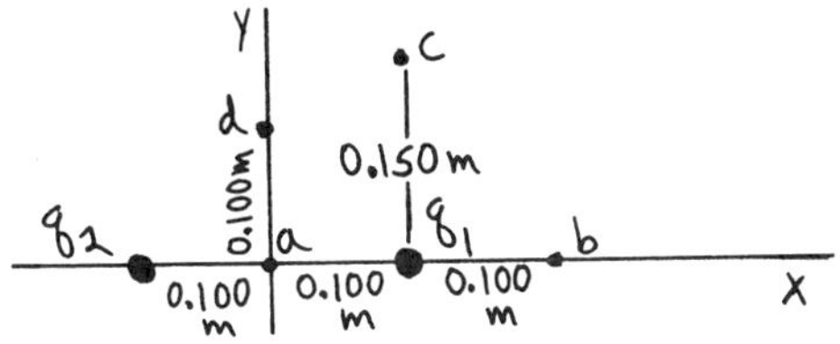

The resultant electric field is the vector sum of the field $\vec{E}_1$ of q_1 and $\vec{E}_2$ of q_2.

a) At point a (the origin) the electric fields $\vec{E}_1$ and $\vec{E}_2$ are:

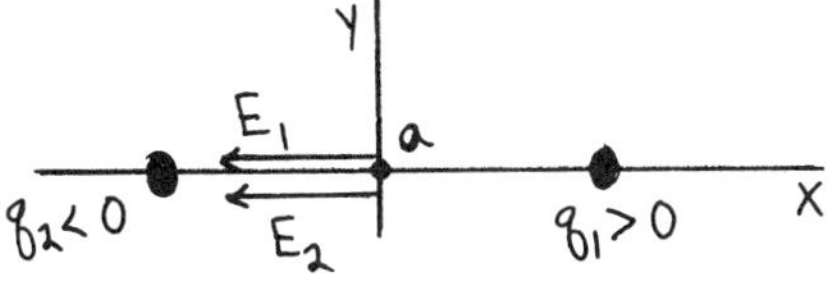

$$E_1 = \frac{1}{4\pi\epsilon_0}\frac{|q_1|}{r_1^2} = (8.988\times10^9\ N\cdot m^2/C^2)\frac{2.00\times10^{-8}\ C}{(0.100\ m)^2}$$
$$E_1 = 1.798\times10^4\ N/C$$
$$E_2 = \frac{1}{4\pi\epsilon_0}\frac{|q_2|}{r_2^2} = (8.988\times10^9\ N\cdot m^2/C^2)\frac{2.00\times10^{-8}\ C}{(0.100\ m)^2}$$
$$E_2 = 1.798\times10^4\ N/C$$

$E_{1x} = -1.798\times10^4\ N/C,\ E_{1y} = 0$ $\qquad$ $E_{2x} = -1.798\times10^4\ N/C,\ E_{2y} = 0$

$$E_x = E_{1x} + E_{2x} = 2(-1.798\times10^4\ N/C) = -3.60\times10^4\ N/C$$

22-37 (cont)

$$E_y = E_{1y} + E_{2y} = 0$$

The resultant field at point a has magnitude $\underline{3.60\times10^4\ \text{N/C}}$ and is in the $\underline{-x\text{-direction}}$.

b) At point b the electric fields $\vec{E}_1$ and $\vec{E}_2$ are:

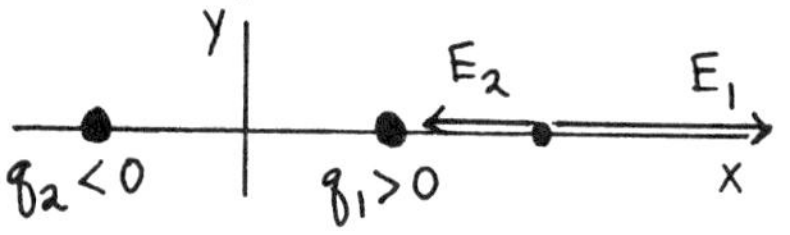

$$E_1 = \frac{1}{4\pi\epsilon_0}\frac{|q_1|}{r_1^2} = (8.988\times10^9\ \text{N}\cdot\text{m}^2/\text{C}^2)\frac{2.00\times10^{-8}\ \text{C}}{(0.100\ \text{m})^2}$$

$$E_1 = 1.798\times10^4\ \text{N/C}$$

$$E_2 = \frac{1}{4\pi\epsilon_0}\frac{|q_2|}{r_2^2} = (8.988\times10^9\ \text{N}\cdot\text{m}^2/\text{C}^2)\frac{2.00\times10^{-8}\ \text{C}}{(0.300\ \text{m})^2}$$

$$E_2 = 1.997\times10^3\ \text{N/C}$$

$E_{1x} = +1.798\times10^4\ \text{N/C},\ E_{1y} = 0$ $\qquad$ $E_{2x} = -1.997\times10^3\ \text{N/C},\ E_{2y} = 0$

$$E_x = E_{1x} + E_{2x} = +1.798\times10^4\ \text{N/C} - 1.997\times10^3\ \text{N/C} = 1.60\times10^4\ \text{N/C}$$

$$E_y = E_{1y} + E_{2y} = 0$$

The resultant electric field at point b has magnitude $\underline{1.60\times10^4\ \text{N/C}}$ and is in the $\underline{+x\text{-direction}}$.

c)

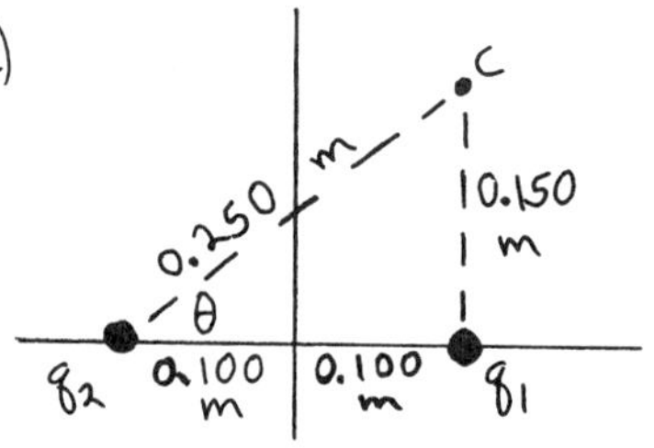

$$E_1 = \frac{1}{4\pi\epsilon_0}\frac{|q_1|}{r_1^2} = (8.988\times10^9\ \text{N}\cdot\text{m}^2/\text{C}^2)\frac{2.00\times10^{-8}\ \text{C}}{(0.150\ \text{m})^2}$$

$$E_1 = 7.989\times10^3\ \text{N/C}$$

$$E_2 = \frac{1}{4\pi\epsilon_0}\frac{|q_2|}{r_2^2} = (8.988\times10^9\ \text{N}\cdot\text{m}^2/\text{C}^2)\frac{2.00\times10^{-8}\ \text{C}}{(0.250\ \text{m})^2}$$

$$E_2 = 2.876\times10^3\ \text{N/C}$$

$$\sin\theta = \frac{0.150\ \text{m}}{0.250\ \text{m}} = 0.600;\quad \cos\theta = \frac{0.200\ \text{m}}{0.250\ \text{m}} = 0.800$$

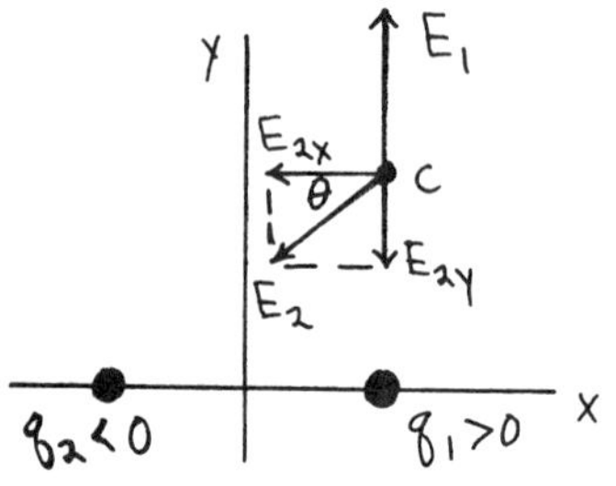

$$E_{1x} = 0,\quad E_{1y} = 7.989\times10^3\ \text{N/C}$$

$$E_{2x} = -E_2\cos\theta = -(2.876\times10^3\ \text{N/C})(0.800) = -2.301\times10^3\ \text{N/C}$$

$$E_{2y} = -E_2\sin\theta = -(2.876\times10^3\ \text{N/C})(0.600) = -1.726\times10^3\ \text{N/C}$$

$$E_x = E_{1x} + E_{2x} = 0 - 2.301\times10^3\ \text{N/C} = -2.301\times10^3\ \text{N/C}$$

$$E_y = E_{1y} + E_{2y} = 7.989\times10^3\ \text{N/C} - 1.726\times10^3\ \text{N/C}$$

$$E_y = 6.263\times10^3\ \text{N/C}$$

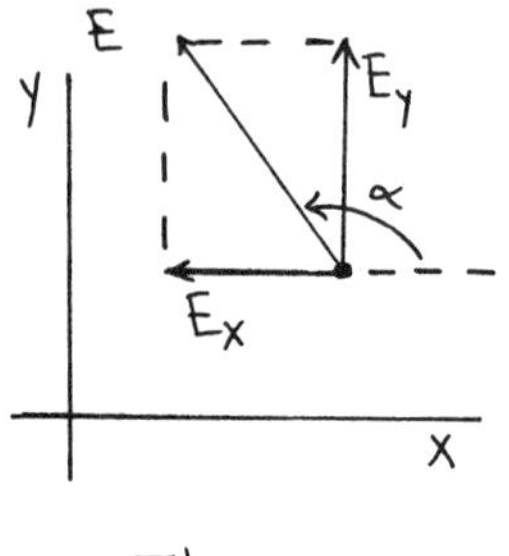

$$E = \sqrt{E_x^2 + E_y^2} = 6.67\times10^3\ \text{N/C}$$

$$\tan\alpha = \frac{E_y}{E_x} = \frac{6.263\times10^3\ \text{N/C}}{-2.301\times10^3\ \text{N/C}} = -2.723 \Rightarrow \alpha = 110^\circ$$

The resultant field at point c has magnitude $\underline{6.67\times10^3\ \text{N/C}}$ and is directed at an angle of $\underline{110^\circ}$ counterclockwise from the +x-axis.

22-37 (cont)

d)

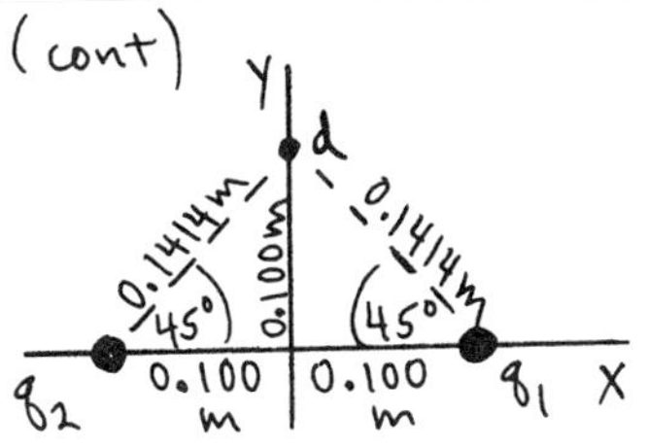

$E_1 = \frac{1}{4\pi\epsilon_0}\frac{|q_1|}{r_1^2} = (8.988\times10^9\ \mathrm{N\cdot m^2/C^2})\frac{2.00\times10^{-8}\ \mathrm{C}}{(0.1414\ \mathrm{m})^2}$

$E_1 = 8.991\times10^3\ \mathrm{N/C}$

$E_2 = \frac{1}{4\pi\epsilon_0}\frac{|q_2|}{r_2^2} = 8.991\times10^3\ \mathrm{N/C}$

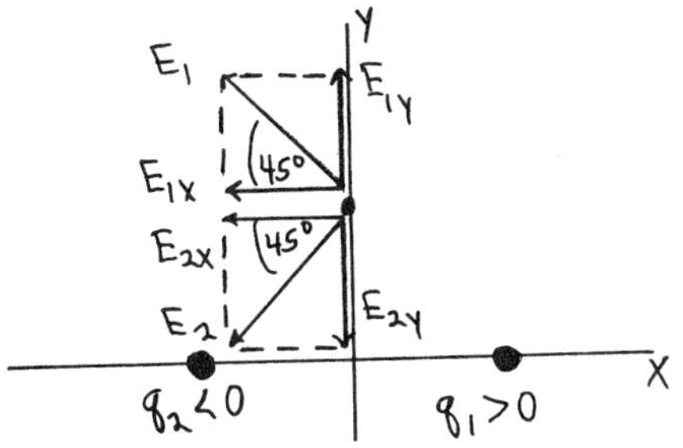

$E_{1x} = -E_1\cos45° = -6.357\times10^3\ \mathrm{N/C}$

$E_{1y} = +E_1\sin45° = +6.357\times10^3\ \mathrm{N/C}$

$E_{2x} = -E_2\cos45° = -6.357\times10^3\ \mathrm{N/C}$

$E_{2y} = -E_2\sin45° = -6.357\times10^3\ \mathrm{N/C}$

$E_x = E_{1x} + E_{2x} = -6.357\times10^3\ \mathrm{N/C} - 6.357\times10^3\ \mathrm{N/C}$

$E_x = -1.271\times10^4\ \mathrm{N/C}$

$E_y = E_{1y} + E_{2y} = 6.357\times10^3\ \mathrm{N/C} - 6.357\times10^3\ \mathrm{N/C} = 0$

The resultant field has magnitude 1.27×10^4 N/C and is in the $-x$-direction.

22-47

a)

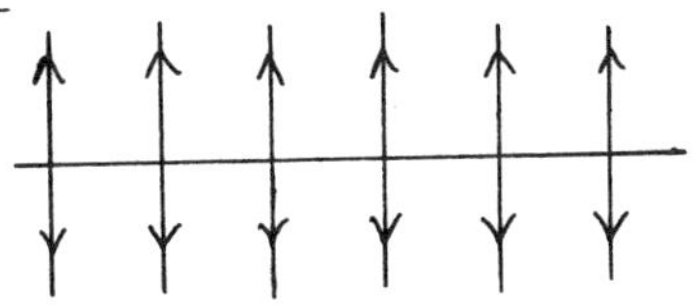

The electric field lines are perpendicular to the line of charge.

b)

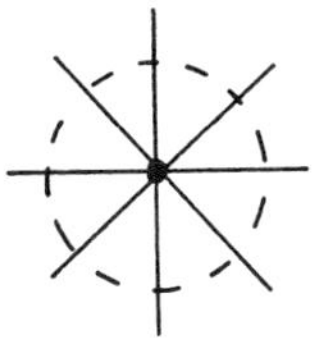

The magnitude of the electric field is inversely proportional to the spacing of the field lines. Consider a circle of radius r with the line of charge passing through the center. The spacing of field lines is the same all around the circle, and in the direction perpendicular to the plane of the circle the lines are equally spaced, so E depends only on the distance r. The number of field lines passing out through the circle is independent of the radius of the circle, so the spacing of the field lines is proportional to the reciprocal of the circumference $2\pi r$ of the circle. Hence E is proportional to $1/r$.

22-49

a) $p = qd$ (Eq. 22-14)

$d = \frac{p}{q} = \frac{8.2\times10^{-12}\ \mathrm{C\cdot m}}{3.5\times10^{-9}\ \mathrm{C}} = 2.3\times10^{-3}\ \mathrm{m} = 2.3\ \mathrm{mm}$

b)

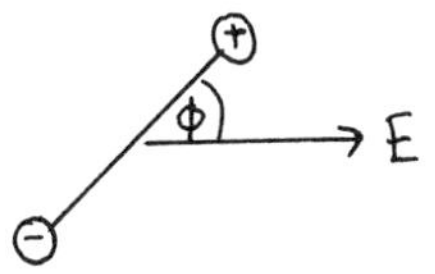

$\tau = pE\sin\phi$

$E = \frac{\tau}{p\sin\phi} = \frac{7.0\times10^{-9}\ \mathrm{N\cdot m}}{(8.2\times10^{-12}\ \mathrm{C\cdot m})\sin35.0°} = 1.5\times10^3\ \mathrm{N/C}$

22-51

a) $U(\phi) = -\vec{p}\cdot\vec{E} = -pE\cos\phi$, where ϕ is the angle between $\vec{p}$ and $\vec{E}$.

parallel $\Rightarrow \phi = 0$ $\quad U(0^\circ) = -pE$

perpendicular $\Rightarrow \phi = 90^\circ$ $\quad U(90^\circ) = 0$

$$\Delta U = U(90^\circ) - U(0^\circ) = +pE = (5.0\times10^{-30}\ \mathrm{C\cdot m})(2.0\times10^{5}\ \mathrm{N/C}) = \underline{1.0\times10^{-24}\ \mathrm{J}}$$

Problems

22-55

a) The number of Na^+ ions in 0.100 mol of NaCl is $N = nN_A$. The charge of one ion is $+e$, so the total charge is

$$q_1 = nN_Ae = (0.100\ \mathrm{mol})(6.022\times10^{23}\ \mathrm{ions/mol})(1.602\times10^{-19}\ \mathrm{C/ion}) = 9.647\times10^{3}\ \mathrm{C}$$

There are the same number of Cl^- ions and each has charge $-e$, so $q_2 = -9.647\times10^{3}\ \mathrm{C}$.

$$F = \frac{1}{4\pi\epsilon_0}\frac{|q_1q_2|}{r^2} = (8.988\times10^{9}\ \mathrm{N\cdot m^2/C^2})\frac{(9.647\times10^{3}\ \mathrm{C})^2}{(0.0200\ \mathrm{m})^2} = \underline{2.09\times10^{21}\ \mathrm{N}}$$

b) $a = \frac{F}{m}$

Need the mass of 0.100 mol of Cl^- ions. For Cl, $M = 35.453\times10^{-30}\ \mathrm{kg/mol}$, so $m = (0.100\ \mathrm{mol})(35.453\times10^{-3}\ \mathrm{kg/mol}) = 35.45\times10^{-4}\ \mathrm{kg}$.

Then $a = \frac{F}{m} = \frac{2.09\times10^{21}\ \mathrm{N}}{35.45\times10^{-4}\ \mathrm{kg}} = 5.90\times10^{23}\ \mathrm{m/s^2}$.

c) Such a huge force being applied to separate all the ions this way is not reasonable.

22-57

a)

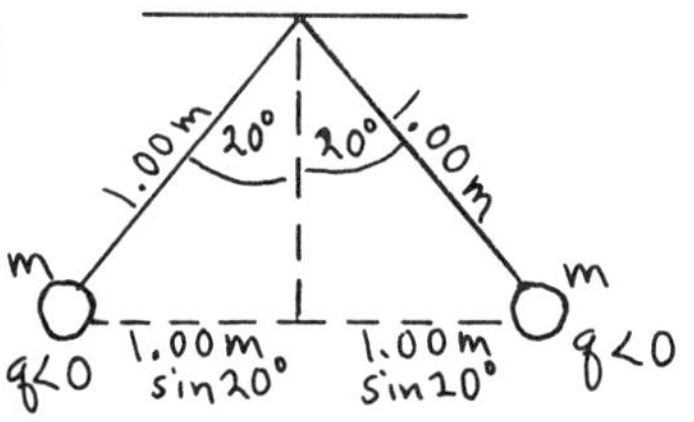

sphere on the left:

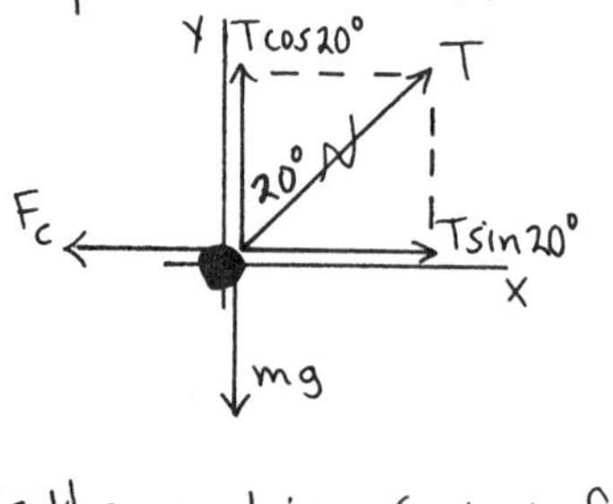

sphere on the right:

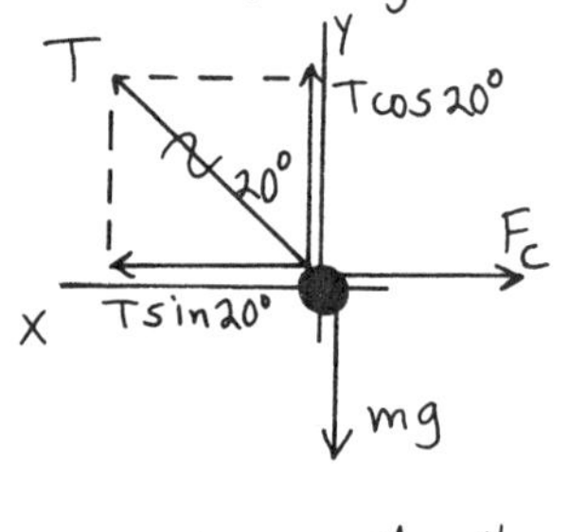

F_c is the repulsive Coulomb force exerted by one sphere on the other.

b) From either force diagram in part (a):

$\sum F_y = ma_y$ $\qquad$ $\sum F_x = ma_x$

$T\cos 20^\circ - mg = 0$ $\qquad$ $T\sin 20^\circ - F_c = 0$

$T = \frac{mg}{\cos 20^\circ}$ $\qquad$ $F_c = T\sin 20^\circ$

22-57 (cont)

Use the first equation to eliminate T in the second

$\Rightarrow \quad F_c = \left(\frac{mg}{\cos 20^\circ}\right)(\sin 20^\circ) = mg\tan 20^\circ$

$$F_c = \frac{1}{4\pi\epsilon_0}\frac{|q_1 q_2|}{r^2} = \frac{1}{4\pi\epsilon_0}\frac{q^2}{r^2} = \frac{1}{4\pi\epsilon_0}\frac{q^2}{[2(1.00\,\text{m})\sin 20^\circ]^2}$$

$$F_c = mg\tan 20^\circ \Rightarrow mg\tan 20^\circ = \frac{1}{4\pi\epsilon_0}\frac{q^2}{[2(1.00\,\text{m})\sin 20^\circ]^2}$$

$$q = (2.00\,\text{m})\sin 20^\circ\sqrt{\frac{mg\tan 20^\circ}{1/4\pi\epsilon_0}} = (2.00\,\text{m})\sin 20^\circ\sqrt{\frac{(12.0\times10^{-3}\,\text{kg})(9.80\,\text{m/s}^2)\tan 20^\circ}{8.988\times10^9\,\text{N}\cdot\text{m}^2/\text{C}^2}} = \underline{1.49\times10^{-6}\,\text{C}}$$

c) The separation between the two spheres is now $2L\sin\theta$. $q = 1.49\,\mu\text{C}$ as found in part (b). $F_c = \frac{1}{4\pi\epsilon_0}\frac{q^2}{(2L\sin\theta)^2}$ and $F_c = mg\tan\theta$.

$$\Rightarrow \frac{1}{4\pi\epsilon_0}\frac{q^2}{(2L\sin\theta)^2} = mg\tan\theta$$

$$(\sin\theta)^2\tan\theta = \frac{1}{4\pi\epsilon_0}\frac{q^2}{4L^2mg} = (8.988\times10^9\,\text{N}\cdot\text{m}^2/\text{C}^2)\frac{(1.49\times10^{-6}\,\text{C})^2}{4(0.400\,\text{m})^2(12.0\times10^{-3}\,\text{kg})(9.80\,\text{m/s}^2)} = 0.2651$$

$$\boxed{(\sin\theta)^2\tan\theta = 0.2651}$$

Solve this equation by trial and error. This will go quicker if we can make a good estimate of the value of θ that solves this equation. θ small gives $\tan\theta \approx \sin\theta$. With this approximation the equation becomes

$\sin^3\theta = 0.2651$ and $\sin\theta = 0.6424$, so $\theta = 40.0^\circ$

Now refine this guess:

θ	$\sin^2\theta\tan\theta$	θ	$\sin^2\theta\tan\theta$	θ	$\sin^2\theta\tan\theta$
40.0°	0.3467	37.0°	0.2729	36.6°	0.2640
39.0°	0.3207	36.0°	0.2510	36.7°	0.2662
38.0°	0.2961	36.5°	0.2618		

$\Rightarrow \underline{\theta = 36.7^\circ}$

22-61

a) 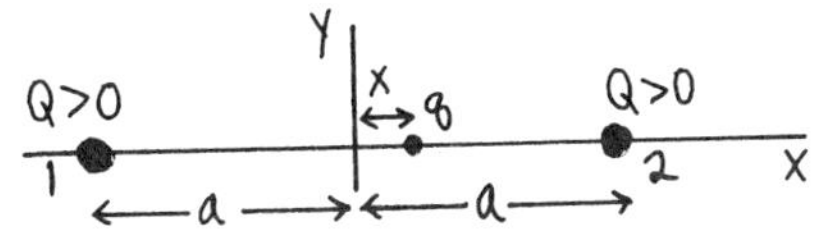

Assume $q>0$, but the result is the same if $q<0$.

Find the net force on q:

$F_x = F_{1x} + F_{2x}$ and $F_{1x} = +F_1$ and $F_{2x} = -F_2$

$$F_1 = \frac{1}{4\pi\epsilon_0}\frac{qQ}{(a+x)^2}, \quad F_2 = \frac{1}{4\pi\epsilon_0}\frac{qQ}{(a-x)^2}$$

$$F_x = F_1 - F_2 = \frac{qQ}{4\pi\epsilon_0}\left[\frac{1}{(a+x)^2} - \frac{1}{(a-x)^2}\right] = \frac{qQ}{4\pi\epsilon_0 a^2}\left[+(1+\tfrac{x}{a})^{-2} - (1-\tfrac{x}{a})^{-2}\right]$$

Since $x \ll a$ we can use the binomial expansion for $(1-\frac{x}{a})^{-2}$ and $(1+\frac{x}{a})^{-2}$ and keep only the first two terms: $(1+z)^n \approx 1 + nz$

For $(1-\frac{x}{a})^{-2}$, $z = -\frac{x}{a}$ and $n=-2$ so $(1-\frac{x}{a})^{-2} \approx 1 + \frac{2x}{a}$

For $(1+\frac{x}{a})^{-2}$, $z = +\frac{x}{a}$ and $n=-2$ so $(1+\frac{x}{a})^{-2} \approx 1 - \frac{3x}{a}$

Then $F \approx \frac{qQ}{4\pi\epsilon_0 a^2}\left[(1-\frac{2x}{a}) - (1+\frac{2x}{a})\right] = -\left(\frac{qQ}{\pi\epsilon_0 a^3}\right)x$

22-61 (cont)

For simple harmonic motion $F=-kx$ and the frequency of oscillation is $f = \frac{1}{2\pi}\sqrt{\frac{k}{m}}$. The net force here is of this form, with $k = \frac{qQ}{\pi \epsilon_0 a^3}$.

Thus $f = \frac{1}{2\pi}\sqrt{\frac{qQ}{\pi \epsilon_0 m a^3}}$.

b)

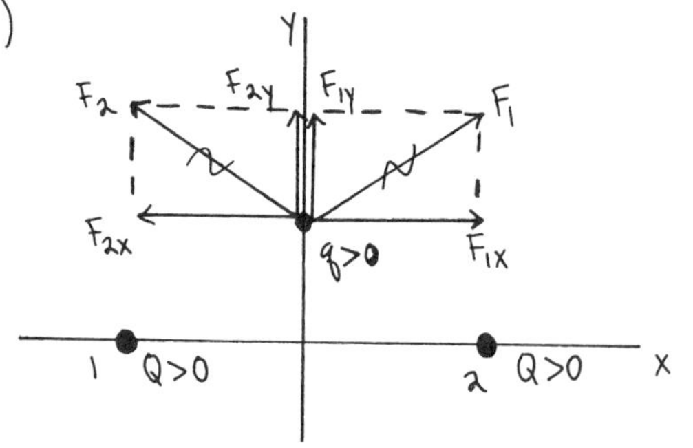

If q is positive the x-components of the forces exerted by the two charges cancel, the y-components add, and the net force is in the +y-direction when $y>0$ and in the −y-direction when $y<0$. The charge moves away from the origin on the y-axis and never returns.

c) If q is negative then the net force is directed toward the origin. In this case the force is a restoring force and the charge oscillates about the origin along the y-axis.

22-63

a)

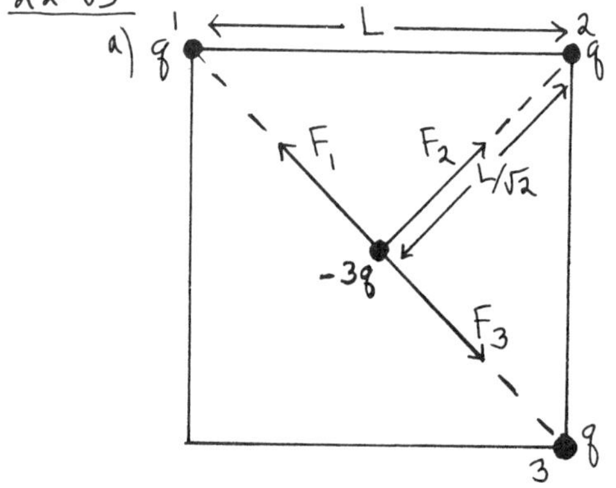

$\vec{F}_1 + \vec{F}_3 = 0$ so the net force is $\vec{F} = \vec{F}_2$.

$F = \frac{1}{4\pi\epsilon_0} \frac{q(3q)}{(L/\sqrt{2})^2} = \frac{6q^2}{4\pi\epsilon_0 L^2}$, away from the vacant corner.

b)

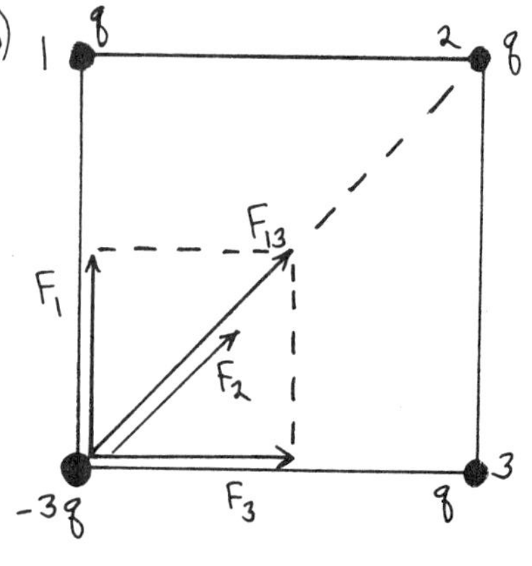

$F_2 = \frac{1}{4\pi\epsilon_0}\frac{q(3q)}{(\sqrt{2}L)^2} = \frac{3q^2}{4\pi\epsilon_0 (2L^2)}$

$F_1 = F_3 = \frac{1}{4\pi\epsilon_0}\frac{q(3q)}{L^2} = \frac{3q^2}{4\pi\epsilon_0 L^2}$

The vector sum of F_1 and F_3 is $F_{13} = \sqrt{F_1^2 + F_3^2}$

$F_{13} = \sqrt{2}\, F_1 = \frac{3\sqrt{2}q^2}{4\pi\epsilon_0 L^2}$; $\vec{F}_{13}$ and $\vec{F}_2$ are in the same direction.

$F = F_{13} + F_2 = \frac{3q^2}{4\pi\epsilon_0 L^2}\left(\sqrt{2} + \frac{1}{2}\right)$, and is directed along the diagonal toward the opposite corner.

22-65

a)

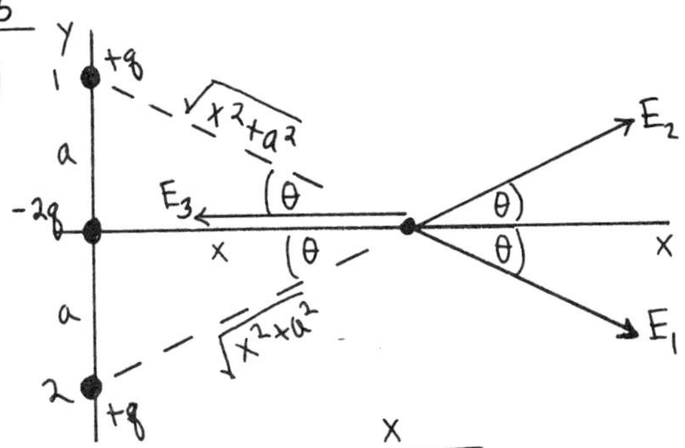

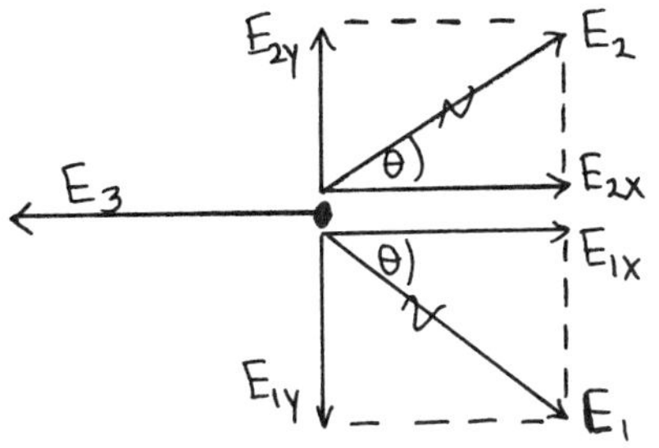

$$\cos\theta = \frac{x}{\sqrt{x^2+a^2}}$$

$$E_1 = E_2 = \frac{1}{4\pi\epsilon_0}\frac{q}{a^2+x^2}$$

$$E_3 = \frac{1}{4\pi\epsilon_0}\frac{2q}{x^2}$$

$E_{1y} = -E_1\sin\theta$, $E_{2y} = +E_1\sin\theta$ so $E_y = E_{1y} + E_{2y} = 0.$

$E_{1x} = E_{2x} = +E_1\cos\theta = \dfrac{1}{4\pi\epsilon_0}\dfrac{q}{a^2+x^2}\dfrac{x}{\sqrt{x^2+a^2}}$, $E_{3x} = -E_3$

$$E_x = E_{1x} + E_{2x} + E_{3x} = 2\left(\frac{1}{4\pi\epsilon_0}\frac{q}{a^2+x^2}\frac{x}{\sqrt{x^2+a^2}}\right) - \frac{2q}{4\pi\epsilon_0 x^2}$$

$$E_x = -\frac{2q}{4\pi\epsilon_0}\left(\frac{1}{x^2} - \frac{x}{(a^2+x^2)^{3/2}}\right) = -\frac{2q}{4\pi\epsilon_0 x^2}\left(1 - \frac{1}{(1+a^2/x^2)^{3/2}}\right)$$

Thus $E = \dfrac{2q}{4\pi\epsilon_0 x^2}\left(1 - \dfrac{1}{(1+a^2/x^2)^{3/2}}\right)$, in the $-x$-direction.

b) $x \gg a \Rightarrow \dfrac{a^2}{x^2} \ll 1$ and $\left(1+\dfrac{a^2}{x^2}\right)^{-3/2} \approx 1 - \dfrac{3a^2}{2x^2}$

$$E \approx \frac{2q}{4\pi\epsilon_0 x^2}\left(1-\left(1-\frac{3a^2}{2x^2}\right)\right) = \frac{3qa^2}{4\pi\epsilon_0 x^4}$$

$E \sim \dfrac{1}{x^4}$

For a point charge $E \sim \dfrac{1}{x^2}$ and for a dipole $E \sim \dfrac{1}{x^3}$.

22-67

a)

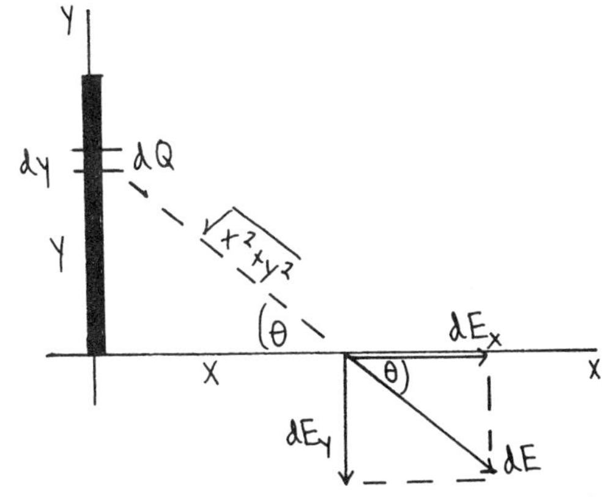

$\sin\theta = \dfrac{y}{\sqrt{x^2+y^2}}$, $\cos\theta = \dfrac{x}{\sqrt{x^2+y^2}}$

Slice the charge distribution up into small pieces of length dy. The charge dQ in each slice is $dQ = Q\dfrac{dy}{a}$. The electric field this produces at a distance x along the x-axis is dE. Calculate the components of $d\vec{E}$ and then integrate over the charge distribution to find the components of the total field.

$$dE = \frac{1}{4\pi\epsilon_0}\frac{dQ}{x^2+y^2} = \frac{Q}{4\pi\epsilon_0 a}\frac{dy}{x^2+y^2}$$

$$dE_x = dE\cos\theta = \frac{Qx}{4\pi\epsilon_0 a}\frac{dy}{(x^2+y^2)^{3/2}}$$

$$dE_y = -dE\sin\theta = -\frac{Q}{4\pi\epsilon_0 a}\frac{y\,dy}{(x^2+y^2)^{3/2}}$$

22-67 (cont)

$$E_x = \int dE_x = \frac{Qx}{4\pi\epsilon_0 a}\int_0^a \frac{dy}{(x^2+y^2)^{3/2}} = \frac{Qx}{4\pi\epsilon_0 a}\left[\frac{1}{x^2}\frac{y}{\sqrt{x^2+y^2}}\Big|_0^a\right] = \frac{Q}{4\pi\epsilon_0 x}\frac{1}{\sqrt{x^2+a^2}}$$

$$E_y = \int dE_y = -\frac{Q}{4\pi\epsilon_0 a}\int_0^a \frac{y\,dy}{(x^2+y^2)^{3/2}} = -\frac{Q}{4\pi\epsilon_0 a}\left(-\frac{1}{(x^2+y^2)^{1/2}}\Big|_0^a\right) = -\frac{Q}{4\pi\epsilon_0 a}\left(\frac{1}{x} - \frac{1}{\sqrt{x^2+a^2}}\right)$$

b) $\vec{F} = q_0\vec{E}$

$F_x = -q\,E_x = -\frac{qQ}{4\pi\epsilon_0 x}\frac{1}{\sqrt{x^2+a^2}}$, $F_y = -qE_y = +\frac{qQ}{4\pi\epsilon_0 a}\left(\frac{1}{x} - \frac{1}{\sqrt{x^2+a^2}}\right)$

c) For $x \gg a$, $\frac{1}{\sqrt{x^2+a^2}} = \frac{1}{x}\left(1+\frac{a^2}{x^2}\right)^{-1/2} = \frac{1}{x}\left(1-\frac{a^2}{2x^2}\right) = \frac{1}{x} + \frac{a^2}{2x^3}$

$F_x \cong -\frac{qQ}{4\pi\epsilon_0 x^2}$, $F_y \cong \frac{qQ}{4\pi\epsilon_0 a}\left(\frac{1}{x} - \frac{1}{x} - \frac{a^2}{2x^3}\right) = -\frac{qQ}{8\pi\epsilon_0 a x^3}$

For $x \gg a$, $F_y \ll F_x$ and $F \cong |F_x| = \frac{qQ}{4\pi\epsilon_0 x^2}$ and $\vec{F}$ is in the $-x$-direction.
For $x \gg a$ the charge distribution Q acts like a point charge.

22-75

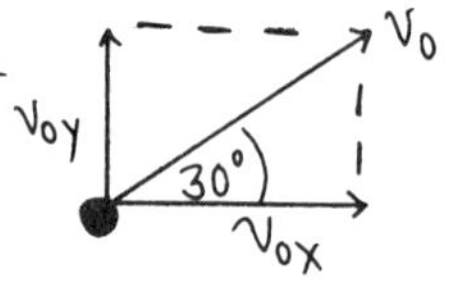

$\vec{F} = q\vec{E}$

$q = -e \Rightarrow \vec{F} = -e\vec{E}$

The electric field is upward so the electric force on the negatively charged electron is downward and has magnitude $F = eE$.

Free-body diagram for the electron:

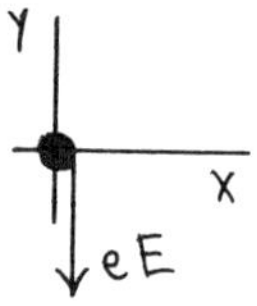

$\Sigma F_y = ma_y$

$-eE = ma$

$a = -\frac{eE}{m} = -\frac{(1.602\times10^{-19}\,C)(400\,N/C)}{9.109\times10^{-31}\,kg} = -7.035\times10^{13}\,m/s^2$

(This is much larger in magnitude than g, so gravity can be neglected.)

a) $v_y = 0$ (at maximum height)

$v_{0y} = v_0\sin30° = (3.00\times10^6\,m/s)\sin30°$

$v_{0y} = 1.50\times10^6\,m/s$

$a_y = -7.035\times10^{13}\,m/s^2$

$y - y_0 = ?$

$v_y^2 = v_{0y}^2 + 2a_y(y-y_0)$

$y - y_0 = \frac{v_y^2 - v_{0y}^2}{2a_y} = \frac{0-(1.50\times10^6\,m/s)^2}{2(-7.035\times10^{13}\,m/s^2)}$

$y - y_0 = 0.0160\,m = \underline{1.60\,cm}$

b) Want $x - x_0$ when $y - y_0 = 0$:

$y - y_0 = 0$

$v_{0y} = 1.50\times10^6\,m/s$

$a_y = -7.035\times10^{13}\,m/s^2$

$t = ?$

$\cancel{y-y_0} = v_{0y}t + \frac{1}{2}a_y t^2$

$-\frac{1}{2}a_y t^2 = v_{0y}t$

$t = -\frac{2v_{0y}}{a_y} = -\frac{2(1.50\times10^6\,m/s)}{-7.035\times10^{13}\,m/s^2} = 4.264\times10^{-8}\,s$

Then

$x - x_0 = ?$

$v_{0x} = v_0\cos30° = (3.00\times10^6\,m/s)\cos30° = 2.598\times10^6\,m/s$

$a_x = 0$

$t = 4.264\times10^{-8}\,s$

$x - x_0 = v_{0x}t + \cancel{\frac{1}{2}a_x t^2}$

$x - x_0 = (2.598\times10^6\,m/s)(4.264\times10^{-8}\,s)$

$x - x_0 = 0.111\,m = \underline{11.1\,cm}$

22-75 (cont)

c) a_y is constant and downward; $a_x = 0$. The motion is just like projectile motion, just with a different numerical value for a_y. Thus the path of the electron is a parabola:

22-77

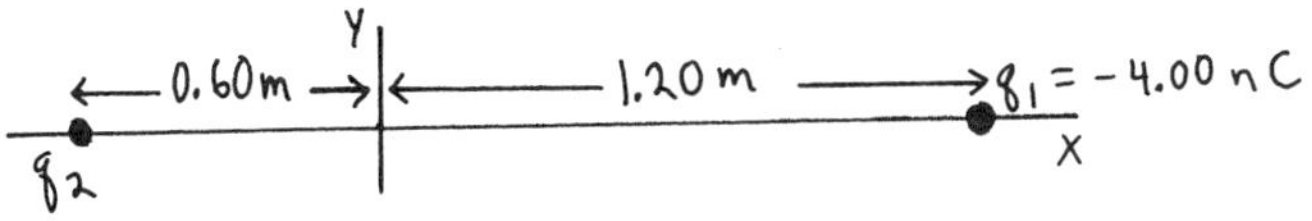

The electric field of q_1 at the origin has magnitude $E_1 = \frac{1}{4\pi\epsilon_0}\frac{|q_1|}{r_1^2} = (8.988 \times 10^9\ N \cdot m^2/C^2)\frac{4.00 \times 10^{-9}\ C}{(1.20\ m)^2}$ $= 25.0\ N/C$ and is in the +x-direction.

The resultant field at the origin is given by $\vec{E} = \vec{E}_1 + \vec{E}_2 \Rightarrow \vec{E}_2 = \vec{E} - \vec{E}_1$ is the field at the origin due to charge q_2.

$\Rightarrow E_{2x} = E_x - E_{1x}$

a) $E_x = +50.0\ N/C$, $E_{1x} = +25.0\ N/C \Rightarrow E_{2x} = 50.0\ N/C - 25.0\ N/C = +25.0\ N/C$

$\vec{E}_2$ is in the +x-direction $\Rightarrow q_2$ is positive.

$$E_2 = \frac{1}{4\pi\epsilon_0}\frac{|q_2|}{r_2^2} \Rightarrow |q_2| = \frac{r_2^2 E_2}{(1/4\pi\epsilon_0)} = \frac{(0.60\ m)^2(25.0\ N/C)}{8.988 \times 10^9\ N \cdot m^2/C^2} = 1.00 \times 10^{-9}\ C$$

$$q_2 = +1.00 \times 10^{-9}\ C$$

b) $E_x = -50.0\ N/C$, $E_{1x} = +25.0\ N/C$

$E_{2x} = E_x - E_{1x} = -50.0\ N/C - 25.0\ N/C = -75.0\ N/C$

$\vec{E}_2$ is in the -x-direction $\Rightarrow q_2$ is negative

$$E_2 = \frac{1}{4\pi\epsilon_0}\frac{|q_2|}{r_2^2} \Rightarrow |q_2| = \frac{r_2^2 E_2}{(1/4\pi\epsilon_0)} = \frac{(0.60\ m)^2(75.0\ N/C)}{8.988 \times 10^9\ N \cdot m^2/C^2} = 3.00 \times 10^{-9}\ C$$

$$q_2 = -3.00 \times 10^{-9}\ C$$

22-79

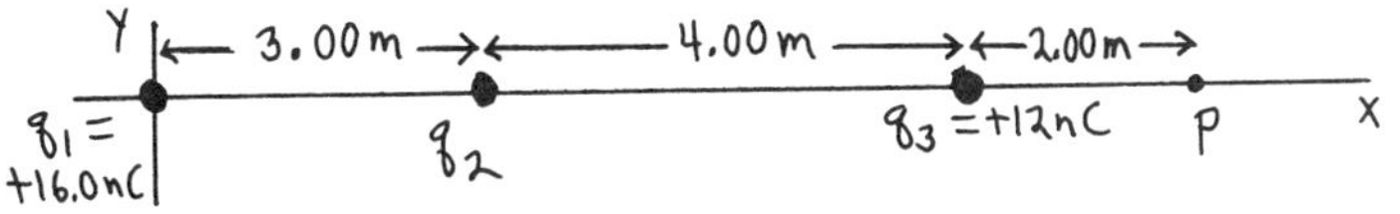

A point P the resultant field has magnitude 18.0 N/C and is in the +x-direction, so $E_x = +18.0\ N/C$.

$\vec{E} = \vec{E}_1 + \vec{E}_2 + \vec{E}_3 \Rightarrow \vec{E}_2 = \vec{E} - \vec{E}_1 - \vec{E}_3$

$E_{2x} = E_x - E_{1x} - E_{3x}$; calculate E_{2x} and from it deduce the sign and magnitude of q_2.

The electric field $\vec{E}_1$ of q_1 at point P has magnitude

$E_1 = \frac{1}{4\pi\epsilon_0}\frac{|q_1|}{r_1^2} = (8.988 \times 10^9\ N \cdot m^2/C^2)\frac{16.0 \times 10^{-9}\ C}{(9.00\ m)^2} = 1.775\ N/C$ and is in the +x-direction

$\Rightarrow E_{1x} = +1.775\ N/C$.

22-79 (cont)

The electric field $\vec{E}_3$ of q_3 at point P has magnitude

$$E_3 = \frac{1}{4\pi\epsilon_0}\frac{|q_3|}{r_3^2} = (8.988\times10^9\ \mathrm{N\cdot m^2/C^2})\frac{12.0\times10^{-9}\ \mathrm{C}}{(2.00\ \mathrm{m})^2} = 26.96\ \mathrm{N/C}$$ and is in the +x-direction

$\Rightarrow E_{3x} = +26.96$ N/C.

Thus $E_{2x} = E_x - E_{1x} - E_{3x} = +18.0\ \mathrm{N/C} - 1.775\ \mathrm{N/C} - 26.96\ \mathrm{N/C} = -10.74\ \mathrm{N/C}$

$\Rightarrow \vec{E}_2$ is in the −x-direction at P $\Rightarrow q_2$ is negative.

$$E_2 = \frac{1}{4\pi\epsilon_0}\frac{|q_2|}{r_2^2} \Rightarrow |q_2| = \frac{r_2^2 E_2}{(1/4\pi\epsilon_0)} = \frac{(6.00\ \mathrm{m})^2(10.74\ \mathrm{N/C})}{(8.988\times10^9\ \mathrm{N\cdot m^2/C^2})} = 4.30\times10^{-8}\ \mathrm{C}$$

$q_2 = -43.0$ nC

22-81

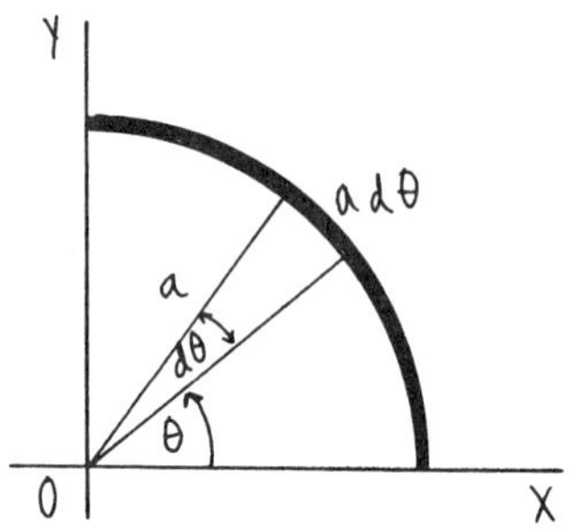

Divide the charge distribution into small segments, use the point charge formula for the electric field due to each small segment, and integrate over the charge distribution to find the x and y components of the total field.

A small segment that subtends angle $d\theta$ has length $a\,d\theta$ and contains charge $dQ = \left(\frac{a\,d\theta}{\frac{1}{2}\pi a}\right)Q = \frac{2Q}{\pi}d\theta$ ($\frac{1}{2}\pi a$ is the total length of the charge distribution).

The charge is negative, so the field at the origin is directed toward the small segment. The small segment is located at angle θ as shown in the sketch.

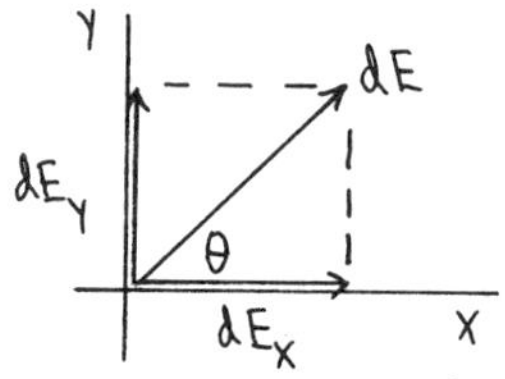

$$dE = \frac{1}{4\pi\epsilon_0}\frac{|dQ|}{r^2} = \frac{1}{4\pi\epsilon_0}\frac{|dQ|}{a^2} = \frac{Q}{2\pi^2\epsilon_0 a^2}d\theta$$

$$dE_x = dE\cos\theta = \frac{Q}{2\pi^2\epsilon_0 a^2}\cos\theta\,d\theta$$

$$dE_y = dE\sin\theta = \frac{Q}{2\pi^2\epsilon_0 a^2}\sin\theta\,d\theta$$

$$E_x = \int dE_x = \frac{Q}{2\pi^2\epsilon_0 a^2}\int_0^{\pi/2}\cos\theta\,d\theta = \frac{Q}{2\pi^2\epsilon_0 a^2}\left(\sin\theta\Big|_0^{\pi/2}\right) = \frac{Q}{2\pi^2\epsilon_0 a^2}$$

$$E_y = \int dE_y = \frac{Q}{2\pi^2\epsilon_0 a^2}\int_0^{\pi/2}\sin\theta\,d\theta = \frac{Q}{2\pi^2\epsilon_0 a^2}\left(-\cos\theta\Big|_0^{\pi/2}\right) = \frac{Q}{2\pi^2\epsilon_0 a^2}$$

(Note that $E_x = E_y$, as expected from symmetry.)

22-83

For an infinite charge sheet $E = \frac{|\sigma|}{2\epsilon_0}$, the same at all points, and is directed away from the sheet if it has positive charge and toward the sheet if it has negative charge.

Obtain the net electric field as the vector sum of the fields due to each sheet.

22-83 (cont)

a)

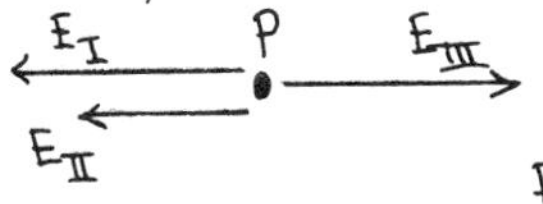

$$E = E_I + E_{II} - E_{III} = \frac{|\sigma_I| + |\sigma_{II}| - |\sigma_{III}|}{2\epsilon_0} = \frac{+0.0200\,C/m^2 + 0.0100\,C/m^2 - 0.0200\,C/m^2}{2\epsilon_0}$$

$$E = \frac{+0.0100\,C/m^2}{2(8.854\times10^{-12}\,C^2/N\cdot m^2)} = \underline{5.65\times10^8\,N/C}, \ \underline{\text{to left}}$$

b)

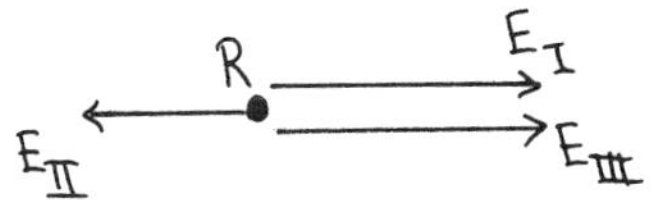

$$E = E_I + E_{III} - E_{II} = \frac{|\sigma_I| + |\sigma_{III}| - |\sigma_{II}|}{2\epsilon_0}$$

$$E = \frac{0.0200\,C/m^2 + 0.0200\,C/m^2 - 0.0100\,C/m^2}{2\epsilon_0} = \frac{0.0300\,C/m^2}{2(8.854\times10^{-12}\,C^2/N\cdot m^2)}$$

$$E = \underline{1.69\times10^9\,N/C}, \ \underline{\text{to right}}$$

c)

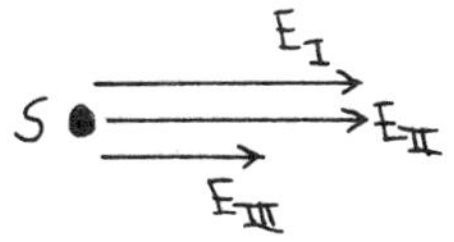

$$E = E_I + E_{II} + E_{III} = \frac{|\sigma_I| + |\sigma_{II}| + |\sigma_{III}|}{2\epsilon_0}$$

$$E = \frac{0.0200\,C/m^2 + 0.0200\,C/m^2 + 0.0100\,C/m^2}{2\epsilon_0} = \frac{0.0500\,C/m^2}{2(8.854\times10^{-12}\,C^2/N\cdot m^2)}$$

$$E = \underline{2.82\times10^9\,N/C}, \ \underline{\text{to right}}$$

d)

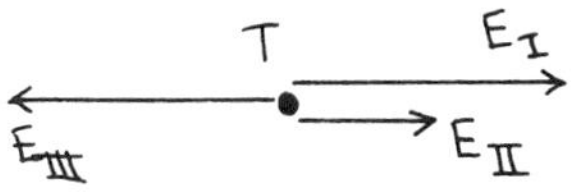

$E = E_I + E_{II} - E_{III}$, the same as at point P

$E = \underline{5.65\times10^8\,N/C}$, $\underline{\text{to right}}$

22-85

The electric field due to the first sheet, which is in the xy-plane, is $\vec{E}_1 = \frac{\sigma}{2\epsilon_0}\hat{k}$ for $z>0$ and $\vec{E}_1 = -\frac{\sigma}{2\epsilon_0}\hat{k}$ for $z<0$.

We can write this as $\vec{E}_1 = \frac{\sigma}{2\epsilon_0}\frac{|z|}{z}\hat{k}$, since $\frac{|z|}{z} = +1$ for $z>0$ and $\frac{|z|}{z} = \frac{-z}{z} = -1$ for $z<0$.

Similarly, we can write the electric field due to the second sheet as $\vec{E}_2 = -\frac{\sigma}{2\epsilon_0}\frac{|x|}{x}\hat{\imath}$, since the charge density is $-\sigma$.

The net field is $\vec{E} = \vec{E}_1 + \vec{E}_2 = \frac{\sigma}{2\epsilon_0}\left(-\frac{|x|}{x}\hat{\imath} + \frac{|z|}{z}\hat{k}\right)$.

CHAPTER 23

Exercises 1, 3, 5, 9, 13, 17

Problems 21, 25, 29, 35, 37, 39, 41, 43, 45, 47

Exercises

23-1

a) $\Phi_E = \int E \cos\phi \, dA$, where ϕ is the angle between the normal to the sheet $\hat{n}$ and the electric field $\vec{E}$. In this problem E and $\cos\phi$ are constant over the surface so $\Phi_E = E\cos\phi \int dA = E\cos\phi\, A = (12.0 \text{ N/C})(\cos 70°)(0.500 \text{ m}^2) = \underline{2.05 \text{ N}\cdot\text{m}^2/\text{C}}$.

b) Φ_E is independent of the shape of the sheet as long as ϕ and E are constant at all points on the sheet.

c) $\Phi_E = E\cos\phi\, A$. Φ_E is largest for $\phi = 0°$, so $\cos\phi = 1$ and $\Phi_E = EA$.

d) Φ_E is smallest for $\phi = 90°$, so $\cos\phi = 0$ and $\Phi_E = 0$.

23-3

a)

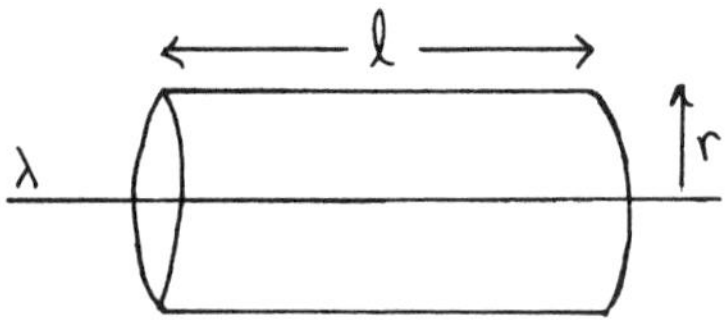

end view

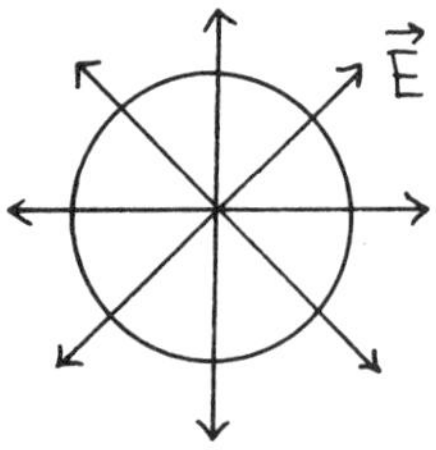

The area of the curved part of the cylinder is $A = 2\pi r l$.

The electric field is parallel to the end caps of the cylinder, so $\vec{E}\cdot\vec{A} = 0$ for the ends and the flux through the cylinder end caps is zero.

The electric field is normal to the curved surface of the cylinder and has the same magnitude $E = \frac{\lambda}{2\pi\epsilon_0 r}$ at all points on this surface. Thus $\phi = 0°$ and

$$\Phi_E = EA\cos\phi = EA = \left(\frac{\lambda}{2\pi\epsilon_0 r}\right)(2\pi r l) = \frac{\lambda l}{\epsilon_0} = \frac{(5.00\times10^{-6} \text{ C/m})(0.300 \text{ m})}{8.854\times10^{-12} \text{ C}^2/\text{N}\cdot\text{m}^2} = \underline{1.69\times10^5 \text{ N}\cdot\text{m}^2/\text{C}}$$

b) In the calculation in part (a) the radius r of the cylinder divided out, so the flux is the same, $\Phi_E = \underline{1.69\times10^5 \text{ N}\cdot\text{m}^2/\text{C}}$.

c) $\Phi_E = \frac{\lambda l}{\epsilon_0} = \frac{(5.00\times10^{-6} \text{ C/m})(0.900 \text{ m})}{8.854\times10^{-12} \text{ C}^2/\text{N}\cdot\text{m}^2} = 5.08\times10^5 \text{ N}\cdot\text{m}^2/\text{C}$ (three times the flux calculated in parts (b) and (c))

23-5

a) $\vec{E} = E_x\hat{i} + E_y\hat{j}$ with $E_x = 2.50$ N/C and $E_y = -4.20$ N/C. $A = L^2 = 0.0400 \text{ m}^2$

face S_1

$\hat{n} = -\hat{j}$

23-5 (cont)

$\Phi_E = \vec{E}\cdot\vec{A} = \vec{E}\cdot(A\hat{n}) = (E_x\hat{\imath} + E_y\hat{\jmath})\cdot(-A\hat{\jmath}) = -E_yA = -(-4.20\text{ N/C})(0.0400\text{ m}^2)$

$\Phi_E = \underline{0.168\text{ N}\cdot\text{m}^2/\text{C}}$

face S_2

$\hat{n} = +\hat{k}$

$\Phi_E = \vec{E}\cdot\vec{A} = \vec{E}\cdot(A\hat{n}) = (E_x\hat{\imath} + E_y\hat{\jmath})\cdot(A\hat{k}) = 0$

face S_3

$\hat{n} = +\hat{\jmath}$

$\Phi_E = \vec{E}\cdot\vec{A} = \vec{E}\cdot(A\hat{n}) = (E_x\hat{\imath} + E_y\hat{\jmath})\cdot(A\hat{\jmath}) = E_yA = (-4.20\text{ N/C})(0.0400\text{ m}^2) = \underline{-0.168\text{ N}\cdot\text{m}^2/\text{C}}$

face S_4

$\hat{n} = -\hat{k}$

$\Phi_E = \vec{E}\cdot\vec{A} = \vec{E}\cdot A\hat{n} = (E_x\hat{\imath} + E_y\hat{\jmath})\cdot(-A\hat{k}) = 0$

face S_5

$\hat{n} = +\hat{\imath}$

$\Phi_E = \vec{E}\cdot\vec{A} = \vec{E}\cdot A\hat{n} = (E_x\hat{\imath} + E_y\hat{\jmath})\cdot(A\hat{\imath}) = E_xA = (2.50\text{ N/C})(0.0400\text{ m}^2) = \underline{0.100\text{ N}\cdot\text{m}^2/\text{C}}$

face S_6

$\hat{n} = -\hat{\imath}$

$\Phi_E = \vec{E}\cdot\vec{A} = \vec{E}\cdot A\hat{n} = (E_x\hat{\imath} + E_y\hat{\jmath})\cdot(-A\hat{\imath}) = -E_xA = -(2.50\text{ N/C})(0.0400\text{ m}^2) = \underline{-0.100\text{ N}\cdot\text{m}^2/\text{C}}$

b) Add the flux through each of the six faces:

$\Phi_E = 0.168\text{ N}\cdot\text{m}^2/\text{C} + 0 - 0.168\text{ N}\cdot\text{m}^2/\text{C} + 0 + 0.100\text{ N}\cdot\text{m}^2/\text{C} - 0.100\text{ N}\cdot\text{m}^2/\text{C} = 0$

The total electric flux through all sides is zero.

23-9

It is rather difficult to calculate the flux directly from $\Phi_E = \oint \vec{E}\cdot d\vec{A}$ since the magnitude of $\vec{E}$ and its angle with $d\vec{A}$ varies over the surface of the cube. A much easier approach is to use Gauss's law to calculate the total flux through the cube:

$$\Phi_E = \frac{Q_{encl}}{\epsilon_0} = \frac{3.60\times10^{-9}\text{ C}}{8.854\times10^{-12}\text{ C}^2/\text{N}\cdot\text{m}^2} = 406.6\text{ N}\cdot\text{m}^2/\text{C}$$

By symmetry the flux is the same through each of the six faces, so the flux through one face is $\frac{1}{6}(406.6\text{ N}\cdot\text{m}^2/\text{C}) = \underline{67.8\text{ N}\cdot\text{m}^2/\text{C}}$.

(Note that the size of the cube did not enter into the calculations.)

23-13

Example 23-5 derived that the electric field just outside the surface of a spherical conductor that has net charge q is $E = \frac{1}{4\pi\epsilon_0}\frac{q}{R^2}$.

$$q = \frac{R^2E}{(1/4\pi\epsilon_0)} = \frac{(0.110\text{ m})^2(1400\text{ N/C})}{8.988\times10^9\text{ N}\cdot\text{m}^2/\text{C}^2} = 1.885\times10^{-9}\text{ C}$$

23-13 (cont)

Each electron has a charge of magnitude $e = 1.602\times10^{-19}$ C, so the number of excess electrons needed is $\dfrac{1.885\times10^{-9}\text{ C}}{1.602\times10^{-19}\text{ C}} = \underline{1.18\times10^{10}}$.

23-17

a) Consider the charge on a length l of the cylinder. This can be expressed as $q = \lambda l$. But since the surface area is $2\pi R l$ it can also be expressed as $q = \sigma 2\pi R l$. These two expressions must be equal, so

$\lambda l = \sigma 2\pi R l$ and $\underline{\lambda = 2\pi R \sigma}$.

b)

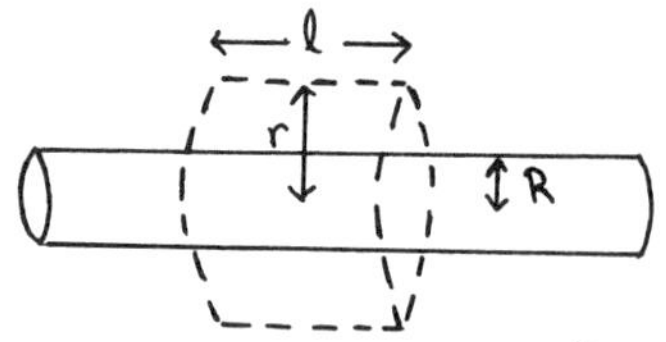

Apply Gauss's law to a Gaussian surface that is a cylinder of length l, radius r and whose axis coincides with the axis of the charge distribution.

$$Q_{encl} = \sigma(2\pi R l), \quad \Phi_E = 2\pi r l E$$

$$\Phi_E = \frac{Q_{encl}}{\epsilon_0} \Rightarrow 2\pi r l E = \frac{\sigma(2\pi R l)}{\epsilon_0}$$

$$\underline{E = \frac{\sigma R}{\epsilon_0 r}}$$

Example 23-6 shows that the electric field of an infinite line of charge is $E = \dfrac{\lambda}{2\pi\epsilon_0 r}$.

c) $\sigma = \dfrac{\lambda}{2\pi R}$, so $E = \dfrac{\sigma R}{\epsilon_0 r} = \dfrac{R}{\epsilon_0 r}\left(\dfrac{\lambda}{2\pi R}\right) = \underline{\dfrac{\lambda}{2\pi\epsilon_0 r}}$, the same as for an infinite line of charge that is along the axis of the cylinder.

Problems

23-21

a) Find the net flux through the parallelepiped surface, and then use that in Gauss's law to find the net charge within. Flux out of the surface is positive and flux into the surface is negative.

$\vec{E}_1$ gives flux out of the surface:

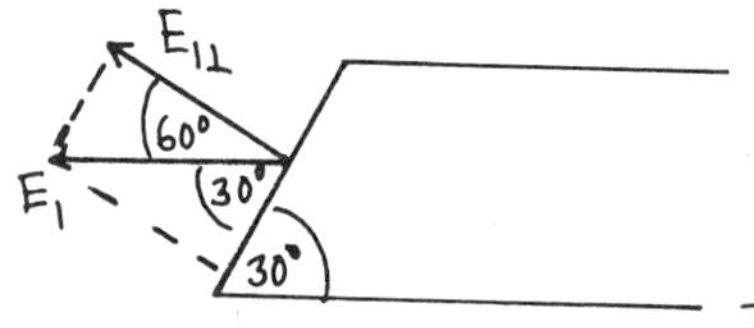

$$\Phi_{E_1} = +E_{1\perp} A$$

$$A = (0.0600\text{ m})(0.0500\text{ m}) = 3.00\times10^{-3}\text{ m}^2$$

$$E_{1\perp} = E_1 \cos 60° = (3.50\times10^4\text{ N/C})\cos 60° = 1.75\times10^4\text{ N/C}$$

$$\Phi_{E_1} = +E_{1\perp} A = +(1.75\times10^4\text{ N/C})(3.00\times10^{-3}\text{ m}^2) = +52.5\text{ N}\cdot\text{m}^2/\text{C}$$

$\vec{E}_2$ gives (negative) flux into the surface:

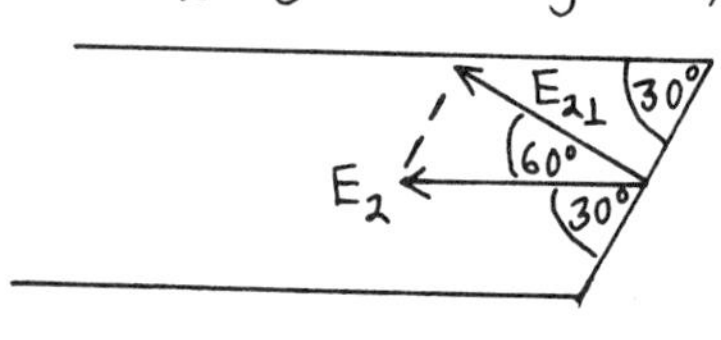

$$\Phi_{E_2} = -E_{2\perp} A$$

$$E_{2\perp} = E_2 \cos 60° = (6.00\times10^4\text{ N/C})\cos 60° = 3.00\times10^4\text{ N/C}$$

$$\Phi_{E_2} = -(3.00\times10^4\text{ N/C})(3.00\times10^{-3}\text{ m}^2) = -90.0\text{ N}\cdot\text{m}^2/\text{C}$$

23-21 (cont)

The net flux is $\Phi_E = \Phi_{E_1} + \Phi_{E_2} = +52.5\ N\cdot m^2/C - 90.0\ N\cdot m^2/C = -37.5\ N\cdot m^2/C$.
The net flux is negative (inward), so the net charge enclosed is negative.
Apply Gauss's law: $\Phi_E = \frac{Q_{encl}}{\epsilon_0}$

$$Q_{encl} = \Phi_E \epsilon_0 = (-37.5\ N\cdot m^2/C)(8.854\times10^{-12}\ C^2/N\cdot m^2) = \underline{-3.32\times10^{-10}\ C}$$

b) If there were no charge within the parallelpiped the net flux would be zero. This is not the case, so there is charge inside. The electric field lines that pass out through the surface of the parallelpiped must terminate on charges, so there must be charges outside the parallelpiped.

23-25

Find the electric field $\vec{E}$ produced by the shell and then use $\vec{F} = q\vec{E}$ to find the force the shell exerts on the point charge.

a)

Apply Gauss's law to a spherical Gaussian surface that has radius $r > R$ and that is concentric with the shell.

$\Phi_E = E(4\pi r^2)$, $Q_{encl} = Q$

$$\Phi_E = \frac{Q_{encl}}{\epsilon_0} \Rightarrow E(4\pi r^2) = \frac{Q}{\epsilon_0}$$

$$E = \frac{Q}{4\pi\epsilon_0 r^2}\text{, radially outward}$$

Then $F = qE = \frac{qQ}{4\pi\epsilon_0 r^2}$, radially outward. (Since q is positive $\vec{E}$ and $\vec{F}$ are in the same direction.)

b)

Apply Gauss's law to a spherical Gaussian surface with radius $r < R$ and concentric with the shell.

$\Phi_E = E(4\pi r^2)$, $Q_{encl} = 0$

$$\Phi_E = \frac{Q_{encl}}{\epsilon_0} \Rightarrow E(4\pi r^2) = 0,\quad E = 0 \text{ so } F = 0.$$

23-29

a) (i) $r < a$

Apply Gauss's law to a spherical Gaussian surface with radius $r < a$.

$\Phi_E = EA = E(4\pi r^2)$

$Q_{encl} = 0$; no charge is enclosed

$$\Phi_E = \frac{Q_{encl}}{\epsilon_0} \Rightarrow E(4\pi r^2) = 0 \text{ and } E = 0$$

(ii) $a < r < b$

Points in this region are in the conductor of the small shell $\Rightarrow E = 0$.

23-29 (cont)

(iii) $b < r < c$

Apply Gauss's law to a spherical Gaussian surface with radius $b < r < c$.

$\Phi_E = EA = E(4\pi r^2)$

The Gaussian surface encloses all of the small shell and none of the larger shell $\Rightarrow Q_{encl} = +2q$.

$$\Phi_E = \frac{Q_{encl}}{\epsilon_0} \Rightarrow E(4\pi r^2) = \frac{+2q}{\epsilon_0}$$

$$E = \frac{2q}{4\pi\epsilon_0 r^2}$$

(iv) $c < r < d$

Points in this region are in the conductor of the larger shell $\Rightarrow E = 0$.

(v) $r > d$

Apply Gauss's law to a spherical Gaussian surface with radius $r > d$.

$\Phi_E = EA = E(4\pi r^2)$

The Gaussian surface encloses all of the small shell and all of the larger shell $\Rightarrow Q_{encl} = 2q + 4q = 6q$.

$$\Phi_E = \frac{Q_{encl}}{\epsilon_0} \Rightarrow E(4\pi r^2) = \frac{6q}{\epsilon_0}$$

$$E = \frac{6q}{4\pi\epsilon_0 r^2}$$

b) (i) charge on inner surface of the small shell

Apply Gauss's law to a spherical Gaussian surface with radius $a < r < b$. This surface lies within the conductor of the small shell, where $E = 0$, so $\Phi_E = 0$. Thus by Gauss's law $Q_{encl} = 0$, so there is zero charge on the inner surface of the small shell.

(ii) charge on outer surface of the small shell

The total charge on the small shell is $+2q$. We found in part (i) that there is zero charge on the inner surface of the shell, so all $+2q$ must reside on the outer surface.

(iii) charge on inner surface of large shell

Apply Gauss's law to a spherical Gaussian surface with radius $c < r < d$. This surface lies within the conductor of the large shell, where $E = 0$, so $\Phi_E = 0$. Thus by Gauss's law $Q_{encl} = 0$. The surface encloses the $+2q$ on the small shell so there must be charge $-2q$ on the inner surface of the larger shell, to make the total enclosed charge zero.

(iv) charge on outer surface of large shell

The total charge on the larger shell is $+4q$. We showed in part (iii) that the charge on the inner surface is $-2q$, so there must be $+6q$ on the outer surface.

23-35

a)

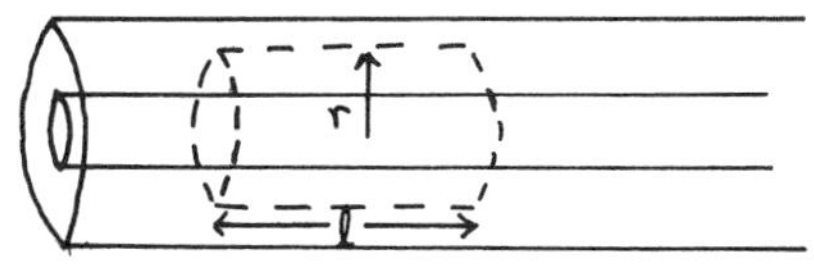

Apply Gauss's law to a Gaussian cylinder of length l and radius r, where $a<r<b$.

$$\Phi_E = \frac{Q_{encl}}{\epsilon_0}$$

$\Phi_E = E(2\pi r l)$

$Q_{encl} = \lambda l$ (the charge on the length l of the inner conductor that is inside the Gaussian surface)

$$\Rightarrow E(2\pi r l) = \frac{\lambda l}{\epsilon_0}$$

$$E = \frac{\lambda}{2\pi\epsilon_0 r}$$

b)

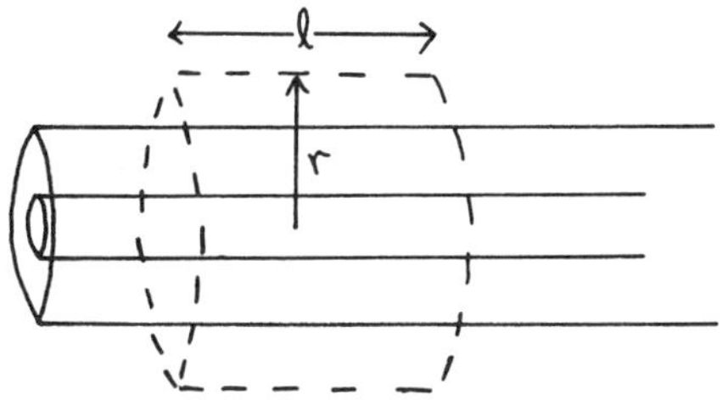

Apply Gauss's law to a Gaussian cylinder of length l and radius r, where $r>c$:

$$\Phi_E = \frac{Q_{encl}}{\epsilon_0}$$

$\Phi_E = E(2\pi r l)$

$Q_{encl} = \lambda l$ (the charge on the length l of the inner conductor that is inside the Gaussian surface; the outer conductor carries no net charge)

$$\Rightarrow E(2\pi r l) = \frac{\lambda l}{\epsilon_0}$$

$$E = \frac{\lambda}{2\pi\epsilon_0 r}$$

c) $E=0$ within a conductor.

Thus $E=0$ for $r<a$; $E=\frac{\lambda}{2\pi\epsilon_0 r}$ for $a<r<b$; $E=0$ for $b<r<c$; $E=\frac{\lambda}{2\pi\epsilon_0 r}$ for $r>c$.

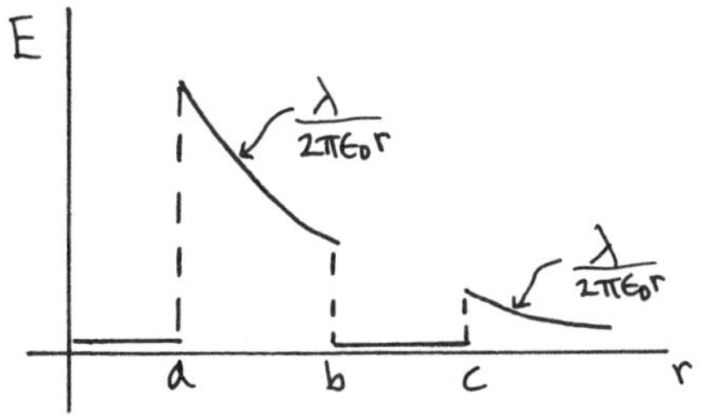

d) inner surface

Apply Gauss's law to a Gaussian cylinder with radius r, where $b<r<c$. This surface lies within the conductor of the outer cylinder, where $E=0$, so $\Phi_E=0$. Thus by Gauss's law $Q_{encl}=0$. The surface encloses charge λl on the inner conductor, so it must enclose charge $-\lambda l$ on the inner surface of the outer conductor. The charge per unit length on the inner surface of the outer cylinder is $-\lambda$.

outer surface

The outer cylinder carries no net charge. So if there is charge per unit length $-\lambda$ on its inner surface there must be charge per unit length $+\lambda$ on the outer surface.

23-37

a)

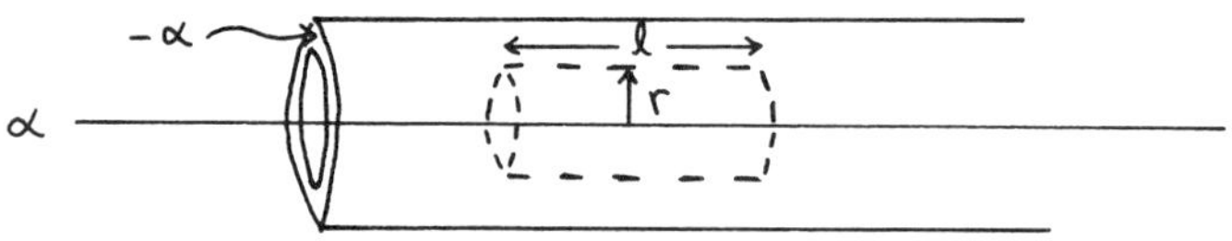

(i) $r<a$ Apply Gauss's law to a cylindrical Gaussian surface of length l and radius r, where $r<a$. $\Phi_E = \frac{Q_{encl}}{\epsilon_0}$

$$\Phi = E(2\pi r l)$$

$Q_{encl} = \alpha l$ (the charge on the length l of the line of charge)

$$\Rightarrow E(2\pi r l) = \frac{\alpha l}{\epsilon_0}$$

$$E = \frac{\alpha}{2\pi\epsilon_0 r}$$

(ii) $a<r<b$

Points in this region are within the conducting tube, so $E=0$.

(iii) $r>b$

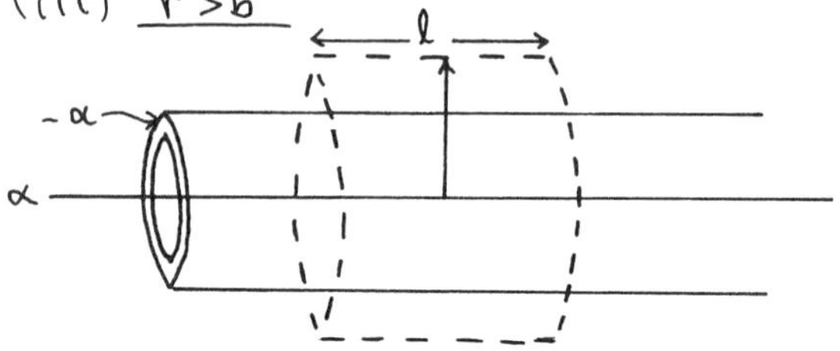

Apply Gauss's law to a cylindrical Gaussian surface of length l and radius r, where $r>b$:

$$\Phi_E = \frac{Q_{encl}}{\epsilon_0}$$

$$\Phi_E = E(2\pi r l)$$

$Q_{encl} = \alpha l$ (charge on length l of the line of charge) $- \alpha l$ (charge on length l of tube)

$\Rightarrow Q_{encl} = 0$

Thus $E(2\pi r l) = 0 \Rightarrow E = 0$.

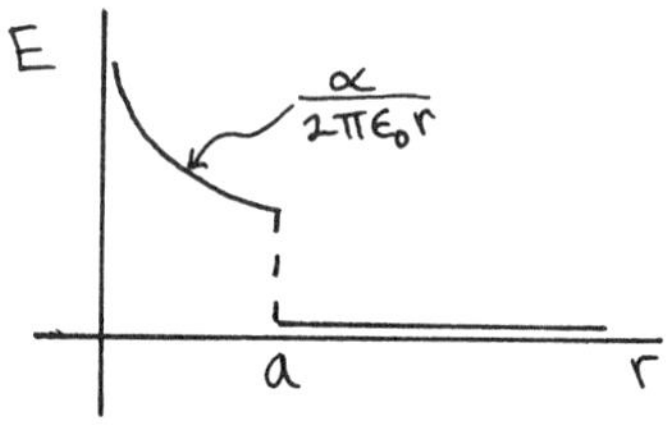

b) (i) inner surface

Apply Gauss's law to a cylindrical Gaussian surface of length l and radius r, where $a<r<b$. This surface lies within the conductor of the tube, where $E=0$, so $\Phi_E = 0$. Then by Gauss's law $Q_{encl}=0$. The surface encloses charge αl on the line of charge so must enclose charge $-\alpha l$ on the inner surface of the tube. The charge per unit length on the inner surface of the tube is $-\alpha$.

(ii) outer surface

The net charge per unit length on the tube is $-\alpha$. We have shown in part (i) that this must all reside on the inner surface, so there is no net charge on the outer surface of the tube.

23-39

Example 23-9 shows that the electric field of a uniformly charged sphere is $E = \frac{Qr}{4\pi\epsilon_0 R^3}$ for $r<R$ and $E = \frac{Q}{4\pi\epsilon_0 r^2}$ for $r>R$.

The force on the electron has magnitude $F = qE = eE$. The acceleration has magnitude $a = \frac{F}{m} = \frac{eE}{m}$, where $\frac{e}{m} = \frac{1.60\times10^{-19}\,C}{9.109\times10^{-31}\,kg} = 1.759\times10^{11}\,C/kg$.

a) $r = 2R$

$$E = \frac{Q}{4\pi\epsilon_0 r^2} = \frac{Q}{4\pi\epsilon_0 (2R)^2} = \frac{1}{4}\left(\frac{Q}{4\pi\epsilon_0 R^2}\right) = \frac{1}{4}\left(\frac{26e}{4\pi\epsilon_0 R^2}\right) = \frac{1}{4}\frac{(8.988\times10^9\,N\cdot m^2/C^2)(26)(1.602\times10^{-19}\,C)}{(4.6\times10^{-15}\,m)^2}$$

$$E = \frac{1}{4}(1.769\times10^{21}\,N/C) = 4.423\times10^{20}\,N/C$$

$$a = \frac{e}{m}E = (1.759\times10^{11}\,C/kg)(4.423\times10^{20}\,N/C) = \underline{7.78\times10^{31}\,m/s^2}$$

b) $r = R$

$$E = \frac{Q}{4\pi\epsilon_0 r^2} = \frac{Q}{4\pi\epsilon_0 R^2} = 1.769\times10^{21}\,N/C$$

$$a = \frac{e}{m}E = (1.759\times10^{11}\,C/kg)(1.769\times10^{21}\,N/C) = \underline{3.11\times10^{32}\,m/s^2}$$ (4 times larger than for $r=2R$)

c) $r = \frac{R}{2}$

$$E = \frac{Qr}{4\pi\epsilon_0 R^3} = \frac{Q\left(\frac{R}{2}\right)}{4\pi\epsilon_0 R^3} = \frac{1}{2}\left(\frac{Q}{4\pi\epsilon_0 R^2}\right) = \frac{1}{2}(1.769\times10^{21}\,N/C) = 8.795\times10^{20}\,N/C$$

$$a = \frac{e}{m}E = (1.759\times10^{11}\,C/kg)(8.795\times10^{20}\,N/C) = \underline{1.55\times10^{22}\,m/s^2}$$ ($\frac{1}{2}$ the value at $r=R$)

23-41

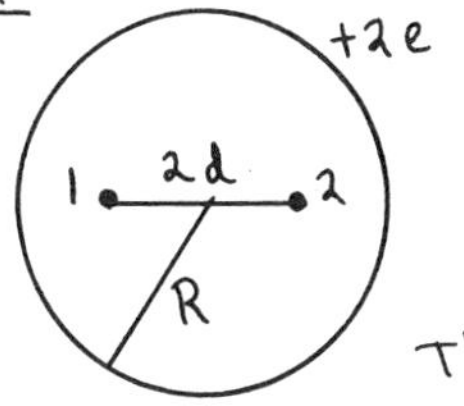

If the electrons are in equilibrium the net force on each one is zero.

Consider the forces on electron 2.
There is a repulsive force F_1 due to the other electron, electron 1.

$$F_1 = \frac{1}{4\pi\epsilon_0}\frac{e^2}{(2d)^2}$$

The electric field inside the uniform distribution of positive charge is $E = \frac{Qr}{4\pi\epsilon_0 R^3}$ (Example 23-9), where $Q = +2e$. At the position of electron 2, $r=d$. The force F_{cd} exerted by the positive charge distribution is $F_{cd} = eE = \frac{e(2e)d}{4\pi\epsilon_0 R^3}$ and is attractive.

The force diagram for electron 2 is

Net force equals zero $\Rightarrow F_1 = F_{cd}$

$$\frac{1}{4\pi\epsilon_0}\frac{e^2}{4d^2} = \frac{2e^2 d}{4\pi\epsilon_0 R^3}$$

$$\frac{1}{4d^2} = \frac{2d}{R^3}$$

$$d^3 = \frac{R^3}{8} \text{ and } \underline{d = R/2}$$

23-43

$\rho(r) = \rho_0\left(1-\frac{r}{R}\right)$ for $r \le R$ where $\rho_0 = \frac{3Q}{\pi R^3}$
$\rho(r) = 0$ for $r \ge R$

a) The charge density varies with r inside the spherical volume.

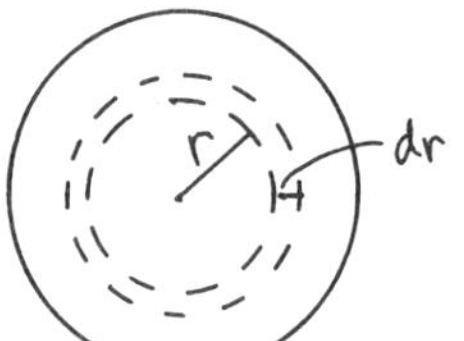

Divide the volume up into thin concentric shells, of radius r and thickness dr. The volume of such a shell is $dV = 4\pi r^2\,dr$.
The charge contained within the shell is

$$dq = \rho(r)\,dV = 4\pi r^2 \rho_0\left(1-\frac{r}{R}\right)dr$$

The total charge Q in the charge distribution is obtained by integrating dq over all such shells into which the sphere can be subdivided:

$$Q = \int dq = \int_0^R 4\pi r^2 \rho_0\left(1-\frac{r}{R}\right)dr = 4\pi\rho_0\int_0^R\left(r^2 - \frac{1}{R}r^3\right)dr = 4\pi\rho_0\left(\frac{r^3}{3} - \frac{r^4}{4R}\right)\Big|_0^R$$

$$Q = 4\pi\rho_0\left(\frac{R^3}{3} - \frac{R^4}{4R}\right) = 4\pi\rho_0\left(\frac{R^3}{12}\right) = 4\pi\left(\frac{3Q}{\pi R^3}\right)\left(\frac{R^3}{12}\right) = Q,$$ as was to be shown.

b) Apply Gauss's law to a spherical surface of radius r, where $r > R$:

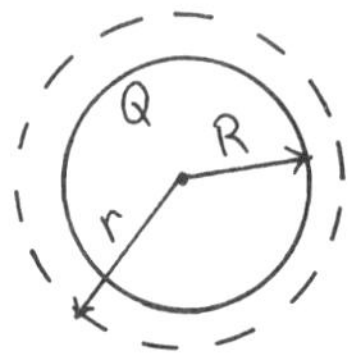

$$\Phi_E = \frac{Q_{encl}}{\epsilon_0}$$

$$E(4\pi r^2) = \frac{Q}{\epsilon_0}$$

$$E = \frac{Q}{4\pi\epsilon_0 r^2}$$; same as for point charge of charge Q.

c) Apply Gauss's law to a spherical surface of radius r, where $r < R$:

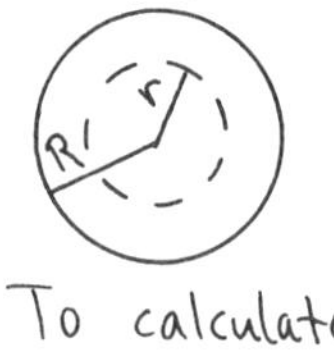

$$\Phi_E = \frac{Q_{encl}}{\epsilon_0}$$

$$\Phi_E = E(4\pi r^2)$$

To calculate the enclosed charge Q_{encl} use the same technique as in part (a), except integrate dq out to r rather than R. (We want only that charge inside radius r.)

$$Q_{encl} = \int_0^r 4\pi r'^2 \rho_0\left(1-\frac{r'}{R}\right)dr' = 4\pi\rho_0\int_0^r\left(r'^2 - \frac{r'^3}{R}\right)dr' = 4\pi\rho_0\left(\frac{r'^3}{3} - \frac{r'^4}{4R}\right)\Big|_0^r$$

$$Q_{encl} = 4\pi\rho_0\left(\frac{r^3}{3} - \frac{r^4}{4R}\right) = 4\pi\rho_0 r^3\left(\frac{1}{3} - \frac{1}{4}\frac{r}{R}\right)$$

$$\rho_0 = \frac{3Q}{\pi R^3} \Rightarrow Q_{encl} = 12Q\frac{r^3}{R^3}\left(\frac{1}{3} - \frac{1}{4}\frac{r}{R}\right) = Q\left(\frac{r^3}{R^3}\right)\left(4 - 3\frac{r}{R}\right)$$

Thus $E(4\pi r^2) = \frac{Q}{\epsilon_0}\left(\frac{r^3}{R^3}\right)\left(4 - 3\frac{r}{R}\right)$

$$E = \frac{Qr}{4\pi\epsilon_0 R^3}\left(4 - 3\frac{r}{R}\right),\quad r \le R$$

d) With $r = R$ in the $r \ge R$ expression (part (b)):

$$E(R) = \frac{Q}{4\pi\epsilon_0 R^2}$$

With $r = R$ in the $r \le R$ expression (part (c)):

$$E(R) = \frac{Q}{4\pi\epsilon_0 R^2}(4-3) = \frac{Q}{4\pi\epsilon_0 R^2}$$

So these two results agree at the surface of the sphere.

23-45

a)

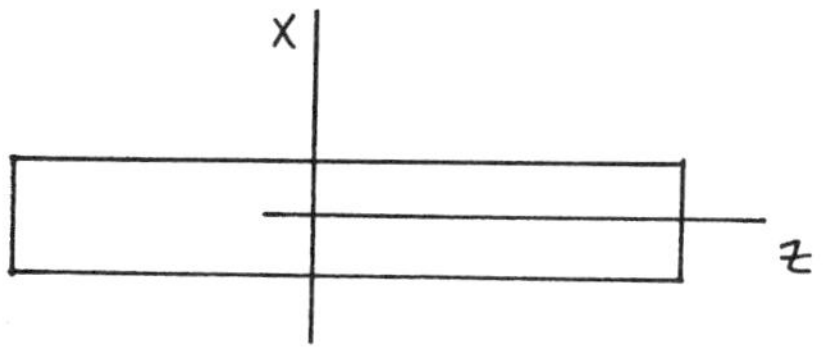

$\rho(x) = \rho_0 \left(\frac{x}{d}\right)^2$

The charge distribution is symmetric about $x=0$, so by symmetry $E(x) = E(-x)$. But for $x>0$ the field is in the $+x$ direction and for $x<0$ the field is in the $-x$ direction. At $x=0$ the field can't be both in the $+x$ and $-x$ directions so must be zero. That is, $E_x(x) = -E_x(-x)$. At $x=0$ this gives $E_x(0) = -E_x(0)$ and this equation is satisfied only for $E_x(0) = 0$.

b) $|x| > d$ (outside the slab)

Apply Gauss's law to a cylindrical Gaussian surface whose axis is perpendicular to the slab and whose end caps have area A and are the same distance $|x| > d$ from $x=0$.

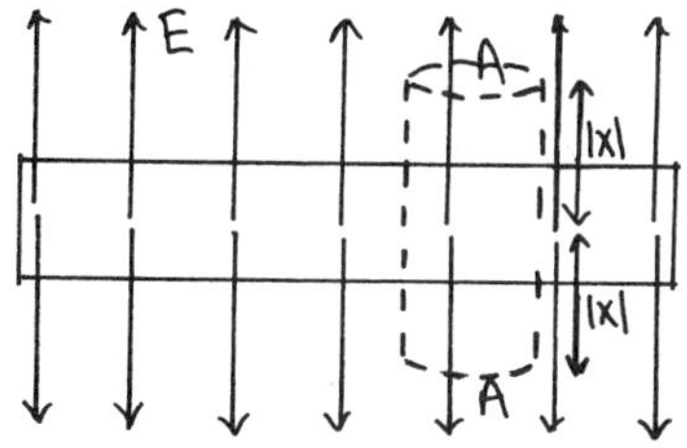

$\Phi_E = 2EA$

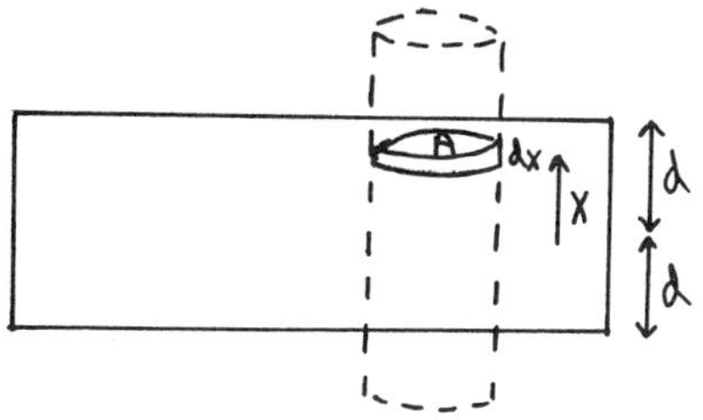

To find Q_{encl} consider a thin disk at coordinate x and with thickness dx. The charge within this disk is

$dq = \rho\, dV = \rho A\, dx = \frac{\rho_0 A}{d^2} x^2\, dx$

The total charge enclosed by the Gaussian cylinder is

$Q_{encl} = 2\int_0^d dq = \frac{2\rho_0 A}{d^2}\int_0^d x^2\, dx = \frac{2\rho_0 A}{d^2}\frac{d^3}{3} = \frac{2}{3}\rho_0 A d$

Then $\Phi_E = \frac{Q_{encl}}{\epsilon_0}$ gives $2EA = \frac{2\rho_0 A d}{3\epsilon_0}$

$E = \frac{\rho_0 d}{3\epsilon_0}$

$\vec{E}$ is directed away from $x=0$, so $\vec{E} = \frac{\rho_0 d}{3\epsilon_0}\frac{x}{|x|}\hat{\imath}$.

$|x| < d$ (inside the slab)

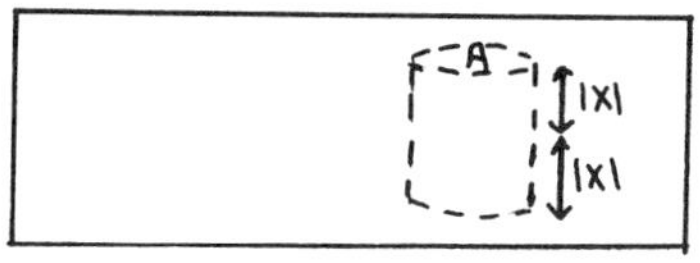

Apply Gauss's law to a cylindrical Gaussian surface whose axis is perpendicular to the slab and whose end caps have area A and are the same distance $|x| < d$ from $x=0$.

23-45 (cont)

$\Phi_E = 2EA$

Q_{encl} is found as above, but now the integral on dx is only from 0 to x instead of 0 to d.

$Q_{encl} = 2\int_0^x dq = \frac{2\rho_0 A}{d^2}\int_0^x x^2\,dx = \frac{2\rho_0 A}{d^2}\frac{x^3}{3}$

Then $\Phi_E = \frac{Q_{encl}}{\epsilon_0}$ gives $2EA = \frac{2\rho_0 A x^3}{3d^2}$

$E = \frac{\rho_0 x^3}{3\epsilon_0 d^2}$

$\vec{E}$ is directed away from $x=0$, so $\vec{E} = \frac{\rho_0 x^3}{3\epsilon_0 d^2}\frac{x}{|x|}\hat{\imath}$.

Note that $E=0$ at $x=0$ as noted in part (a).
Note also that the expressions for $|x|>d$ and $|x|<d$ agree for $x=d$.

23-47

a) For an insulating sphere of uniform charge density ρ and centered at the origin, the electric field inside the sphere is given by $E = \frac{Qr}{4\pi\epsilon_0 R^3}$ (Example 23-9), where $\vec{r}$ is the vector from the center of the sphere to the point where E is calculated.

But $\rho = \frac{3Q}{4\pi R^3}$ so this may be written as $E = \frac{\rho r}{3\epsilon_0}$. And $\vec{E}$ is radially outward, in the direction of $\vec{r}$, so $\vec{E} = \frac{\rho\vec{r}}{3\epsilon_0}$.

For a sphere centered at $\vec{r}=\vec{b}$, a point inside the sphere located by $\vec{r}$ is located by the vector $\vec{r}-\vec{b}$ relative to the center of the sphere.

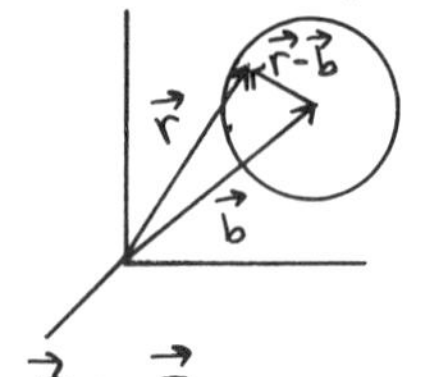

Thus $\vec{E} = \frac{\rho(\vec{r}-\vec{b})}{3\epsilon_0}$

b) $\vec{E} = \vec{E}_{uniform} + \vec{E}_{hole}$, where $\vec{E}_{uniform}$ is the field of a uniformly charged sphere with charge density ρ and $\vec{E}_{hole}$ is the field of a sphere located at the hole and with charge density $-\rho$. (Within the spherical hole the net charge density is $+\rho-\rho=0$.)

$\vec{E}_{uniform} = \frac{\rho\vec{r}}{3\epsilon_0}$, where $\vec{r}$ is a vector from the center of the sphere.

$\vec{E}_{hole} = \frac{-\rho(\vec{r}-\vec{b})}{3\epsilon_0}$, at points inside the hole.

Then $\vec{E} = \frac{\rho\vec{r}}{3\epsilon_0} + \left(\frac{-\rho(\vec{r}-\vec{b})}{3\epsilon_0}\right) = \frac{\rho}{3\epsilon_0}(\vec{r}-(\vec{r}-\vec{b})) = \frac{\rho\vec{b}}{3\epsilon_0}$.

$\vec{E}$ is independent of $\vec{r}$ so is uniform inside the hole. The direction of $\vec{E}$ inside the hole is in the direction of the vector $\vec{b}$, the direction from the center of the insulating sphere to the center of the hole.

CHAPTER 24

Exercises 1, 3, 5, 7, 13, 15, 17, 25, 27, 29, 35, 37, 39

Problems 51, 53, 57, 59, 63, 65, 67, 71, 73, 77, 79, 81

Exercises

24-1

a) $U = \frac{1}{4\pi\epsilon_0}\frac{qq'}{r} = (8.988\times10^9\ \text{N}\cdot\text{m}^2/\text{C}^2)\frac{(+9.10\times10^{-6}\ \text{C})(-0.420\times10^{-6}\ \text{C})}{0.960\ \text{m}} = \underline{-0.0358\ \text{J}}$

(Note: The electrical potential energy of a pair of point charges can be either positive or negative, depending on the signs of the charges.)

b)

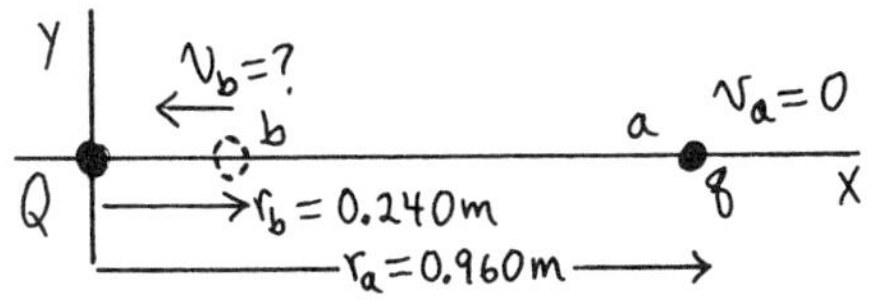

Use conservation of energy:

$K_a + U_a + W_{other} = K_b + U_b$

Only the electric force does work $\Rightarrow W_{other} = 0$ and $U = \frac{1}{4\pi\epsilon_0}\frac{qQ}{r}$.

$K_a = 0$ (released from rest)

$U_a = -0.0358$ J (from part (a))

$K_b = \frac{1}{2}mv_b^2$

$U_b = \frac{1}{4\pi\epsilon_0}\frac{qQ}{r_b} = (8.988\times10^9\ \text{N}\cdot\text{m}^2/\text{C}^2)\frac{(+9.10\times10^{-6}\ \text{C})(-0.420\times10^{-6}\ \text{C})}{0.240\ \text{m}} = -0.1431\ \text{J}$

Then $K_a + U_a + W_{other} = K_b + U_b$ gives

$U_a = K_b + U_b$

$K_b = U_a - U_b$

$\frac{1}{2}mv_b^2 = U_a - U_b \Rightarrow v_b = \sqrt{\frac{2(U_a - U_b)}{m}} = \sqrt{\frac{2(-0.0358\ \text{J} - (-0.1431\ \text{J}))}{3.20\times10^{-4}\ \text{kg}}} = \underline{25.9\ \text{m/s}}$

24-3

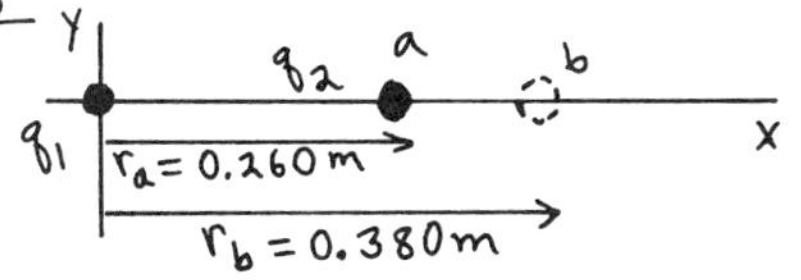

$W_{a\to b} = U_a - U_b$

$U_a = \frac{1}{4\pi\epsilon_0}\frac{q_1q_2}{r_a} = (8.988\times10^9\ \text{N}\cdot\text{m}^2/\text{C}^2)\frac{(-5.80\times10^{-6}\ \text{C})(+4.30\times10^{-6}\ \text{C})}{0.260\ \text{m}}$

$U_a = -0.8622\ \text{J}$

$U_b = \frac{1}{4\pi\epsilon_0}\frac{q_1q_2}{r_b} = (8.988\times10^9\ \text{N}\cdot\text{m}^2/\text{C}^2)\frac{(-5.80\times10^{-6}\ \text{C})(+4.30\times10^{-6}\ \text{C})}{0.380\ \text{m}} = -0.5899\ \text{J}$

$W_{a\to b} = U_a - U_b = -0.8622\ \text{J} - (-0.5899\ \text{J}) = \underline{-0.272\ \text{J}}$

The attractive force on q_2 is in the $-x$-direction, so it does negative work on q_2 when q_2 moves to larger x.

24-5

a)

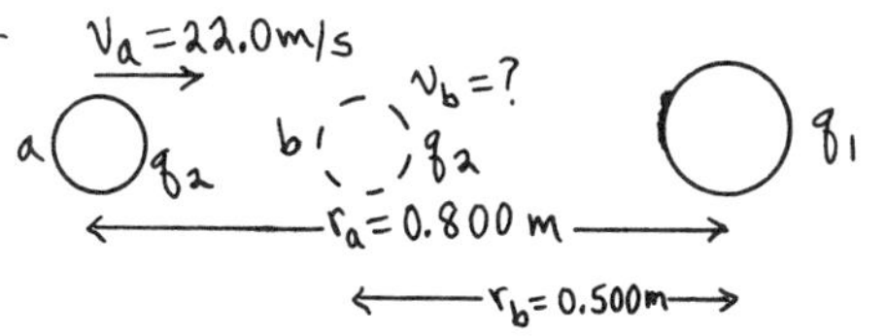

Use conservation of energy:

$K_a + U_a + W_{other} = K_b + U_b$

Let point a be where q_2 is 0.800 m from q_1 and point b be where q_2 is 0.500 m from q_1.

24-5 (cont)

Only the electric force does work $\Rightarrow W_{other} = 0$ and $U = \frac{1}{4\pi\epsilon_0}\frac{q_1 q_2}{r}$

$K_a = \frac{1}{2}mv_a^2 = \frac{1}{2}(2.00\times10^{-3}\text{ kg})(22.0\text{ m/s})^2 = 0.4840\text{ J}$

$U_a = \frac{1}{4\pi\epsilon_0}\frac{q_1 q_2}{r_a} = (8.988\times10^9\text{ N}\cdot\text{m}^2/\text{C}^2)\frac{(+7.50\times10^{-6}\text{ C})(+3.00\times10^{-6}\text{ C})}{0.800\text{ m}} = +0.2528\text{ J}$

$K_b = \frac{1}{2}mv_b^2$

$U_b = \frac{1}{4\pi\epsilon_0}\frac{q_1 q_2}{r_b} = (8.988\times10^9\text{ N}\cdot\text{m}^2/\text{C}^2)\frac{(+7.50\times10^{-6}\text{ C})(+3.00\times10^{-6}\text{ C})}{0.500\text{ m}} = +0.4045\text{ J}$

The conservation of energy equation then gives

$K_b = K_a + (U_a - U_b)$

$\frac{1}{2}mv_b^2 = +0.4840\text{ J} + (0.2528\text{ J} - 0.4045\text{ J}) = 0.3323\text{ J}$

$v_b = \sqrt{\frac{2(0.3323\text{ J})}{2.00\times10^{-3}\text{ kg}}} = \underline{18.2\text{ m/s}}$

(The potential energy increases when the two positively charged spheres get closer together, so the kinetic energy and speed decrease.)

b) Let point c be where q_2 has its speed momentarily reduced to zero. Apply conservation of energy to points a and c: $K_a + U_a + W_{other} = K_c + U_c$.

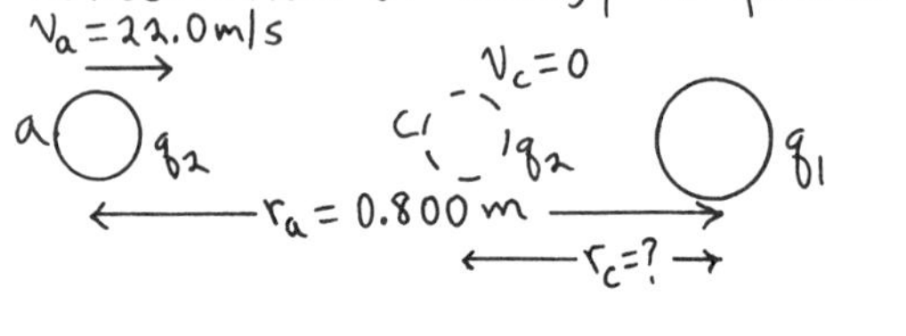

$K_a = +0.4840\text{ J}$ (from part (a))

$U_a = +0.2528\text{ J}$ (from part (a))

$K_c = 0$ (at distance of closest approach the speed is zero)

$U_c = \frac{1}{4\pi\epsilon_0}\frac{q_1 q_2}{r_c}$

$\Rightarrow K_a + U_a = U_c$

$\frac{1}{4\pi\epsilon_0}\frac{q_1 q_2}{r_c} = +0.4840\text{ J} + 0.2528\text{ J} = 0.7368\text{ J}$

$r_c = \frac{1}{4\pi\epsilon_0}\frac{q_1 q_2}{0.7368\text{ J}} = (8.988\times10^9\text{ N}\cdot\text{m}^2/\text{C}^2)\frac{(+7.50\times10^{-6}\text{ C})(+3.00\times10^{-6}\text{ C})}{+0.7368\text{ J}} = \underline{0.274\text{ m}}$

24-7

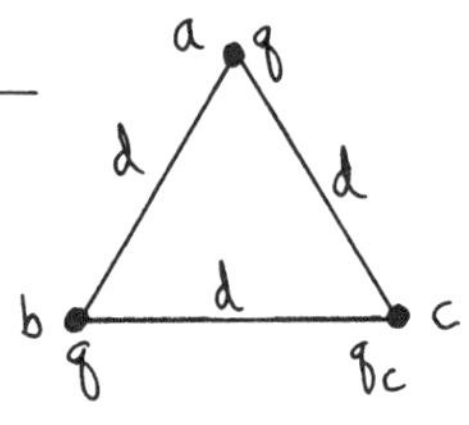

$W = -\Delta U = -(U_2 - U_1)$

Let 1 be where all charges are infinitely far apart.

Let 2 be where charges are at the corners of the triangle.

$U_1 = 0$

$U_2 = U_{ab} + U_{ac} + U_{bc} = \frac{1}{4\pi\epsilon_0 d}(q^2 + 2qq_c)$

Want $W = 0$, so $W = -(U_2 - U_1)$ gives $0 = -U_2$

$0 = -\frac{1}{4\pi\epsilon_0 d}(q^2 + 2qq_c)$

$q^2 + 2qq_c = 0$

$\underline{q_c = -q/2}$

24-13

a) The direction of $\vec{E}$ is always from high potential to low potential
$\Rightarrow$ point a is at higher potential.

b) $V_a - V_b = -\int_b^a \vec{E}\cdot d\vec{l} = -\int_b^a E\,dx = -E(x_a - x_b)$

$E = -\dfrac{V_a - V_b}{x_a - x_b} = -\dfrac{+730\,V}{0.80\,m - 1.20\,m} = \underline{1820\,V/m}$

c) $W_{b\to a} = q(V_b - V_a) = (-0.200\times10^{-6}\,C)(-730\,V) = \underline{+1.46\times10^{-4}\,J}$

The electric force does positive work on a negative charge when the negative charge moves from low potential (point b) to high potential (point a).

24-15

$W_{a\to b} = q'\int_a^b \vec{E}\cdot d\vec{l}$

Use coordinates where +y is upward and +x is to the right. Then $\vec{E} = E\hat{j}$ with $E = 5.00\times10^4\,N/C$.

a) 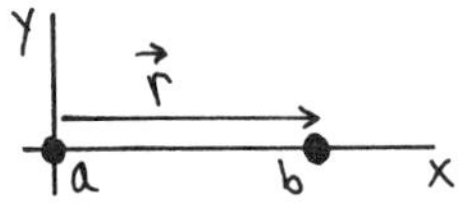

$d\vec{l} = dx\,\hat{i}$

$\vec{E}\cdot d\vec{l} = (E\hat{j})\cdot(dx\,\hat{i}) = 0 \Rightarrow W_{a\to b} = q'\int_a^b \vec{E}\cdot d\vec{l} = 0$

The electric force on the positive charge is upward (in the direction of the electric field) and does no work for a horizontal displacement of the charge.

b) 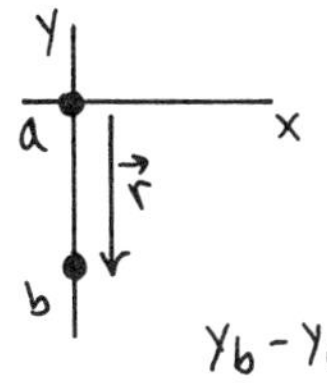

$d\vec{l} = dy\,\hat{j}$

$\vec{E}\cdot d\vec{l} = (E\hat{j})\cdot(dy\,\hat{j}) = E\,dy$

$W_{a\to b} = q'\int_a^b \vec{E}\cdot d\vec{l} = q'E\int_a^b dy = q'E(y_b - y_a)$

$y_b - y_a = -0.670\,m$, negative since the displacement is downward and we have taken +y to be upward.

$W_{a\to b} = q'E(y_b - y_a) = (+3.70\times10^{-8}\,C)(5.00\times10^4\,N/C)(-0.670\,m) = \underline{-1.24\times10^{-3}\,J}$

The electric force on the positive charge is upward so it does negative work for a downward displacement of the charge.

c) $d\vec{l} = dx\,\hat{i} + dy\,\hat{j}$ (The displacement has both horizontal and vertical components.)

$\vec{E}\cdot d\vec{l} = (E\hat{j})\cdot(dx\,\hat{i} + dy\,\hat{j}) = E\,dy$ (only the vertical component of the displacement contributes to the work)

$W_{a\to b} = q'\int_a^b \vec{E}\cdot d\vec{l} = q'E\int_a^b dy = q'E(y_b - y_a)$

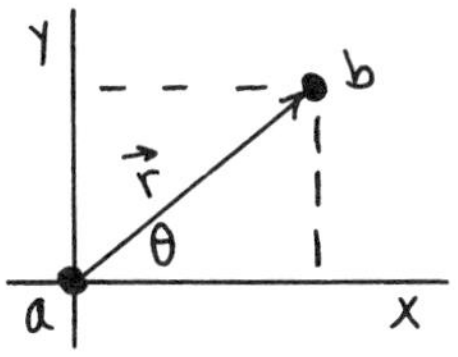

$y_a = 0$

$y_b = r\sin\theta = (2.60\,m)\sin 45^\circ = 1.838\,m$

(The vertical component of the 2.60 m displacement is 1.838 m upward.)

$W_{a\to b} = q'E(y_b - y_a) = (+3.70\times10^{-8}\,C)(5.00\times10^4\,N/C)(+1.838\,m) = \underline{3.40\times10^{-3}\,J}$

24-17

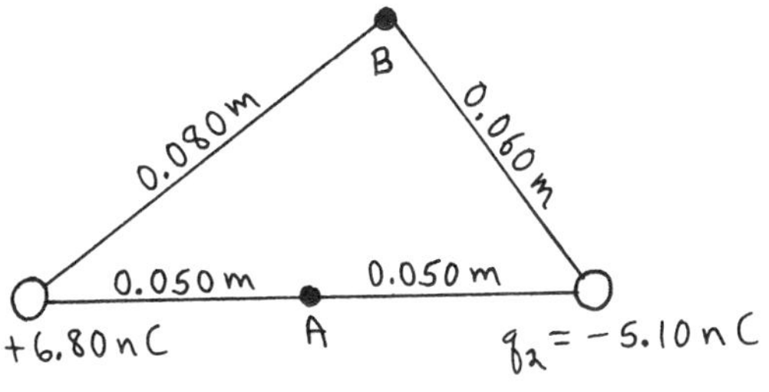

$$V = \frac{1}{4\pi\epsilon_0}\sum_i \frac{q_i}{r_i}$$

a) $V_A = \frac{1}{4\pi\epsilon_0}\left(\frac{q_1}{r_{A1}} + \frac{q_2}{r_{A2}}\right) = (8.988\times10^9\ \text{N}\cdot\text{m}^2/\text{C}^2)\left(\frac{+6.80\times10^{-9}\ \text{C}}{0.050\ \text{m}} + \frac{-5.10\times10^{-9}\ \text{C}}{0.050\ \text{m}}\right)$

$V_A = \underline{+306\ \text{V}}$

b) $V_B = \frac{1}{4\pi\epsilon_0}\left(\frac{q_1}{r_{B1}} + \frac{q_2}{r_{B2}}\right) = (8.988\times10^9\ \text{N}\cdot\text{m}^2/\text{C}^2)\left(\frac{+6.80\times10^{-9}\ \text{C}}{0.080\ \text{m}} + \frac{-5.10\times10^{-9}\ \text{C}}{0.060\ \text{m}}\right) = \underline{0}$

c) $W_{B\to A} = q'(V_B - V_A) = (2.50\times10^{-9}\ \text{C})(0 - 306\ \text{V}) = \underline{-7.65\times10^{-7}\ \text{J}}$

The electric force does negative work on the positive charge when it moves from low potential (point B) to high potential (point A).

24-25

a) From Example 24-9 $\quad E = \frac{V_{ab}}{d} = \frac{500\ \text{V}}{0.0850\ \text{m}} = \underline{5880\ \text{V/m}}$

b) $F = |q|E = (2.40\times10^{-9}\ \text{C})(5880\ \text{V/m}) = \underline{1.41\times10^{-5}\ \text{N}}$

c)

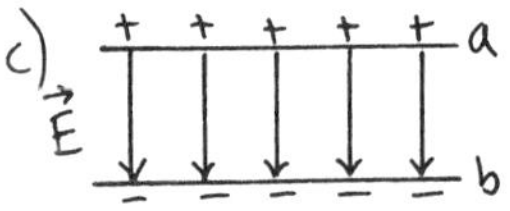

The plate with positive charge (plate a) is at higher potential. The electric field is directed from high potential toward low potential (or $\vec{E}$ is from + charge to – charge) $\Rightarrow$ $\vec{E}$ points from a to b. Hence the force that $\vec{E}$ exerts on the positive charge is from a to b, so it does positive work.

$W = \int_a^b \vec{F}\cdot d\vec{l} = Fd$, where d is the separation between the plates.

$W = Fd = (1.41\times10^{-5}\ \text{N})(0.0850\ \text{m}) = \underline{+1.20\times10^{-6}\ \text{J}}$

d) $V_a - V_b = +500\ \text{V}$ (plate a is at higher potential)

$\Delta U = U_a - U_b = q(V_a - V_b) = (2.40\times10^{-9}\ \text{C})(500\ \text{V}) = \underline{+1.20\times10^{-6}\ \text{J}}$

We see that $W_{a\to b} = U_a - U_b$.

24-27

a) $E = \frac{V_{ab}}{d} \Rightarrow d = \frac{V_{ab}}{E} = \frac{3.75\times10^3\ \text{V}}{3.00\times10^6\ \text{N/C}} = 1.25\times10^{-3}\ \text{m} = \underline{1.25\ \text{mm}}$

b) $E = \frac{\sigma}{\epsilon_0}$

$\sigma = E\epsilon_0 = (3.00\times10^6\ \text{V/m})(8.854\times10^{-12}\ \text{C}^2/\text{N}\cdot\text{m}^2) = \underline{2.66\times10^{-5}\ \text{C/m}^2}$

24-29

a) When the electron is on either side of the center of the ring, the ring exerts an attractive force directed toward the center of the ring. This restoring force produces oscillatory motion of the electron along the axis of

24-29 (cont)

the ring, with amplitude 25.0 cm. The force on the electron is <u>not</u> of the form $F = -kx$ so the oscillatory motion is not simple harmonic motion.

b) $K_a + U_a = K_b + U_b$ with a at the initial position of the electron and b at the center of the ring. From Example 24-11, $V = \frac{1}{4\pi\epsilon_0}\frac{Q}{\sqrt{x^2+a^2}}$. $x_a = 25.0$ cm, $x_b = 0$.

$K_a = 0$ (released from rest), $K_b = \frac{1}{2}mv^2$

$\frac{1}{2}mv^2 = U_a - U_b$

$U = qV = -eV \Rightarrow v = \sqrt{\frac{2e(V_b - V_a)}{m}}$

$$V_a = \frac{1}{4\pi\epsilon_0}\frac{Q}{\sqrt{x_a^2+a^2}} = (8.988\times10^9\ N\cdot m^2/C^2)\frac{12.0\times10^{-9}C}{\sqrt{(0.250m)^2+(0.100m)^2}} = 400.6\ V$$

$$V_b = \frac{1}{4\pi\epsilon_0}\frac{Q}{\sqrt{x_b^2+a^2}} = (8.988\times10^9\ N\cdot m^2/C^2)\frac{12.0\times10^{-9}C}{0.100m} = 1078.6\ V$$

$$v = \sqrt{\frac{2e(V_b-V_a)}{m}} = \sqrt{\frac{2(1.602\times10^{-19}C)(1078.6V - 400.6V)}{9.109\times10^{-31}kg}} = \underline{1.54\times10^7\ m/s}$$

24-35

$V = axy + by^2 + cy$

a) $E_x = -\frac{\partial V}{\partial x} = -ay$

$E_y = -\frac{\partial V}{\partial y} = -ax - 2by - c = -(ax + 2by + c)$

$E_z = -\frac{\partial V}{\partial z} = 0$

b) $E = 0 \Rightarrow E_x = E_y = E_z = 0$

$E_x = 0 \Rightarrow y = 0$ and any x and z

$E_y = 0 \Rightarrow -(ax + 2by + c) = 0$

But already know that must have $y = 0$ so $ax + c = 0 \Rightarrow x = -\frac{c}{a}$ and any z

$x = -\frac{c}{a} = -\frac{5.00\ V/m}{3.00\ V/m^2} = -1.67m$

Thus $E = 0$ for $x = -1.67$ m, $y = 0$, and any value of z.

24-37

a) V is independent of x and $z \Rightarrow E_x = -\frac{\partial V}{\partial x} = 0$ and $E_z = -\frac{\partial V}{\partial z} = 0$.

$E_y = -\frac{\partial V}{\partial y}$

For $y < 0$, $V = 0$ so $E_y = 0$. $E_x = E_y = E_z = 0$ so $E = 0$.

For $0 < y < d$, $V = Cy$ so $E_y = -C$. $\vec{E} = -C\hat{j}$.

For $y > d$, $V = Cd$ and $E_y = -\frac{\partial(Cd)}{\partial y} = 0$. $E_x = E_y = E_z = 0$ so $E = 0$.

b) There is a uniform electric field in the $-y$-direction for y between 0 and d and zero field elsewhere. This is the field of two infinite sheets of charge

24-37 (cont)

parallel to the xz-plane, one with charge $+\sigma$ at $y=d$ and one with charge $-\sigma$ at $y=0$.

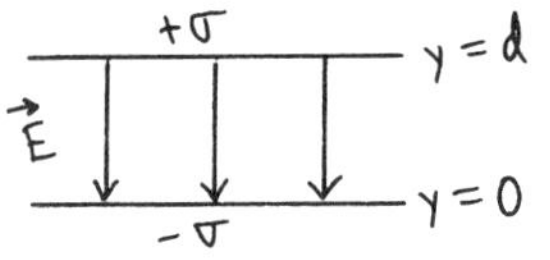

For this charge distribution $E=\frac{\sigma}{\epsilon_0}$ between the plates, from the positive plate toward the negative plate, and $E=0$ outside the plates.

To have $E=C$ between the plates need $C=\frac{\sigma}{\epsilon_0}$, so $\sigma=C\epsilon_0$.

24-39

a) Calculate the acceleration of the electron produced by the electric force: $\vec{F}_E=q\vec{E}$. Since $q=-e$ is negative $\vec{F}_E$ and $\vec{E}$ are in opposite directions; $\vec{E}$ is upward so $\vec{F}_E$ is downward. The magnitude of F_E is $F_E=|q|E=eE$.

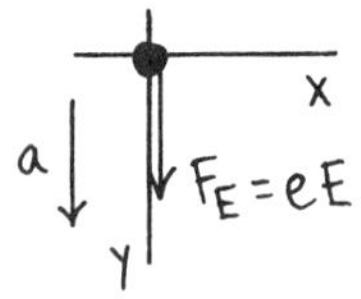

$\sum F_y=ma_y$

$F_E=ma$

$eE=ma$

$a=\frac{eE}{m}=\frac{(1.602\times10^{-19}\,C)(1.40\times10^{3}\,N/C)}{9.109\times10^{-31}\,kg}$

$a=2.462\times10^{14}\,m/s^2$

(This is much larger than $g=9.80\,m/s^2$, so the gravity force on the electron can be neglected.)

The acceleration is constant and downward, so the motion is like that of a projectile. Use the horizontal motion to find the time and then use the time to find the vertical displacement.

x-component

$v_{ox}=v_0=8.00\times10^6\,m/s$

$a_x=0$

$x-x_0=0.060\,m$

$t=?$

$x-x_0=v_{ox}t+\cancel{\tfrac{1}{2}a_xt^2}$

$t=\frac{x-x_0}{v_{ox}}=\frac{0.060\,m}{8.00\times10^6\,m/s}=7.5\times10^{-9}\,s$

y-component

$v_{oy}=0$

$a_y=2.462\times10^{14}\,m/s^2$

$t=7.5\times10^{-9}\,s$

$y-y_0=?$

$y-y_0=\cancel{v_{oy}t}+\tfrac{1}{2}a_yt^2$

$y-y_0=\tfrac{1}{2}(2.462\times10^{14}\,m/s^2)(7.5\times10^{-9}\,s)^2=0.00692\,m=\underline{0.692\,cm}$

b)

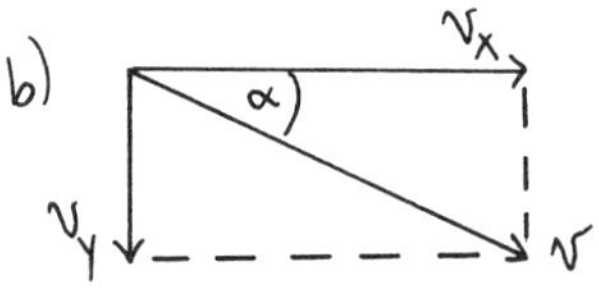

$v_x=v_{ox}=8.00\times10^6\,m/s$ (since $a_x=0$)

$v_y=v_{oy}+a_yt=0+(2.462\times10^{14}\,m/s^2)(7.5\times10^{-9}\,s)=1.846\times10^6\,m/s$

$\tan\alpha=\frac{v_y}{v_x}=\frac{1.846\times10^6\,m/s}{8.00\times10^6\,m/s}=0.2308\Rightarrow\underline{\alpha=13.0^\circ}$

c) Consider the motion of the electron after it leaves the region between the plates. Outside the plates there is no electric field, so $a=0$. (Gravity can still be neglected.) Use the horizontal motion to find the time it takes the electron to travel 0.120 m horizontally to the screen. From this time find the distance downward that the electron travels.

24-39 (cont)

X-component

$a_x = 0$

$v_{0x} = 8.00\times10^6$ m/s

$x - x_0 = 0.120$ m

$t = ?$

$x - x_0 = v_{0x}t + \cancel{\frac{1}{2}a_x t^2}$

$t = \frac{x - x_0}{v_{0x}} = \frac{0.120\text{ m}}{8.00\times10^6\text{ m/s}} = 1.50\times10^{-8}$ s

Y-component

$t = 1.50\times10^{-8}$ s

$a_y = 0$

$v_{0y} = 1.846\times10^6$ m/s (from part (b))

$y - y_0 = ?$

$y - y_0 = v_{0y}t + \cancel{\frac{1}{2}a_y t^2}$

$y - y_0 = (1.846\times10^6\text{ m/s})(1.50\times10^{-8}\text{ s}) = 0.0277$ m

$y - y_0 = 2.77$ cm

The electron travels downward a distance 0.69 cm while it is between the plates and a distance 2.77 cm while traveling from the edge of the plates to the screen. The total downward deflection is 0.69 cm + 2.77 cm = 3.46 cm.

Problems

24-51

a) Example 23-9 shows that the electric field outside a uniformly charged sphere with positive charge is $E = \frac{1}{4\pi\epsilon_0}\frac{Q}{r^2}$, directed radially outward.

Apply $V_a - V_b = \int_a^b \vec{E}\cdot d\vec{l}$ with $b = \infty$ so $V_b = V_\infty = 0$ and $a = r > R$, where R is the radius of the charge distribution.

$V_r = \int_r^\infty E\,dr = \frac{Q}{4\pi\epsilon_0}\int_r^\infty \frac{dr}{r^2} = \frac{Q}{4\pi\epsilon_0}\left(-\frac{1}{r}\Big|_r^\infty\right) = \frac{Q}{4\pi\epsilon_0 r}$, the same as if all the charge were concentrated at the center of the sphere.

At the surface of the nucleus $r = R$ and

$V = \frac{Q}{4\pi\epsilon_0 R} = (8.988\times10^9\text{ N}\cdot\text{m}^2/\text{C}^2)\frac{92(1.602\times10^{-19}\text{ C})}{7.40\times10^{-15}\text{ m}} = 1.79\times10^7$ V

b)

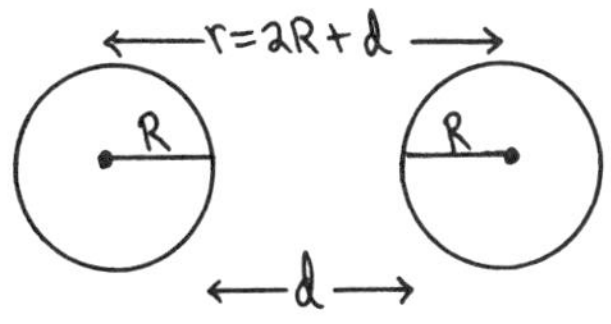

$d = 12.0\times10^{-15}$ m

Outside each sphere the potential is the same as if all the charge of that sphere is at the center of the sphere, so $U = \frac{1}{4\pi\epsilon_0}\frac{Q^2}{r}$, where $r = 2R + d$ is the separation between the centers of the spheres.

$r = 2R + d = 2(7.40\times10^{-15}\text{ m}) + 12.0\times10^{-15}\text{ m} = 26.8\times10^{-15}$ m

$U = 92e$

$U = \frac{1}{4\pi\epsilon_0}\frac{Q^2}{r} = (8.988\times10^9\text{ N}\cdot\text{m}^2/\text{C}^2)\frac{[(92)(1.602\times10^{-19}\text{C})]^2}{26.8\times10^{-15}\text{ m}} = 7.285\times10^{-11}$ J

$U = 7.285\times10^{-11}\text{ J}\left(\frac{1\text{ eV}}{1.602\times10^{-19}\text{ J}}\right) = 4.55\times10^8\text{ eV} = 455$ MeV

24-53

a) 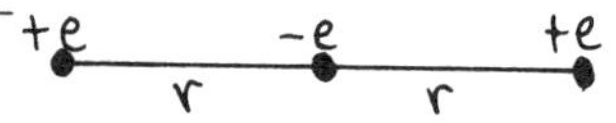

$r = \frac{1.07\times10^{-10}\,m}{2} = 0.535\times10^{-10}\,m$

The potential energy of interaction of the electron with each nucleus is $U = \frac{1}{4\pi\epsilon_0}\frac{(-e^2)}{r}$, so the total potential energy is

$$U = -\frac{2e^2}{4\pi\epsilon_0 r} = -\frac{2(8.988\times10^9\,N\cdot m^2/C^2)(1.60\times10^{-19}\,C)^2}{0.535\times10^{-10}\,m} = \underline{-8.60\times10^{-18}\,J}$$

$$U = -8.60\times10^{-18}\,\cancel{J}\left(\frac{1\,eV}{1.602\times10^{-19}\,\cancel{J}}\right) = \underline{-53.7\,eV}$$

b) 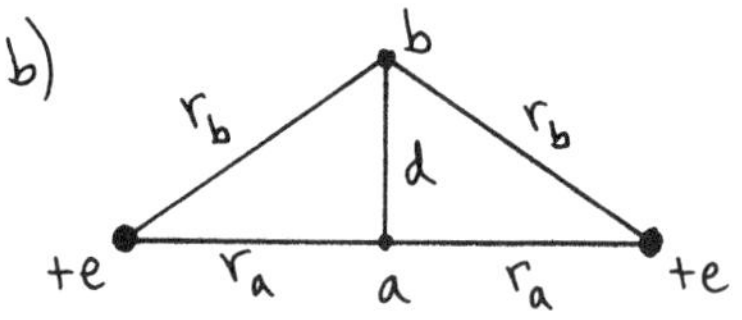

$r_b = \sqrt{r_a^2 + d^2}$

$r_a = r = 0.535\times10^{-10}\,m$

Apply $K_a + U_a = K_b + U_b$ with point a midway between the protons and point b where the electron instantaneously has $v = 0$ (at its maximum displacement d from point a).

$U_a = -\frac{2e^2}{4\pi\epsilon_0 r}$ as in part (a)

$K_a = \frac{1}{2}mv_a^2 = \frac{1}{2}(9.11\times10^{-31}\,kg)(1.70\times10^5\,m/s)^2 = 1.316\times10^{-20}\,J$

$K_b = 0$

$U_b = 2\frac{(-e^2)}{4\pi\epsilon_0 r_b} = -\frac{2e^2}{4\pi\epsilon_0\sqrt{r^2+d^2}}$

Then $K_a + U_a = K_b + U_b$ gives

$$K_a - \frac{2e^2}{4\pi\epsilon_0}\frac{1}{r} = -\frac{2e^2}{4\pi\epsilon_0}\frac{1}{\sqrt{r^2+d^2}}$$

$$K_a = \frac{2e^2}{4\pi\epsilon_0}\left(\frac{1}{r} - \frac{1}{\sqrt{r^2+d^2}}\right) = \frac{2e^2}{4\pi\epsilon_0 r}\left(1 - \frac{1}{\sqrt{1+(d/r)^2}}\right)$$

$$\frac{1}{\sqrt{1+(d/r)^2}} = -\frac{K_a r}{2e^2(1/4\pi\epsilon_0)} + 1 = -\frac{(1.316\times10^{-20}\,J)(0.535\times10^{-10}\,m)}{2(1.60\times10^{-19}\,C)^2(8.988\times10^9\,N\cdot m^2/C^2)} + 1 = 0.99847$$

$$1 + \left(\frac{d}{r}\right)^2 = \frac{1}{(0.99847)^2} = 1.003067$$

$$\frac{d}{r} = \sqrt{0.003067} = 0.05538$$

$$d = 0.05538r = 0.05538(0.535\times10^{-10}\,m) = \underline{2.96\times10^{-12}\,m}$$

24-57

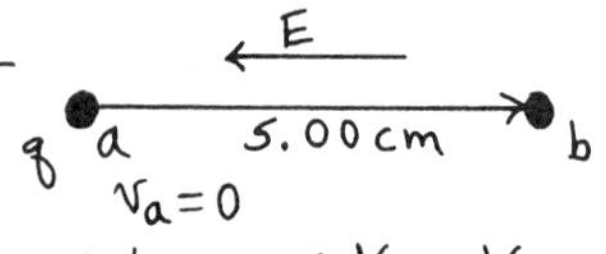

a) $W_{tot} = \Delta K = K_b - \cancel{K_a} = K_b = 6.20\times10^{-5}\,J$

The electric force F_E and the additional force F both do work, so that $W_{tot} = W_{F_E} + W_F$

$$W_{F_E} = W_{tot} - W_F = 6.20\times10^{-5}\,J - 8.45\times10^{-5}\,J = \underline{-2.25\times10^{-5}\,J}$$

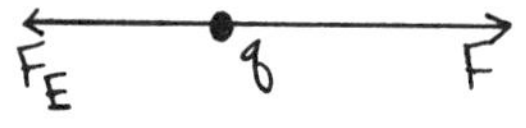

The electric force is to the left (in the direction of the electric field since the particle has positive charge). The displacement is to the right, so the electric force does negative work. The

24-57 (cont)

additional force F is in the direction of the displacement, so it does positive work.

b) For the work done by the electric force,

$W_{a\to b} = q(V_a - V_b) \Rightarrow V_a - V_b = \dfrac{W_{a\to b}}{q} = -\dfrac{2.25\times10^{-5}\,J}{4.30\times10^{-9}\,C} = \underline{-5.23\times10^{3}\,V}$

The starting point (point a) is at 5.23×10^3 V lower potential than the ending point (point b). We know that $V_b > V_a$ because the electric field always points from high potential toward low potential.

c) Since the electric field is uniform and directed opposite to the displacement $W_{a\to b} = -F_E d = -qEd$, where $d = 5.00$ cm is the displacement of the particle.

$E = -\dfrac{W_{a\to b}}{qd} = -\dfrac{V_a - V_b}{d} = -\dfrac{-5.23\times10^3\,V}{0.0500\,m} = \underline{1.05\times10^5\,V/m}$

24-59

a) $V = Cx^{4/3}$

$C = \dfrac{V}{x^{4/3}} = \dfrac{160\,V}{(14.6\times10^{-3}m)^{4/3}} = \underline{4.48\times10^4\,V/m^{4/3}}$

b) $E_x = -\dfrac{\partial V}{\partial x} = -\frac{4}{3}Cx^{1/3}$

(The minus sign means that E_x is in the $-x$-direction, which says that $\vec{E}$ points from the positive anode toward the negative cathode.)

c) $\vec{F} = q\vec{E} \Rightarrow F_x = -eE_x = +\frac{4}{3}eCx^{1/3}$

Halfway between the electrodes $\Rightarrow x = 7.30\times10^{-3}$ m

$F_x = \frac{4}{3}(1.602\times10^{-19}C)(4.48\times10^4 V/m^{4/3})(7.30\times10^{-3}m)^{1/3} = \underline{1.86\times10^{-15}\,N}$

F_x is positive $\Rightarrow$ the force is directed toward the positive anode.

24-63

a) Problem 24-61 derived that $E = \dfrac{V_{ab}}{\ln(b/a)}\dfrac{1}{r}$, where a is the radius of the inner cylinder (wire) and b is the radius of the outer hollow cylinder. The potential difference between the two cylinders is V_{ab}. Midway between the wire and the cylinder wall $\Rightarrow r = \dfrac{a+b}{2} = \dfrac{80.0\times10^{-6}\,m + 0.120\,m}{2} = 0.06004$ m.

$E = \dfrac{V_{ab}}{\ln(b/a)\,r} = \dfrac{60.0\times10^3\,V}{\ln\left(\dfrac{0.120\,m}{80.0\times10^{-6}\,m}\right)(0.06004\,m)} = \underline{1.37\times10^5\,V/m}$

b) $F_E = 10mg$

$|q|E = 10mg$

$|q| = \dfrac{10mg}{E} = \dfrac{10(30.0\times10^{-9}kg)(9.80\,m/s^2)}{1.37\times10^5\,V/m} = \underline{2.15\times10^{-11}\,C}$

24-65

a)

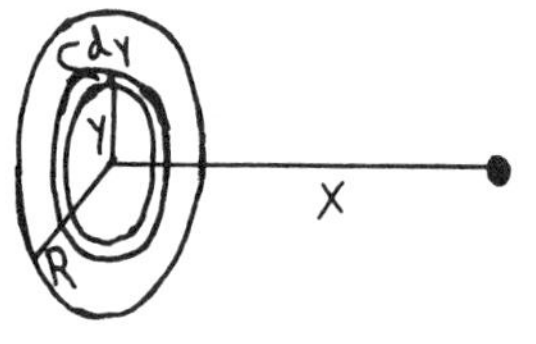

Consider a thin ring of radius y and width dy. The ring has area $2\pi y\,dy$ so the charge on the ring is $dq = \sigma(2\pi y\,dy)$.

The result of Example 24-11 then says that the potential due to this thin ring at the point on the axis at a distance x from the disk is

$$dV = \frac{1}{4\pi\epsilon_0}\frac{dq}{\sqrt{x^2+y^2}} = \frac{2\pi\sigma}{4\pi\epsilon_0}\frac{y\,dy}{\sqrt{x^2+y^2}}$$

$$V = \int dV = \frac{\sigma}{2\epsilon_0}\int_0^R \frac{y\,dy}{\sqrt{x^2+y^2}} = \frac{\sigma}{2\epsilon_0}\left[\sqrt{x^2+y^2}\Big|_0^R\right] = \frac{\sigma}{2\epsilon_0}\left(\sqrt{x^2+R^2}-x\right)$$

Note: For $x \gg R$ this result should reduce to the potential of a point charge with $Q = \sigma\pi R^2$.

$$\sqrt{x^2+R^2} = x\left(1+\frac{R^2}{x^2}\right)^{1/2} \cong x\left(1+\frac{R^2}{2x^2}\right)$$

$$\sqrt{x^2+R^2}-x \cong \frac{R^2}{2x}$$

$$V \cong \frac{\sigma}{2\epsilon_0}\frac{R^2}{2x} = \frac{\sigma\pi R^2}{4\pi\epsilon_0 x} = \frac{Q}{4\pi\epsilon_0 x}$$, as expected.

b) $E_x = -\frac{\partial V}{\partial x} = -\frac{\sigma}{2\epsilon_0}\left(\frac{x}{\sqrt{x^2+R^2}}-1\right) = \frac{\sigma x}{2\epsilon_0}\left(\frac{1}{x}-\frac{1}{\sqrt{x^2+R^2}}\right)$, which agrees with Eq. (22-11).

24-67

a) From Problem 23-34, $E(r) = \frac{\lambda r}{2\pi\epsilon_0 R^2}$ for $r \le R$ (inside the cylindrical charge distribution)

and $E(r) = \frac{\lambda}{2\pi\epsilon_0 r}$ for $r \ge R$. Let $V=0$ at $r=R$ (at the surface of the cylinder).
Use $V_a - V_b = \int_a^b \vec{E}\cdot d\vec{l}$.

$r \ge R$

Take point a to be at R and point b to be at r, where $r>R$. Let $d\vec{l} = d\vec{r}$. $\vec{E}$ and $d\vec{r}$ are both radially outward $\Rightarrow \vec{E}\cdot d\vec{r} = E\,dr$. Thus $V_R - V_r = \int_R^r E\,dr$. $V_R = 0$ gives $V_r = -\int_R^r E\,dr$. In this interval $(r>R)$, $E(r) = \frac{\lambda}{2\pi\epsilon_0 r}$, so

$$V_r = -\int_R^r \frac{\lambda}{2\pi\epsilon_0 r}dr = -\frac{\lambda}{2\pi\epsilon_0}\int_R^r \frac{dr}{r} = -\frac{\lambda}{2\pi\epsilon_0}\ln\left(\frac{r}{R}\right).$$

Note: This expression gives $V_r = 0$ when $r=R$ and the potential decreases (becomes a negative number of larger magnitude) with increasing distance from the cylinder.

$r \le R$

Take point a at r, where $r<R$, and point b at R. $\vec{E}\cdot d\vec{r} = E\,dr$ as before.
$\Rightarrow V_r - V_R = \int_r^R E\,dr$. $V_R = 0$ then gives $V_r = \int_r^R E\,dr$. In this interval $(r<R)$ $E(r) = \frac{\lambda r}{2\pi\epsilon_0 R^2}$ and $V_r = \int_r^R \frac{\lambda r}{2\pi\epsilon_0 R^2}dr = \frac{\lambda}{2\pi\epsilon_0 R^2}\int_r^R r\,dr = \frac{\lambda}{2\pi\epsilon_0 R^2}\left(\frac{R^2}{2}-\frac{r^2}{2}\right)$

$$V_r = \frac{\lambda}{4\pi\epsilon_0}\left(1-\left(\frac{r}{R}\right)^2\right)$$

Note: This expression also gives $V_r = 0$ when $r=R$. The potential is $+\frac{\lambda}{4\pi\epsilon_0}$ at $r=0$ and decreases with increasing r.

24-67 (cont)

b)

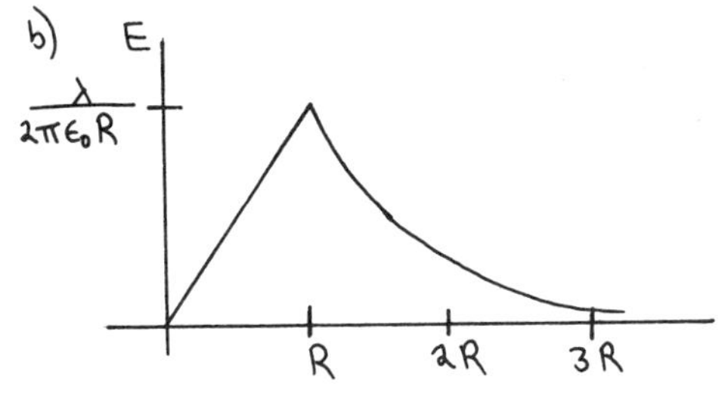

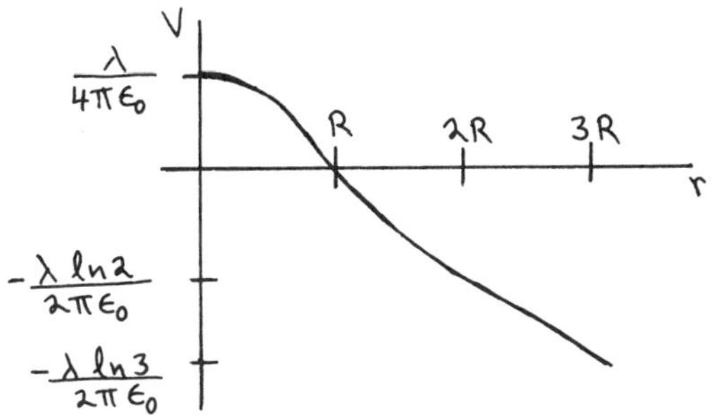

24-71

$$\vec{E} = (3.00\,\text{N/C}\cdot\text{m})x\hat{i} + (4.00\,\text{N/C}\cdot\text{m})y\hat{j}$$

a) S_1 and S_3 are parallel to the xz-plane; S_1 is at $y_1 = 0$ and S_3 is at $y_3 = L$.

$V_1 - V_3 = \int_1^3 \vec{E}\cdot d\vec{l}$; $d\vec{l} = dy\,\hat{j}$, $\vec{E}\cdot d\vec{l} = (4.00\,\text{N/C}\cdot\text{m})\,y\,dy$

$V_1 - V_3 = (4.00\,\text{N/C}\cdot\text{m})\int_{y_1}^{y_3} y\,dy = (4.00\,\text{N/C}\cdot\text{m})\left(\tfrac{1}{2}y^2\big|_0^L\right) = (2.00\,\text{N/C}\cdot\text{m})\,L^2$

$V_1 - V_3 = (2.00\,\text{N/C}\cdot\text{m})(0.200\,\text{m})^2 = \underline{0.0800\,\text{V}}$; $V_1 - V_3$ is positive.

Thus S_1 is at higher potential. ($\vec{E}$ points from high potential toward low potential.)

b) S_2 and S_4 are parallel to the xy-plane; S_2 is at $z_2 = 0$ and S_4 is at $z_4 = L$.

$V_2 - V_4 = \int_2^4 \vec{E}\cdot d\vec{l}$; $d\vec{l} = dz\,\hat{k}$; $\vec{E}\cdot d\vec{l} = 0$ so $V_2 - V_4 = 0$.

c) S_5 and S_6 are parallel to the yz-plane; S_6 is at $x_6 = 0$ and S_5 is at $x_5 = L$.

$V_6 - V_5 = \int_6^5 \vec{E}\cdot d\vec{l}$; $d\vec{l} = dx\,\hat{i}$; $\vec{E}\cdot d\vec{l} = (3.00\,\text{N/C}\cdot\text{m})\,x\,dx$

$V_6 - V_5 = (3.00\,\text{N/C}\cdot\text{m})\int_{x_6}^{x_5} x\,dx = (3.00\,\text{N/C}\cdot\text{m})\left(\tfrac{1}{2}x^2\big|_0^L\right) = (1.50\,\text{N/C}\cdot\text{m})L^2 = (1.50\,\text{N/C}\cdot\text{m})(0.200\,\text{m})^2$

$V_6 - V_5 = \underline{0.0600\,\text{V}}$; $V_6 - V_5$ is positive so S_6 is at higher potential.

($\vec{E}$ points from high to low potential.)

24-73

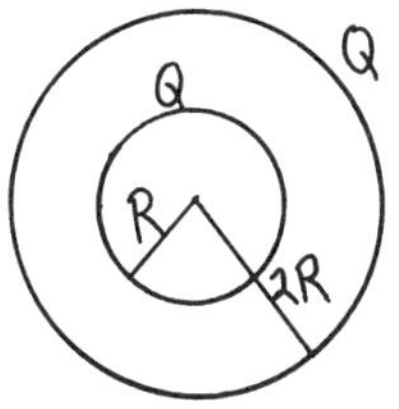

For $R < r < 2R$, $E = \frac{Q}{4\pi\epsilon_0 r^2}$

Let a be at $r_a = R$ and b be at $r_b = 2R$.

$$V_a - V_b = \int_a^b \vec{E}\cdot d\vec{l} = \int_{r_a}^{r_b} \frac{Q}{4\pi\epsilon_0 r^2}\,dr = \frac{Q}{4\pi\epsilon_0}\left(-\frac{1}{r}\Big|_R^{2R}\right) = \frac{Q}{4\pi\epsilon_0}\left(\frac{1}{R} - \frac{1}{2R}\right)$$

$$\underline{V_a - V_b = \frac{Q}{8\pi\epsilon_0 R}}$$

24-77

a)

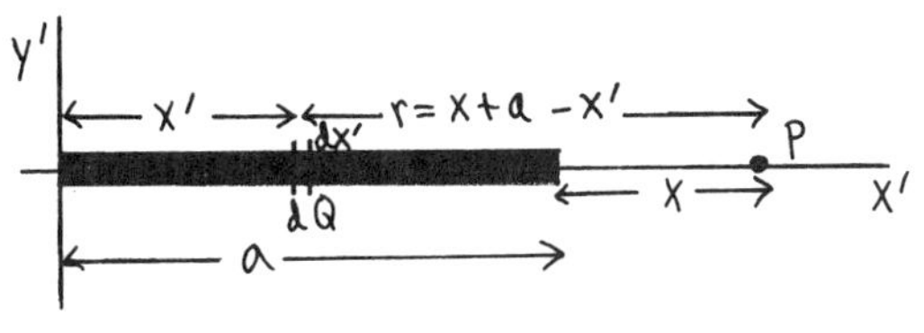

Use coordinates with the origin at the left-hand end of the rod and one axis along the rod. Call the axes x' and y' so as not to confuse them with the distance x given in the problem.

24-77 (cont)

Slice the charged rod up into thin slices of width dx'. Each slice has charge $dQ = Q\frac{dx'}{a}$ and is a distance $r = x+a-x'$ from point P. The potential at P due to the small slice dQ is $dV = \frac{1}{4\pi\epsilon_0}\frac{dQ}{r} = \frac{1}{4\pi\epsilon_0}\frac{Q}{a}\frac{dx'}{x+a-x'}$.

Compute the total V at P due to the whole rod by integrating dV over the length of the rod ($x'=0$ to $x'=a$):

$$V = \int dV = \frac{Q}{4\pi\epsilon_0 a}\int_0^a \frac{dx'}{(x+a-x')} = \frac{Q}{4\pi\epsilon_0 a}\left(-\ln(x+a-x')\Big|_0^a\right) = \frac{Q}{4\pi\epsilon_0 a}\ln\left(\frac{x+a}{x}\right)$$

Note: As $x \to \infty$, $V \to \frac{Q}{4\pi\epsilon_0 a}\ln\left(\frac{x}{x}\right) = 0$.

b)

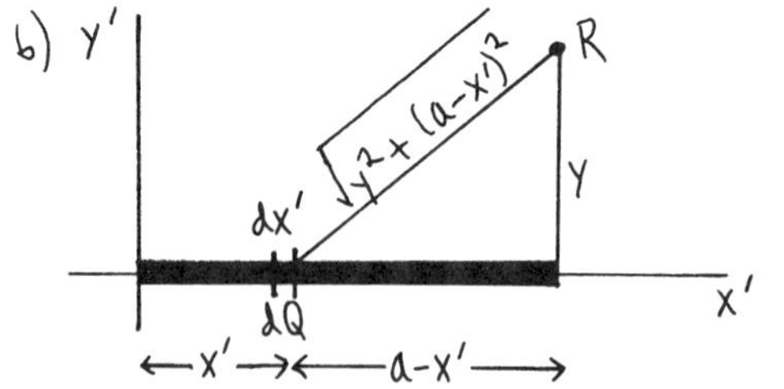

$dQ = \frac{Q}{a}dx'$ as in part (a)
Each slice dQ is a distance $r = \sqrt{y^2+(a-x')^2}$ from point R.

The potential dV at R due to the small slice dQ is $dV = \frac{1}{4\pi\epsilon_0}\frac{dQ}{r} = \frac{1}{4\pi\epsilon_0}\frac{Q}{a}\frac{dx'}{\sqrt{y^2+(a-x')^2}}$

$$V = \int dV = \int_0^a \frac{1}{4\pi\epsilon_0}\frac{Q}{a}\frac{dx'}{\sqrt{y^2+(a-x')^2}} = \frac{Q}{4\pi\epsilon_0 a}\int_0^a \frac{dx'}{y^2+(a-x')^2}$$

In the integral make the change of variable $u = a-x'$; $du = -dx'$

$$V = -\frac{Q}{4\pi\epsilon_0 a}\int_a^0 \frac{du}{\sqrt{y^2+u^2}} = -\frac{Q}{4\pi\epsilon_0 a}\left[\ln(u+\sqrt{y^2+u^2})\Big|_a^0\right] = -\frac{Q}{4\pi\epsilon_0 a}\left[\ln y - \ln(a+\sqrt{y^2+a^2})\right]$$

$V = \frac{Q}{4\pi\epsilon_0 a}\left[\ln\frac{a+\sqrt{a^2+y^2}}{y}\right]$ (The expression for the integral was found in Appendix B.)

Note: As $y\to\infty$, $V \to \frac{Q}{4\pi\epsilon_0 a}\ln\left(\frac{y}{y}\right) = 0$.

c) part (a): $V = \frac{Q}{4\pi\epsilon_0 a}\ln\left(\frac{x+a}{x}\right) = \frac{Q}{4\pi\epsilon_0 a}\ln\left(1+\frac{a}{x}\right)$

From Appendix B, $\ln(1+u) = u - \frac{u^2}{2} + \ldots \Rightarrow \ln\left(1+\frac{a}{x}\right) = \frac{a}{x} - \frac{a^2}{2x^2} \to \frac{a}{x}$ when x is large.

Thus $V \to \frac{Q}{4\pi\epsilon_0 a}\left(\frac{a}{x}\right) = \frac{Q}{4\pi\epsilon_0 x}$.

For large x, V becomes the potential of a point charge.

part (b): $V = \frac{Q}{4\pi\epsilon_0 a}\ln\left[\frac{a+\sqrt{a^2+y^2}}{y}\right] = \frac{Q}{4\pi\epsilon_0 a}\ln\left(\frac{a}{y}+\sqrt{1+\frac{a^2}{y^2}}\right)$

From the binomial theorem (Appendix B), $\sqrt{1+a^2/y^2} = (1+a^2/y^2)^{1/2} = 1+\frac{a^2}{2y^2}+\ldots$

Thus $\frac{a}{y}+\sqrt{1+\frac{a^2}{y^2}} \to 1+\frac{a}{y}+\frac{a^2}{2y^2}+\ldots \to 1+\frac{a}{y}$. Using $\ln(1+u)\approx u$ gives

$V \to \frac{Q}{4\pi\epsilon_0 a}\ln\left(1+\frac{a}{y}\right) \to \frac{Q}{4\pi\epsilon_0 a}\left(\frac{a}{y}\right) = \frac{Q}{4\pi\epsilon_0 y}$.

For large y, V becomes the potential of a point charge.

24-79

a) The potential at the surface of a charged conducting sphere is given by Example 24-8: $V = \frac{1}{4\pi\epsilon_0}\frac{q}{R}$. For spheres A and B this gives $V_A = \frac{Q_A}{4\pi\epsilon_0 R_A}$ and $V_B = \frac{Q_B}{4\pi\epsilon_0 R_B}$.

24-79 (cont)

$V_A = V_B \Rightarrow \dfrac{Q_A}{4\pi\epsilon_0 R_A} = \dfrac{Q_B}{4\pi\epsilon_0 R_B}$.

$$\frac{Q_B}{Q_A} = \frac{R_B}{R_A}$$

And then $R_A = 3R_B \Rightarrow \dfrac{Q_B}{Q_A} = \dfrac{1}{3}$

b) The electric field at the surface of a charged conducting sphere is given in Example 23-5: $E = \dfrac{1}{4\pi\epsilon_0}\dfrac{|q|}{R^2}$. For spheres A and B this gives $E_A = \dfrac{|Q_A|}{4\pi\epsilon_0 R_A^2}$ and $E_B = \dfrac{|Q_B|}{4\pi\epsilon_0 R_B^2}$. $\dfrac{E_B}{E_A} = \dfrac{|Q_B|}{4\pi\epsilon_0 R_B^2}\,\dfrac{4\pi\epsilon_0 R_A^2}{|Q_A|} = \left|\dfrac{Q_B}{Q_A}\right|\left(\dfrac{R_A}{R_B}\right)^2 = \frac{1}{3}(3)^2 = 3.$

24-81

a) $E = \dfrac{1}{4\pi\epsilon_0}\dfrac{Q}{R^2} = (8.988\times10^9\ \text{N}\cdot\text{m}^2/\text{C}^2)\dfrac{3.50\times10^{-8}\ \text{C}}{(0.240\ \text{m})^2} = \underline{5.46\times10^3\ \text{V/m}}$

$V = \dfrac{1}{4\pi\epsilon_0}\dfrac{Q}{R} = RE = (0.240\ \text{m})(5.46\times10^3\ \text{V/m}) = \underline{1.31\times10^3\ \text{V}}$

b) Two conditions must be met:
1) Let q_1 and q_2 be the final charges on each sphere. Then $q_1 + q_2 = Q$ (charge conservation)
2) Let V_1 and V_2 be the final potentials of each sphere. All points of a conductor are at the same potential, so $V_1 = V_2$.

$$V_1 = V_2 \Rightarrow \frac{1}{4\pi\epsilon_0}\frac{q_1}{R_1} = \frac{1}{4\pi\epsilon_0}\frac{q_2}{R_2}$$

$$q_1 = \left(\frac{R_1}{R_2}\right)q_2 = \left(\frac{0.240\ \text{m}}{0.062\ \text{m}}\right)q_2 = 3.871q_2$$

Use this in the first equation:

$$3.871q_2 + q_2 = Q_1$$

$$q_2 = \frac{3.50\times10^{-8}\ \text{C}}{4.871} = \underline{7.19\times10^{-9}\ \text{C}}$$

And then $q_1 = 3.871q_2 = \underline{2.78\times10^{-8}\ \text{C}}$

c) $V_1 = \dfrac{1}{4\pi\epsilon_0}\dfrac{q_1}{R_1} = (8.988\times10^9\ \text{N}\cdot\text{m}^2/\text{C}^2)\dfrac{2.78\times10^{-8}\ \text{C}}{0.240\ \text{m}} = \underline{1.04\times10^3\ \text{V}}$

$V_2 = \dfrac{1}{4\pi\epsilon_0}\dfrac{q_2}{R_2} = (8.988\times10^9\ \text{N}\cdot\text{m}^2/\text{C}^2)\dfrac{7.19\times10^{-9}\ \text{C}}{0.062\ \text{m}} = \underline{1.04\times10^3\ \text{V}}$ ($V_1 = V_2$, as it should)

d) $E_1 = \dfrac{V_1}{R_1} = \dfrac{1.04\times10^3\ \text{V}}{0.240\ \text{m}} = \underline{4.33\times10^3\ \text{V/m}}$

$E_2 = \dfrac{V_2}{R_2} = \dfrac{1.04\times10^3\ \text{V}}{0.062\ \text{m}} = \underline{1.68\times10^4\ \text{V/m}}$

CHAPTER 25

Exercises 1, 5, 7, 9, 11, 19, 23, 25, 27, 29, 31, 35

Problems 41, 43, 45, 47, 49, 51, 53, 57

Exercises

25-1

a) $C = \dfrac{Q}{V_{ab}} \Rightarrow V_{ab} = \dfrac{Q}{C} = \dfrac{0.346\times10^{-6}\,C}{500\times10^{-12}\,F} = \underline{692\,V}$

b) $C = \dfrac{\epsilon_0 A}{d} \Rightarrow A = \dfrac{Cd}{\epsilon_0} = \dfrac{(500\times10^{-12}F)(0.453\times10^{-3}m)}{8.854\times10^{-12}\,C^2/N\cdot m^2} = \underline{0.0256\,m^2}$

c) $V_{ab} = Ed \Rightarrow E = \dfrac{V_{ab}}{d} = \dfrac{692\,V}{0.453\times10^{-3}m} = \underline{1.53\times10^6\,V/m}$

d) $E = \dfrac{\sigma}{\epsilon_0} \Rightarrow \sigma = E\epsilon_0 = (1.53\times10^6 V/m)(8.854\times10^{-12}C^2/N\cdot m^2) = \underline{1.35\times10^{-5}\,C/m^2}$

or
$\sigma = \dfrac{Q}{A} = \dfrac{0.346\times10^{-6}C}{0.0256\,m^2} = 1.35\times10^{-5}\,C/m^2$, which checks

25-5

a) From Example 25-3, $C = 4\pi\epsilon_0 \dfrac{r_a r_b}{r_b - r_a} = 4\pi(8.854\times10^{-12}F/m)\dfrac{(0.120m)(0.150m)}{0.150m - 0.120m}$

$C = 6.68\times10^{-11}F = \underline{66.8\,pF}$

b) From Example 25-3, $E = \dfrac{Q}{4\pi\epsilon_0 r^2}$ between the spheres.

$C = \dfrac{Q}{V} \Rightarrow Q = CV = (66.8\times10^{-12}F)(140V) = 9.352\times10^{-9}C$

$E = \dfrac{Q}{4\pi\epsilon_0 r^2} = \dfrac{(8.988\times10^9 N\cdot m^2/C^2)(9.352\times10^{-9}C)}{r^2} = \dfrac{84.06\,N\cdot m^2/C}{r^2}$

$r = 0.121m \Rightarrow E = \dfrac{84.06\,N\cdot m^2/C}{(0.121\,m)^2} = \underline{5.74\times10^3\,V/m}$

c) $r = 0.149\,m \Rightarrow E = \dfrac{84.06\,N\cdot m^2/C}{(0.149m)^2} = \underline{3.79\times10^3 V/m}$

d) No, the results of parts (b) and (c) show that E is not uniform in the region between the spheres.

25-7

a) From Example 25-4, $C = \dfrac{2\pi\epsilon_0 L}{\ln(r_b/r_a)} = \dfrac{2\pi(8.854\times10^{-12}N\cdot m^2/C^2)(3.5m)}{\ln(4.0mm/2.5mm)} = 4.143\times10^{-10}\,F$

$Q = CV = (4.143\times10^{-10}F)(350\times10^{-3}V) = \underline{1.4\times10^{-10}C}$

b) $\dfrac{C}{L} = \dfrac{4.143\times10^{-10}F}{3.5m} = \underline{1.2\times10^{-10}\,F/m}$

25-9

Do parts (a) and (b) together.

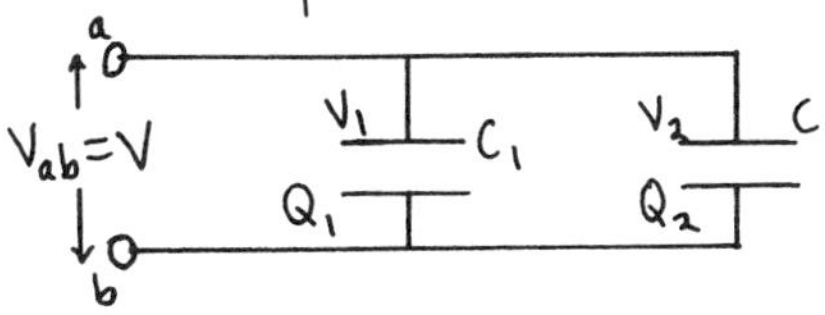

$$V_1 = V_2 = V$$

$$\underline{V_1 = 67.5\text{ V}} \; ; \; \underline{V_2 = 67.5\text{ V}}$$

$$C = \frac{Q}{V} \Rightarrow Q = CV$$

$$Q_1 = C_1V_1 = (4.00\times10^{-6}\text{ F})(67.5\text{ V}) = \underline{2.70\times10^{-4}\text{ C}}$$

$$Q_2 = C_2V_2 = (6.00\times10^{-6}\text{ F})(67.5\text{ V}) = \underline{4.05\times10^{-4}\text{ C}}$$

25-11

Do parts (a) and (b) together.

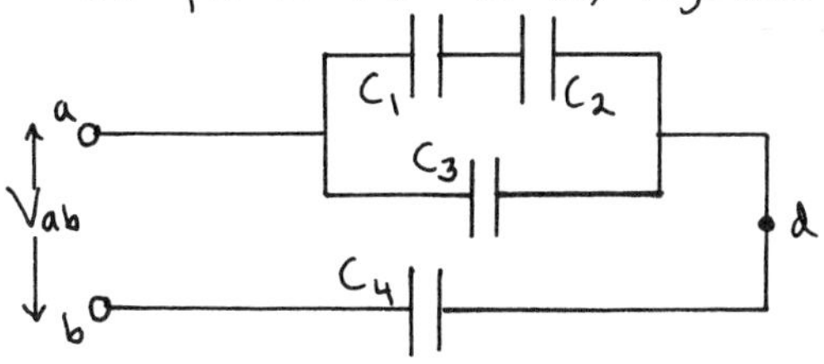

$$C_1 = C_2 = C_3 = C_4 = 2.00\,\mu\text{F}$$

$$V_{ab} = 40.4\text{ V}$$

Simplify the circuit by replacing the capacitor combinations by their equivalents:

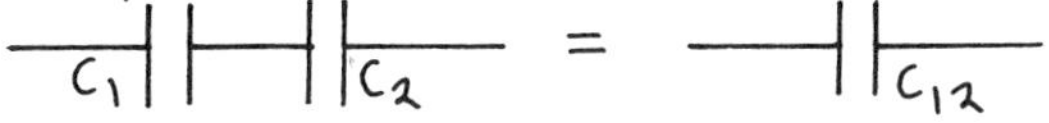

$$\frac{1}{C_{12}} = \frac{1}{C_1} + \frac{1}{C_2} = \frac{C_1+C_2}{C_1C_2}$$

$$C_{12} = \frac{C_1C_2}{C_1+C_2} = \frac{(2.00\times10^{-6}\text{ F})(2.00\times10^{-6}\text{ F})}{2.00\times10^{-6}\text{ F} + 2.00\times10^{-6}\text{ F}} = 1.00\times10^{-6}\text{ F}$$

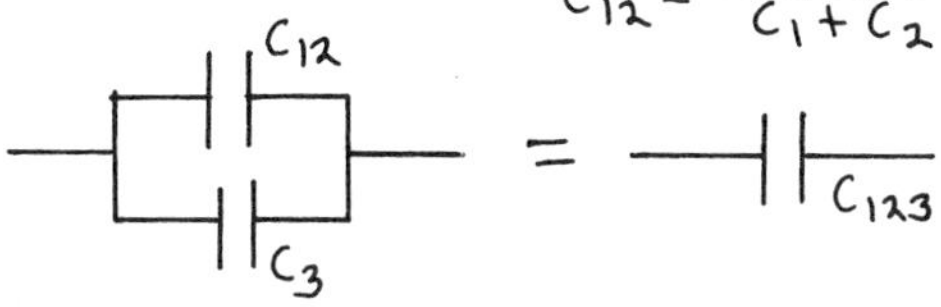

$$C_{123} = C_{12} + C_3 = 1.00\times10^{-6}\text{ F} + 2.00\times10^{-6}\text{ F} = 3.00\times10^{-6}\text{ F}$$

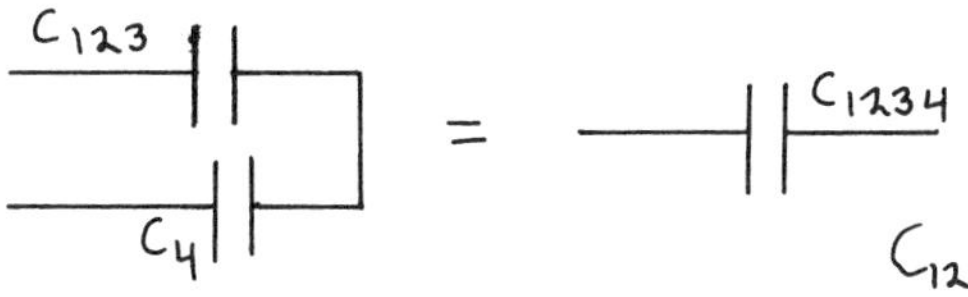

$$\frac{1}{C_{1234}} = \frac{1}{C_{123}} + \frac{1}{C_4} = \frac{C_{123}+C_4}{C_{123}C_4}$$

$$C_{1234} = \frac{C_{123}C_4}{C_{123}+C_4} = \frac{(3.00\times10^{-6}\text{ F})(2.00\times10^{-6}\text{ F})}{3.00\times10^{-6}\text{ F} + 2.00\times10^{-6}\text{ F}} = 1.20\times10^{-6}\text{ F}$$

The circuit is equivalent to

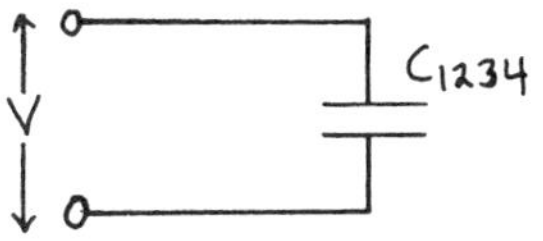

$$V_{1234} = V = 40.4\text{ V}$$

$$Q_{1234} = C_{1234}V = (1.20\times10^{-6}\text{ F})(40.4\text{ V}) = 48.5\,\mu\text{C}$$

Now build back up the original circuit, step by step:

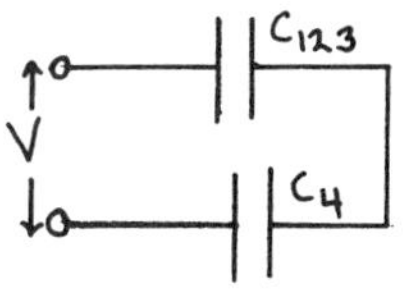

$Q_{123} = Q_4 = Q_{1234} = 48.5\,\mu\text{C}$ (charge same for capacitors in series)

The $V_{123} = \dfrac{Q_{123}}{C_{123}} = \dfrac{48.5\,\mu\text{C}}{3.00\,\mu\text{F}} = 16.2\text{ V}$

$$V_4 = \frac{Q_4}{C_4} = \frac{48.5\,\mu\text{C}}{2.00\,\mu\text{F}} = 24.2\text{ V}$$

Note that $V_4 + V_{123} = 24.2\text{ V} + 16.2\text{ V} = 40.4\text{ V}$, as it should.

25-11 (cont)

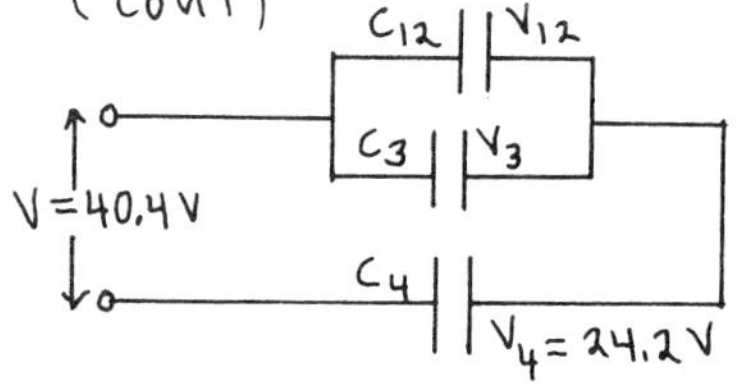

$V_3 = V_{12} = 40.4\text{ V} - V_4 = 16.2\text{ V}$

$Q_3 = C_3 V_3 = (2.00\times10^{-6}\text{ F})(16.2\text{ V}) = 32.4\ \mu\text{C}$
$Q_{12} = C_{12} V_{12} = (1.00\times10^{-6}\text{ F})(16.2\text{ V}) = 16.2\ \mu\text{C}$

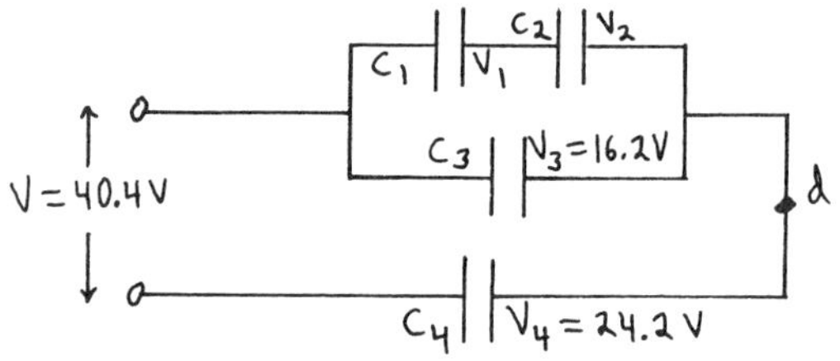

$Q_1 = Q_2 = Q_{12} = 16.2\ \mu\text{C}$ (charge same for capacitors in series)

$V_1 = \dfrac{Q_1}{C_1} = \dfrac{16.2\times10^{-6}\text{ C}}{2.00\times10^{-6}\text{ F}} = 8.1\text{ V}$

$V_2 = \dfrac{Q_2}{C_2} = \dfrac{16.2\times10^{-6}\text{ C}}{2.00\times10^{-6}\text{ F}} = 8.1\text{ V}$

Note that $V_1 + V_2 = 16.2\text{ V}$, which equals V_3 as it should.

Summary: $Q_1 = 16.2\ \mu\text{C}$, $V_1 = 8.1\text{ V}$
$Q_2 = 16.2\ \mu\text{C}$, $V_2 = 8.1\text{ V}$
$Q_3 = 32.4\ \mu\text{C}$, $V_3 = 16.2\text{ V}$
$Q_4 = 48.5\ \mu\text{C}$, $V_4 = 24.2\text{ V}$

c) $V_{ad} = V_3 = \underline{16.2\text{ V}}$

25-19

The energy density is given by Eq. (25-11): $u = \frac{1}{2}\epsilon_0 E^2$

Calculate E: $E = \dfrac{V}{d} = \dfrac{500\text{ V}}{4.00\times10^{-3}\text{ m}} = 1.25\times10^5\text{ V/m}$.

Then $u = \frac{1}{2}\epsilon_0 E^2 = \frac{1}{2}(8.854\times10^{-12}\text{ C}^2/\text{N}\cdot\text{m}^2)(1.25\times10^5\text{ V/m})^2 = \underline{0.0692\text{ J/m}^3}$

25-23

a) From Exercise 25-5, $E = 5.74\times10^3\text{ V/m}$ at $r = 12.1\text{ cm}$.
$u = \frac{1}{2}\epsilon_0 E^2 = \frac{1}{2}(8.854\times10^{-12}\text{ N}\cdot\text{m}^2/\text{C}^2)(5.74\times10^3\text{ V/m})^2 = \underline{1.46\times10^{-4}\text{ J/m}^3}$

b) From Exercise 25-5, $E = 3.78\times10^3\text{ V/m}$ at $r = 14.9\text{ cm}$.
$u = \frac{1}{2}\epsilon_0 E^2 = \frac{1}{2}(8.854\times10^{-12}\text{ N}\cdot\text{m}^2/\text{C}^2)(3.78\times10^3\text{ V/m})^2 = \underline{6.33\times10^{-5}\text{ J/m}^3}$

c) No, the results of parts (a) and (b) show that the energy density is not uniform in the region between the spheres.

25-25

a) $U = \dfrac{Q^2}{2C}$; $Q = q$ and $C = \dfrac{\epsilon_0 A}{x} \Rightarrow U = \dfrac{xq^2}{2\epsilon_0 A}$

b) $x \to x + dx \Rightarrow U = \dfrac{(x+dx)q^2}{2\epsilon_0 A}$

25-25 (cont)

c) $$dW = \frac{(x+dx)q^2}{2\epsilon_0 A} - \frac{xq^2}{2\epsilon_0 A} = \frac{dx\, q^2}{2\epsilon_0 A}$$

But also $dW = F\,dx \Rightarrow F = \frac{q^2}{2\epsilon_0 A}$

d) 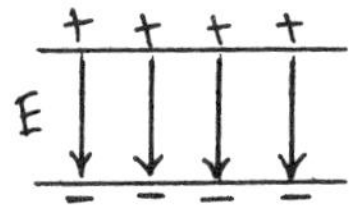

$$E = \frac{\sigma}{\epsilon_0} = \frac{q}{\epsilon_0 A} \Rightarrow F = \tfrac{1}{2} q E, \text{ not } qE$$

The reason for the difference is that E is the field due to both plates. If we consider the positive plate only and calculate its electric field using Gauss's law:

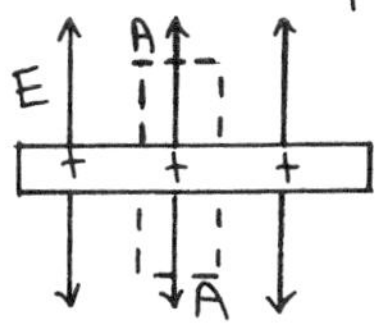

$$\oint \vec{E} \cdot d\vec{A} = \frac{Q_{encl}}{\epsilon_0}$$

$$2EA = \frac{\sigma A}{\epsilon_0} \Rightarrow E = \frac{\sigma}{2\epsilon_0} = \frac{q}{2\epsilon_0 A}$$

The force this field exerts on the other plate, that has charge $-q$, is $F = \frac{q^2}{2\epsilon_0 A}$.

25-27

$C = K\epsilon_0 \frac{A}{d}$. Minimum $A \Rightarrow$ smallest possible d.

d is limited by the requirement that E be less than 2.00×10^7 V/m when V is as large as 6000 V.

$$V = Ed \Rightarrow d = \frac{V}{E} = \frac{6000\text{ V}}{2.00 \times 10^7\text{ V/m}} = 3.00 \times 10^{-4}\text{ m}$$

Then $$A = \frac{Cd}{K\epsilon_0} = \frac{(1.37 \times 10^{-9}\text{ F})(3.00 \times 10^{-4}\text{ m})}{(3.40)(8.854 \times 10^{-12}\text{ C}^2/\text{N}\cdot\text{m}^2)} = \underline{0.0137\text{ m}^2}.$$

25-29

$$E = \frac{E_0}{K} \Rightarrow K = \frac{E_0}{E} = \frac{3.60 \times 10^5\text{ V/m}}{1.80 \times 10^5\text{ V/m}} = 2.00$$

a) $$\sigma_i = \sigma\left(1 - \frac{1}{K}\right)$$

$$\sigma = \epsilon_0 E_0 = (8.854 \times 10^{-12}\text{ C}^2/\text{N}\cdot\text{m}^2)(3.60 \times 10^5\text{ N/C}) = 3.187 \times 10^{-6}\text{ C/m}^2$$

$$\sigma_i = (3.187 \times 10^{-6}\text{ C/m}^2)\left(1 - \frac{1}{2.00}\right) = \underline{1.59 \times 10^{-6}\text{ C/m}^2}$$

b) As calculated above, $\underline{K = 2.00}$.

25-31

a) Since the capacitor remains connected to the power supply the potential difference doesn't change when the dielectric is inserted. Before the dielectric is inserted

$$U_0 = \tfrac{1}{2} C_0 V^2 \Rightarrow V = \sqrt{\frac{2U_0}{C_0}} = \sqrt{\frac{2(1.99 \times 10^{-5}\text{ J})}{255 \times 10^{-9}\text{ F}}} = \underline{12.5\text{ V}}$$

b) $$K = \frac{C}{C_0}$$

$$U_0 = \tfrac{1}{2} C_0 V^2,\quad U = \tfrac{1}{2} C V^2 \Rightarrow \frac{C}{C_0} = \frac{U}{U_0}$$

$$K = \frac{U}{U_0} = \frac{1.99 \times 10^{-5}\text{ J} + 2.69 \times 10^{-5}\text{ J}}{1.99 \times 10^{-5}\text{ J}} = \underline{2.35}$$

25-35

a)

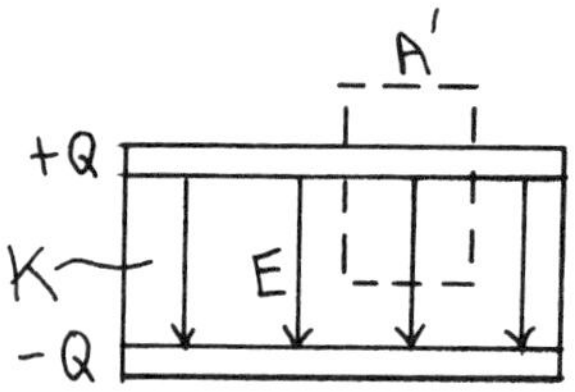

Apply Eq. (25-22) to the dotted surface:

$$\oint K\vec{E}\cdot d\vec{A} = \frac{Q_{encl\text{-}free}}{\epsilon_0}$$

$\oint K\vec{E}\cdot d\vec{A} = KEA'$ since $E=0$ outside the plates

$$Q_{encl\text{-}free} = \sigma A' = \frac{Q}{A}A'$$

Thus $KEA' = \dfrac{\frac{Q}{A}A'}{\epsilon_0} \Rightarrow E = \dfrac{Q}{\epsilon_0 A K}$

b) $V = Ed = \dfrac{Qd}{\epsilon_0 A K}$

c) $C = \dfrac{Q}{V} = \dfrac{Q}{\frac{Qd}{\epsilon_0 AK}} = K\dfrac{\epsilon_0 A}{d} = KC_0 \Rightarrow K = \dfrac{C}{C_0}$, which is Eq. (25-12).

Problems

25-41

a) $C = \dfrac{\epsilon_0 A}{d} = \dfrac{(8.854\times10^{-12}\,C^2/N\cdot m^2)(0.18\,m)^2}{1.16\times10^{-2}\,m} = 2.47\times10^{-11}\,F = \underline{24.7\,pF}$

b) Remains connected to the battery $\Rightarrow$ V stays 50.0 V

$Q = CV = (2.47\times10^{-11}\,F)(50.0\,V) = \underline{1.24\times10^{-9}\,C}$

c) $E = \dfrac{V}{d} = \dfrac{50.0\,V}{1.16\times10^{-2}\,m} = \underline{4.31\times10^{3}\,V/m}$

d) $U = \frac{1}{2}QV = \frac{1}{2}(1.24\times10^{-9}\,C)(50.0\,V) = \underline{3.10\times10^{-8}\,J}$

25-43

a)

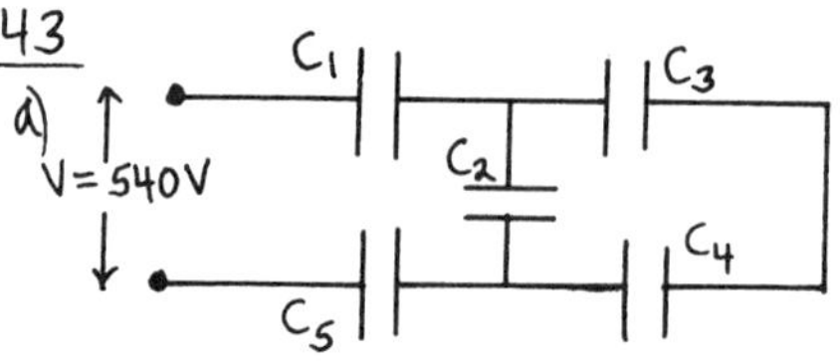

$C_1 = C_5 = 4.6\,\mu F$

$C_2 = C_3 = C_4 = 2.3\,\mu F$

Simplify the circuit by replacing the capacitor combinations by their equivalents:

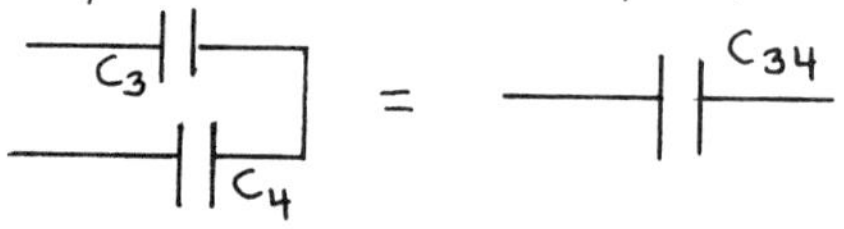

$$\frac{1}{C_{34}} = \frac{1}{C_3} + \frac{1}{C_4} = \frac{C_3+C_4}{C_3C_4}$$

$$C_{34} = \frac{C_3C_4}{C_3+C_4} = \frac{(2.3\times10^{-6}\,F)(2.3\times10^{-6}\,F)}{2.3\times10^{-6}\,F + 2.3\times10^{-6}\,F} = 1.15\times10^{-6}\,F$$

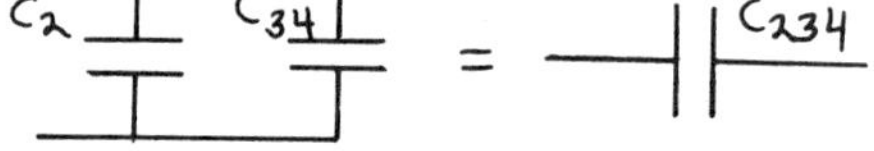

$$C_{234} = C_2 + C_{34} = 2.3\,\mu F + 1.15\,\mu F = 3.45\,\mu F$$

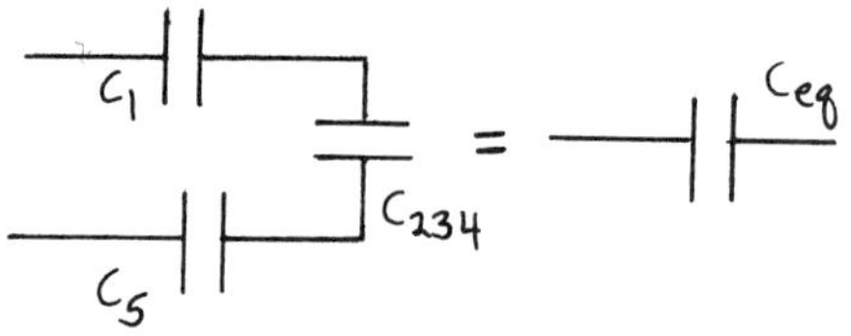

$$\frac{1}{C_{eq}} = \frac{1}{C_1} + \frac{1}{C_5} + \frac{1}{C_{234}} = \frac{2}{4.6\,\mu F} + \frac{1}{3.45\,\mu F}$$

$C_{eq} = \underline{1.38\,\mu F}$

25-43 (cont)

b) Now build the original circuit back up, piece by piece:

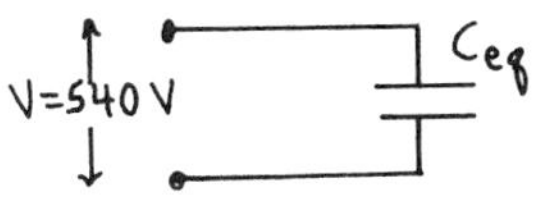

$$Q_{eq} = C_{eq}V = (1.38\,\mu F)(540V) = 745\,\mu C$$

$$Q_1 = Q_5 = Q_{234} = 745\,\mu C \text{ (capacitors in series have same charge)}$$

$$V_1 = \frac{Q_1}{C_1} = \frac{745\,\mu C}{4.6\,\mu F} = 162\,V$$

$$V_5 = \frac{Q_5}{C_5} = \frac{745\,\mu C}{4.6\,\mu F} = 162\,V$$

$$V_{234} = \frac{Q_{234}}{C_{234}} = \frac{745\,\mu C}{3.45\,\mu F} = 216\,V$$

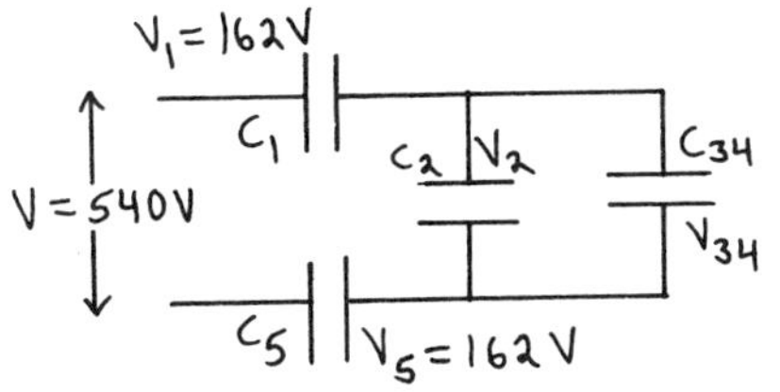

$V_2 = V_{34} = V_{234} = 216V$ (capacitors in parallel have the same potential)

$$Q_2 = C_2V_2 = (2.3\,\mu F)(216\,V) = 497\,\mu C$$

$$Q_{34} = C_{34}V_{34} = (1.15\,\mu F)(216\,V) = 248\,\mu C$$

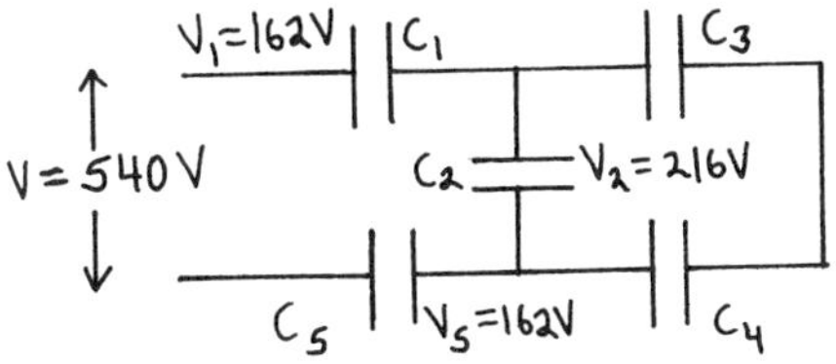

$Q_3 = Q_4 = Q_{34} = 248\,\mu C$ (capacitors in series have same charge)

$$V_3 = \frac{Q_3}{C_3} = \frac{248\,\mu C}{2.3\,\mu F} = 108\,V$$

$$V_4 = \frac{Q_4}{C_4} = \frac{248\,\mu C}{2.3\,\mu F} = 108\,V$$

Summary: $Q_1 = 745\,\mu C$, $V_1 = 162V$
$Q_2 = 497\,\mu C$, $V_2 = 216V$
$Q_3 = 248\,\mu C$, $V_3 = 108V$
$Q_4 = 248\,\mu C$, $V_4 = 108V$
$Q_5 = 745\,\mu C$, $V_5 = 162V$

Note: $V_3 + V_4 = V_2$
$V_1 + V_2 + V_5 = 540V$
$Q_1 = Q_2 + Q_3$
$Q_5 = Q_2 + Q_4$

25-45

a)

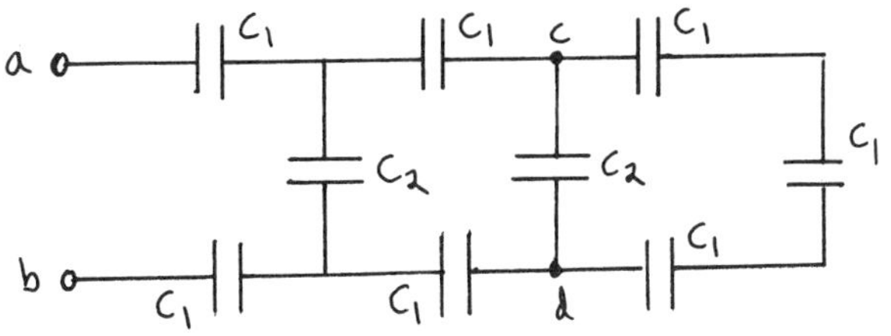

$C_1 = 9.3\,\mu F$
$C_2 = 6.2\,\mu F$

Simplify the network by replacing the capacitor combinations by their equivalents:

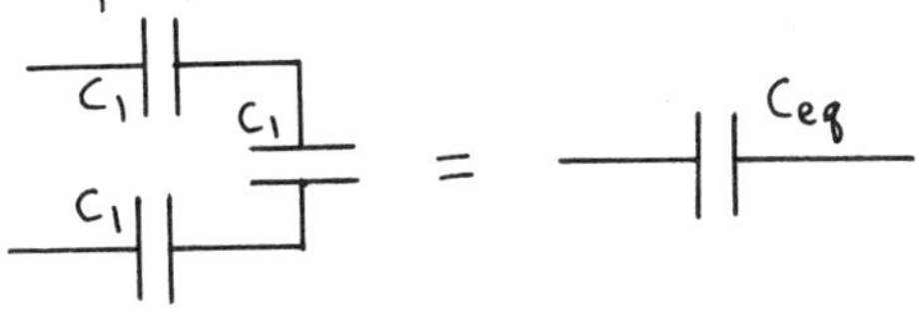

$$\frac{1}{C_{eq}} = \frac{3}{C_1}$$

$$C_{eq} = \frac{C_1}{3} = \frac{9.3\,\mu F}{3} = 3.1\,\mu F$$

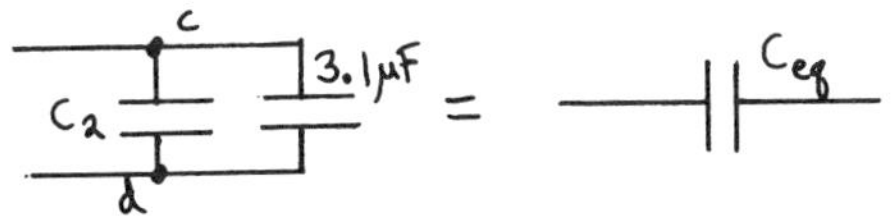

$$C_{eq} = 3.1\,\mu F + C_2 = 3.1\,\mu F + 6.2\,\mu F = 9.3\,\mu F$$

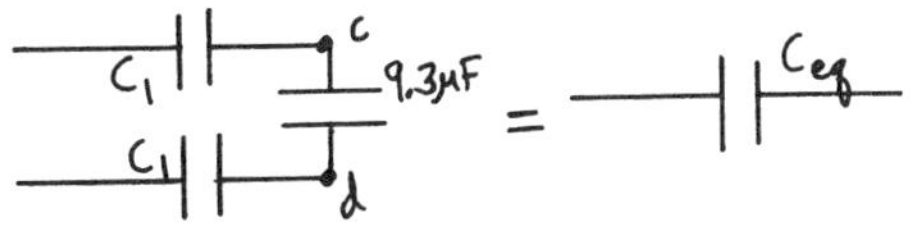

$$\frac{1}{C_{eq}} = \frac{2}{C_1} + \frac{1}{9.3\,\mu F} = \frac{3}{9.3\,\mu F}$$

$$C_{eq} = 3.1\,\mu F$$

25-45 (cont)

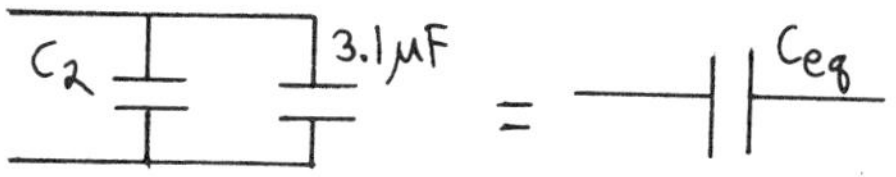

$C_{eq} = C_2 + 3.1\mu F = 6.2\mu F + 3.1\mu F$

$C_{eq} = 9.3\mu F$

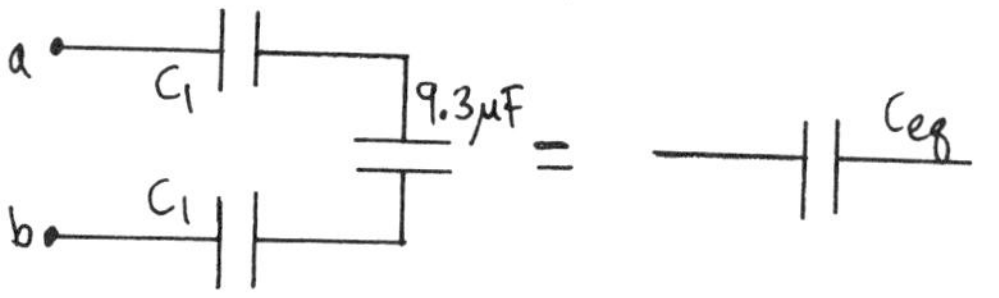

$\frac{1}{C_{eq}} = \frac{2}{C_1} + \frac{1}{9.3\mu F} = \frac{3}{9.3\mu F}$

$C_{eq} = \underline{3.1\mu F}$

b)

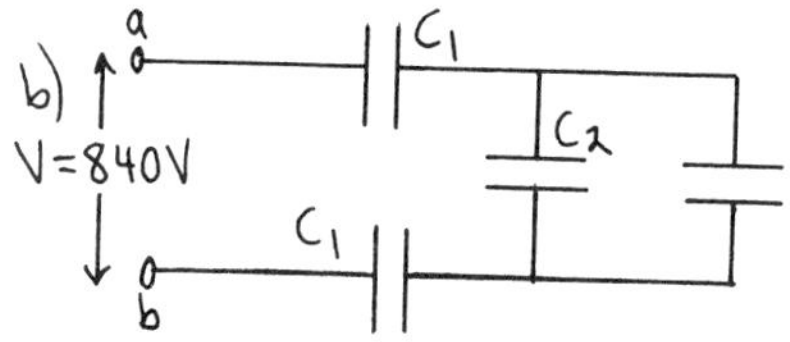

(From part (a) this is the equivalent capacitance of the rest of the network.)

The equivalent network is

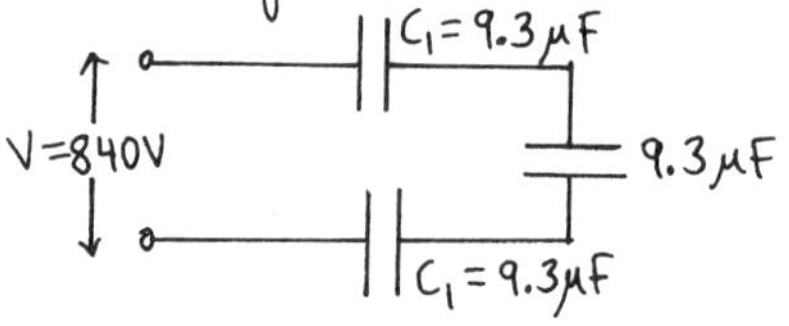

series ⇒ all three capacitors have same Q. But here all three have same C, so by $V = \frac{Q}{C}$ all three must have same V. The three V's must add to 840 V, so each capacitor has V = 280 V. The 9.3μF capacitor is the equivalent of C_2 and the 3.1μF capacitor in parallel, so $V_2 = 280$ V (capacitors in parallel have same potential difference).

$Q_1 = C_1 V_1 = (9.3\mu F)(280 V) = \underline{2.6 \times 10^{-3} C}$

$Q_2 = C_2 V_2 = (6.2\mu F)(280 V) = \underline{1.7 \times 10^{-3} C}$

c) From the potentials deduced in part (b) we have

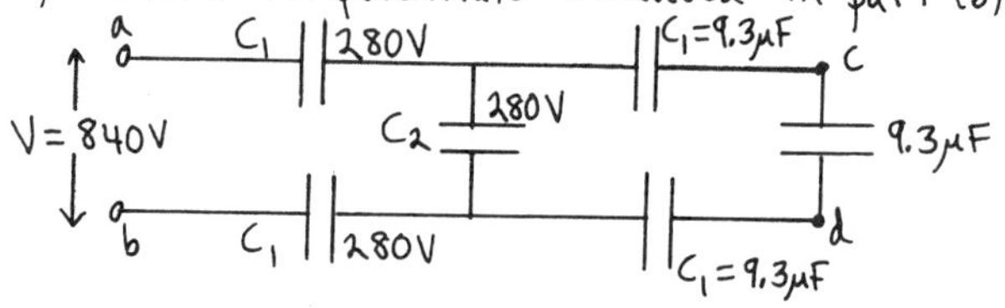

(From part (a) this is the equivalent capacitance of the rest of the network.)

The three right-most capacitors are in series and therefore have the same charge. But their capacitances also are equal, so by $V = \frac{Q}{C}$ they each have the same potential difference. Their potentials must sum to 280 V, so the potential across each is 93.3 V; $V_{cd} = \underline{93.3 V.}$

25-47

a)

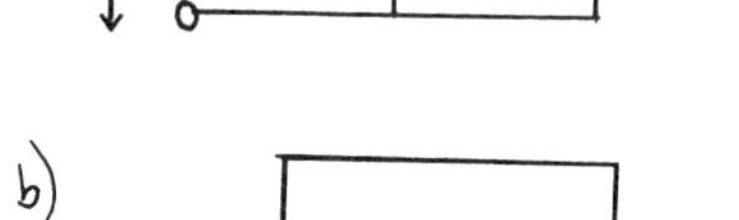

$V_1 = V_2 = \underline{1200 V}$

$Q_1 = C_1 V_1 = (1.00 \times 10^{-6} F)(1200 V) = \underline{1.20 \times 10^{-3} C}$

$Q_2 = C_2 V_2 = (2.00 \times 10^{-6} F)(1200 V) = \underline{2.40 \times 10^{-3} C}$

b)

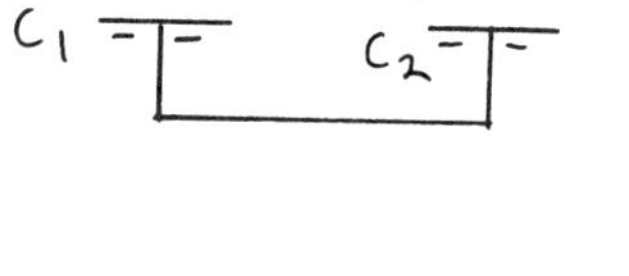

$q_1 + q_2 = Q_2 - Q_1$ (since terminals of unlike sign are connected)

$q_1 + q_2 = 2.40 \times 10^{-3} C - 1.20 \times 10^{-3} C = 1.20 \times 10^{-3} C$

25-47 (cont)

From the circuit diagram can see that $V_1 = V_2$.

$$V_1 = V_2 \Rightarrow \frac{q_1}{C_1} = \frac{q_2}{C_2}$$

$$q_2 = \frac{C_2}{C_1} q_1 = \frac{2.00\,\mu F}{1.00\,\mu F} q_1 = 2.00\, q_1$$

Use this result in the equation $q_1 + q_2 = 1.20\times10^{-3}\,C \Rightarrow 3.00 q_1 = 1.20\times10^{-3}\,C$

$q_1 = \underline{0.400\times10^{-3}\,C}$, $q_2 = \underline{0.800\times10^{-3}\,C}$

$V_1 = \frac{q_1}{C_1} = \frac{0.400\times10^{-3}\,C}{1.00\times10^{-6}\,F} = \underline{400\,V}$ $V_2 = \frac{q_2}{C_2} = \frac{0.800\times10^{-3}\,C}{2.00\times10^{-6}\,F} = \underline{400\,V}$ (so do have $V_1 = V_2$)

25-49

a)

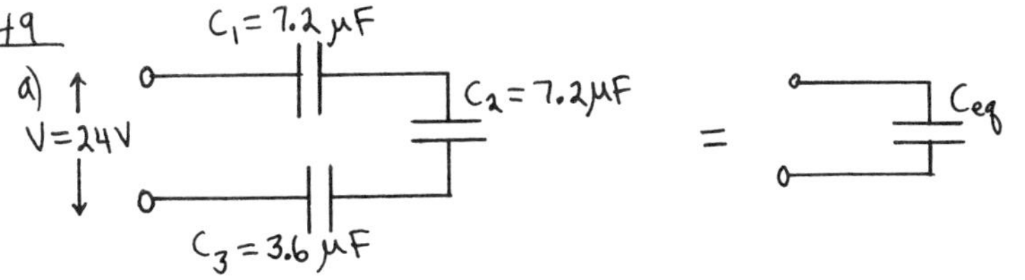

$$\frac{1}{C_{eq}} = \frac{1}{C_1} + \frac{1}{C_2} + \frac{1}{C_3} = \frac{4}{7.2\,\mu F}$$

$$C_{eq} = \frac{7.2\,\mu F}{4} = 1.8\,\mu F$$

$$Q = C_{eq} V = (1.8\,\mu F)(24\,V) = 43\,\mu C$$

The three capacitors are in series so they each have the same charge:

$Q_1 = Q_2 = Q_3 = \underline{43\,\mu C}$

b) $U = \frac{1}{2}Q_1V_1 + \frac{1}{2}Q_2V_2 + \frac{1}{2}Q_3V_3$

But $Q_1 = Q_2 = Q_3 = Q \Rightarrow U = \frac{1}{2}Q(V_1 + V_2 + V_3)$

But also $V_1 + V_2 + V_3 = V = 24\,V$

$\Rightarrow U = \frac{1}{2}QV = \frac{1}{2}(43\times10^{-6}\,C)(24\,V) = \underline{5.2\times10^{-4}\,J}$

c)

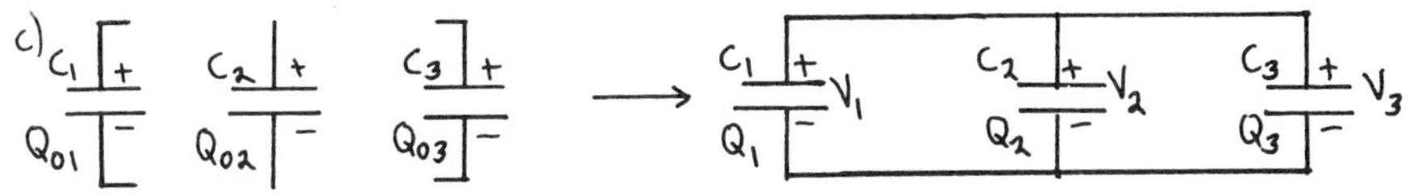

The total positive charge that is available to be distributed on the upper plates of the three capacitors is $Q_0 = Q_{01} + Q_{02} + Q_{03} = 3(43\,\mu C) = 129\,\mu C$.

Thus $\boxed{Q_1 + Q_2 + Q_3 = 129\,\mu C.}$

After the circuit is completed the charge distributes to make $V_1 = V_2 = V_3$.

$V = \frac{Q}{C}$ and $V_1 = V_2 \Rightarrow \frac{Q_1}{C_1} = \frac{Q_2}{C_2}$ and then $C_1 = C_2$ says that $Q_1 = Q_2$.

$V_1 = V_3 \Rightarrow \frac{Q_1}{C_1} = \frac{Q_3}{C_3} \Rightarrow Q_1 = Q_3\left(\frac{C_1}{C_3}\right) = Q_3\left(\frac{7.2\,\mu F}{3.6\,\mu F}\right) = 2Q_3$

Use $Q_2 = Q_1$ and $Q_1 = 2Q_3$ in the above equation $\Rightarrow 2Q_3 + 2Q_3 + Q_3 = 129\,\mu C$

$5Q_3 = 129\,\mu C \Rightarrow Q_3 = 26\,\mu C$

$Q_1 = Q_2 = 52\,\mu C$

Then $V_1 = \frac{Q_1}{C_1} = \frac{52\,\mu C}{7.2\,\mu F} = 7.2\,V$

$V_2 = \frac{Q_2}{C_2} = \frac{52\,\mu C}{7.2\,\mu F} = 7.2\,V$

$V_3 = \frac{Q_3}{C_3} = \frac{26\,\mu C}{3.6\,\mu F} = 7.2\,V$

The voltage across each capacitor in the parallel combination is $\underline{7.2\,V}$.

25-49 (cont)

d) $U = \frac{1}{2}Q_1V_1 + \frac{1}{2}Q_2V_2 + \frac{1}{2}Q_3V_3$

But $V_1 = V_2 = V_3 \Rightarrow U = \frac{1}{2}V_1(Q_1+Q_2+Q_3) = \frac{1}{2}(7.2\text{ V})(129\,\mu\text{C}) = \underline{4.6\times10^{-4}\text{ J}}$

This is less than the original energy of 5.2×10^{-4} J.

25-51

a) The conductor can be at some potential V, where V=0 far from the conductor. This potential depends on the charge Q on the conductor so can define $C = \frac{Q}{V}$ where C will not depend on V or Q.

b) For a solid conducting sphere $V = \frac{Q}{4\pi\epsilon_0 R}$ if $V=0$ at $r\to\infty$.

$C = \frac{Q}{V} = Q\left(\frac{4\pi\epsilon_0 R}{Q}\right) = 4\pi\epsilon_0 R$

c) $C = 4\pi\epsilon_0 R = 4\pi(8.854\times10^{-12}\text{ F/m})(6.38\times10^6\text{ m}) = 7.10\times10^{-4}\text{ F} = 710\,\mu\text{F}$.

The capacitance of the earth is about seven times larger than the largest capacitances in this range.

25-53

From Example 23-9, $E = \frac{1}{4\pi\epsilon_0}\frac{Qr}{R^3}$ for $r<R$ and $E = \frac{1}{4\pi\epsilon_0}\frac{Q}{r^2}$ for $r>R$.

a) $u = \frac{1}{2}\epsilon_0E^2 = \frac{1}{2}\epsilon_0\left(\frac{1}{4\pi\epsilon_0}\frac{Qr}{R^3}\right)^2 = \frac{1}{4\pi\epsilon_0}\frac{Q^2r^2}{8\pi R^6}$

b) $u = \frac{1}{2}\epsilon_0E^2 = \frac{1}{2}\epsilon_0\left(\frac{1}{4\pi\epsilon_0}\frac{Q}{r^2}\right)^2 = \frac{1}{4\pi\epsilon_0}\frac{Q^2}{8\pi r^4}$

c) $dU = u\,dV$ where $dV = 4\pi r^2dr$

For $r<R$, $U = \int_0^R u\,4\pi r^2dr = \frac{1}{4\pi\epsilon_0}\frac{Q^2}{2R^6}\int_0^R r^4dr = \frac{1}{4\pi\epsilon_0}\frac{Q^2}{2R^6}\left(\frac{R^5}{5}\right) = \frac{1}{4\pi\epsilon_0}\frac{Q^2}{10R}$

For $r>R$, $U = \int_R^\infty u\,4\pi r^2dr = \frac{1}{4\pi\epsilon_0}\frac{Q^2}{2}\int_R^\infty\frac{dr}{r^2} = \frac{1}{4\pi\epsilon_0}\frac{Q^2}{2}\left(-\frac{1}{r}\Big|_R^\infty\right) = \frac{1}{4\pi\epsilon_0}\frac{Q^2}{2R}$.

The total electric field energy is $U = \frac{1}{4\pi\epsilon_0}\frac{Q^2}{10R} + \frac{1}{4\pi\epsilon_0}\frac{Q^2}{2R} = \frac{1}{4\pi\epsilon_0}\frac{Q^2}{R}\left(\frac{1}{10}+\frac{1}{2}\right) = \frac{1}{4\pi\epsilon_0}\frac{3Q^2}{5R}$.

25-57

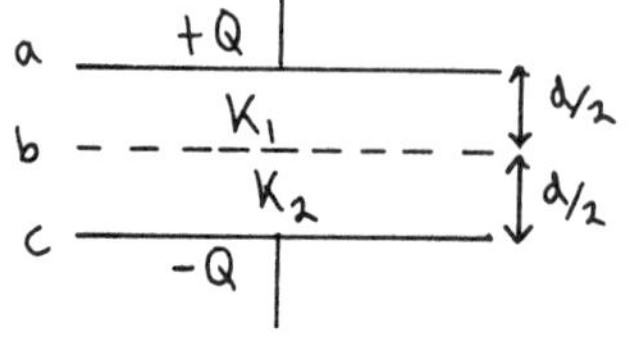

$C = \frac{Q}{V}$, so we need to calculate the effect of the dielectrics on the potential difference between the plates. Let the potential of the positive plate be V_a, the potential of the negative plate be V_c, and the potential midway between the plates where the dielectrics meet be V_b.

Then $C = \frac{Q}{V_a - V_c} = \frac{Q}{V_{ac}}$.

$V_{ac} = V_{ab} + V_{bc}$

The electric field in the absence of any dielectric is $E_0 = \frac{Q}{\epsilon_0 A}$. In the first dielectric the electric field is reduced to

25-57 (cont) $E_1 = \frac{E_0}{K_1} = \frac{Q}{K_1 \epsilon_0 A}$

Thus $V_{ab} = E_1\left(\frac{d}{2}\right) = \frac{Qd}{K_1 2\epsilon_0 A}$.

In the second dielectric the electric field is reduced to $E_2 = \frac{E_0}{K_2} = \frac{Q}{K_2 \epsilon_0 A}$.

Thus $V_{bc} = E_2\left(\frac{d}{2}\right) = \frac{Qd}{K_2 2\epsilon_0 A}$.

Then $V_{ac} = V_{ab} + V_{bc} = \frac{Qd}{K_1 2\epsilon_0 A} + \frac{Qd}{K_2 2\epsilon_0 A} = \frac{Qd}{2\epsilon_0 A}\left(\frac{1}{K_1} + \frac{1}{K_2}\right)$

$V_{ac} = \frac{Qd}{2\epsilon_0 A}\left(\frac{K_1 + K_2}{K_1 K_2}\right)$

This gives $C = \frac{Q}{V_{ac}} = Q\,\frac{2\epsilon_0 A}{Qd}\left(\frac{K_1 K_2}{K_1 + K_2}\right) = \frac{2\epsilon_0 A}{d}\left(\frac{K_1 K_2}{K_1 + K_2}\right)$.

CHAPTER 26

Exercises 3, 5, 11, 13, 15, 19, 21, 23, 27, 31, 37, 39, 41, 43

Problems 45, 49, 51, 55, 57, 59, 63, 67

Exercises

26-3

a) $dQ = I\,dt$

$Q = \int_0^{10.0\,s} (3.0\,A + (0.73\,A/s^2)t^2)\,dt = \left[(3.0\,A)t + (0.243\,A/s^2)t^3\right]\Big|_0^{10.0s}$

$Q = (3.0\,A)(10.0s) + (0.243\,A/s^2)(10.0s)^3 = 30\,C + 243\,C = \underline{273\,C}$

b) $I = \frac{Q}{t} = \frac{273\,C}{10.0\,s} = \underline{27.3\,A}$

26-5

a) Calculate the drift speed v_d: $\quad v_d = \frac{J}{n|q|}$

$J = \frac{I}{A} = \frac{I}{\pi r^2} = \frac{3.55A}{\pi(0.1275\times10^{-3}m)^2} = 6.951\times10^7\,A/m^2$

$v_d = \frac{J}{n|q|} = \frac{6.951\times10^7\,A/m^2}{(8.5\times10^{28}/m^3)(1.602\times10^{-19}C)} = 5.105\times10^{-3}\,m/s$

$t = \frac{L}{v_d} = \frac{0.550\,m}{5.105\times10^{-3}\,m/s} = \underline{1.1\times10^2\,s}$

b) $v_d = \frac{J}{\pi r^2 n|q|}$

$t = \frac{L}{v_d} = \frac{\pi r^2 n|q|L}{J}$

t is proportional to r^2 and hence to d^2 where $d = 2r$ is the wire diameter.

$t = (1.1\times10^2 s)\left(\frac{1.45\,mm}{0.255\,mm}\right)^2 = \underline{3.5\times10^3 s}$

c) The time is proportional to the square of the diameter of the wire.

26-11

a) Eq. (26-5): $\rho = \frac{E}{J} \Rightarrow J = \frac{E}{\rho}$

From Table 26-1 the resistivity for aluminum is $2.75\times10^{-8}\Omega\cdot m$.

$J = \frac{E}{\rho} = \frac{0.49\,V/m}{2.75\times10^{-8}\Omega\cdot m} = 1.782\times10^7\,A/m^2$

$I = JA = J\pi r^2 = (1.782\times10^7 A/m^2)\pi(0.42\times10^{-3}m)^2 = \underline{9.9\,A}$

b) $V = EL = (0.49\,V/m)(12.0\,m) = \underline{5.9\,V}$

c) We can use Ohm's law (Eq. 26-11): $V = IR$

$R = \frac{V}{I} = \frac{5.9\,V}{9.9\,A} = \underline{0.60\,\Omega}$

26-11 (cont)

Or, we can calculate R from the resistivity and the dimensions of the wire (Eq. 26-10)

$$R = \frac{\rho L}{A} = \frac{\rho L}{\pi r^2} = \frac{(2.75\times10^{-8}\,\Omega\cdot m)(12.0\,m)}{\pi\,(0.42\times10^{-3}\,m)^2} = 0.60\,\Omega, \text{ which checks.}$$

26-13

a) $E = \frac{V}{L} = \frac{8.52\,V}{8.00\,m} = \underline{1.06\,V/m}$

b) $E = \rho J \Rightarrow \rho = \frac{E}{J} = \frac{1.06\,V/m}{3.40\times10^{7}\,A/m^2} = \underline{3.13\times10^{-8}\,\Omega\cdot m}$

26-15

$E = \frac{V}{L} = \rho J$, so $\frac{V}{L} = \rho\frac{I}{\pi r^2}$ where $\rho = 20\times10^{-8}\,\Omega\cdot m$ (Table 26-1)

$$r = \sqrt{\frac{\rho I L}{\pi V}} = \sqrt{\frac{(20\times10^{-8}\,\Omega\cdot m)(6.00\,A)(1.70\,m)}{\pi\,(1.50\,V)}} = \underline{0.658\,mm}$$

26-19

$R = R_0\left[1 + \alpha(T - T_0)\right]$

a) $\alpha = 0.0004\,(C°)^{-1}$ (Table 26-2)

Let T_0 be 0.0°C and T be 17.3°C.

$$R_0 = \frac{R}{1+\alpha(T-T_0)} = \frac{100.0\,\Omega}{1+(0.0004\,(C°)^{-1})(17.3\,C°)} = \underline{99.3\,\Omega}$$

b) $\alpha = -0.0005\,(C°)^{-1}$ (Table 26-2)

$$R = R_0\left[1+\alpha(T-T_0)\right] = 0.0190\,\Omega\left[1+(-0.0005\,(C°)^{-1}(32.4\,C°)\right] = \underline{0.0187\,\Omega}$$

26-21

The voltmeter reads the potential difference V_{ab} between the terminals of the battery.

open circuit $\Rightarrow I = 0$

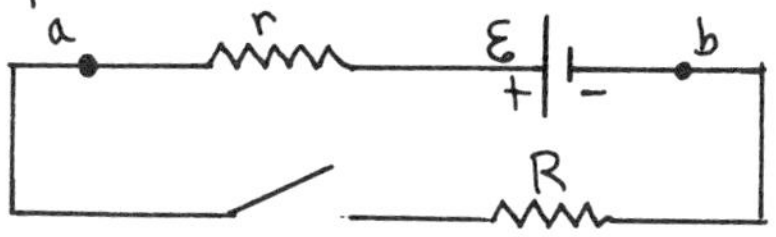

$V_{ab} = \mathcal{E} = \underline{1.56\,V}$

switch closed

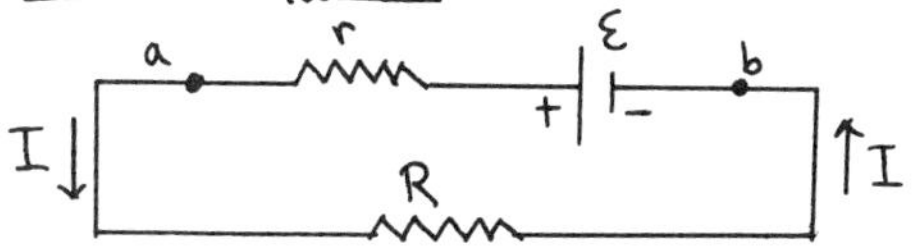

$V_{ab} = \mathcal{E} - Ir = 1.45\,V$

$$r = \frac{\mathcal{E} - 1.45\,V}{I} = \frac{1.56\,V - 1.45\,V}{1.30\,A} = \underline{0.085\,\Omega}$$

26-23

a)

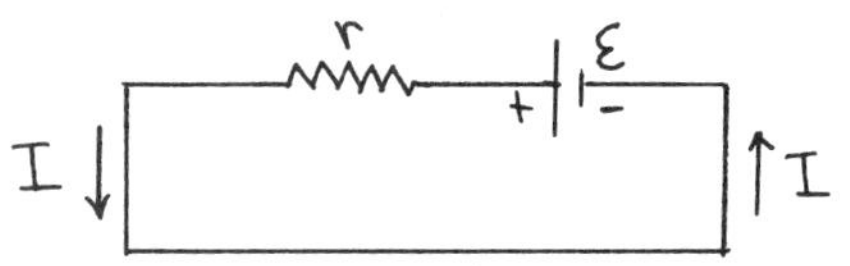

$$\mathcal{E} - Ir = 0$$

$$r = \frac{\mathcal{E}}{I} = \frac{1.50\ \text{V}}{14.8\ \text{A}} = \underline{0.101\ \Omega}$$

b) $r = \frac{\mathcal{E}}{I} = \frac{1.50\ \text{V}}{5.2\ \text{A}} = \underline{0.29\ \Omega}$

c) $r = \frac{\mathcal{E}}{I} = \frac{12.0\ \text{V}}{1000\ \text{A}} = \underline{0.0120\ \Omega}$

26-27

a)

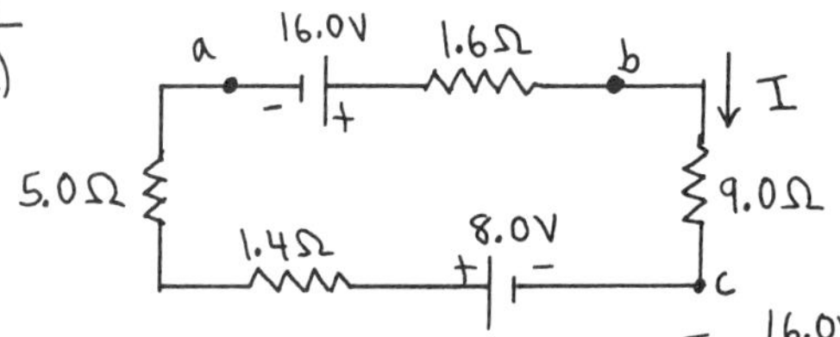

Assume that the current is clockwise. Add up the potential rises and drops as travel clockwise around the circuit:

$$16.0\text{V} - I(1.6\Omega) - I(9.0\Omega) + 8.0\text{V} - I(1.4\Omega) - I(5.0\Omega) = 0$$

$$I = \frac{16.0\text{V} + 8.0\text{V}}{9.0\Omega + 1.4\Omega + 5.0\Omega + 1.6\Omega} = \frac{24.0\ \text{V}}{17.0\ \Omega} = \underline{1.41\ \text{A}},\ \underline{\text{clockwise}}$$

b) $V_a + 16.0\text{V} - I(1.6\Omega) = V_b$

$V_a - V_b = -16.0\text{V} + (1.41\text{A})(1.6\Omega)$

$V_{ab} = -16.0\text{V} + 2.3\text{V} = \underline{-13.7\text{V}}$ (point a is at lower potential; it is the negative terminal)

c) $V_a + 16.0\text{V} - I(1.6\Omega) - I(9.0\Omega) = V_c$

$V_a - V_c = -16.0\text{V} + (1.41\text{A})(1.6\Omega + 9.0\Omega)$

$V_{ac} = -16.0\text{V} + 15.0\text{V} = \underline{-1.0\text{V}}$ (point a is at lower potential than point c)

d) Call the potential zero at point a. Travel clockwise around the circuit.

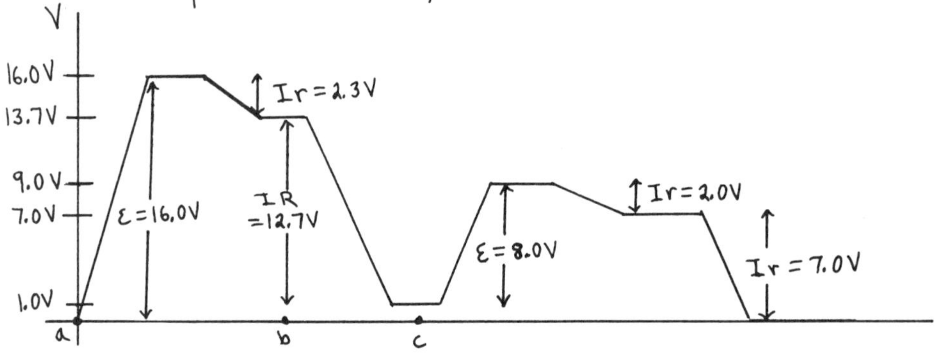

26-31

By definition $p = \frac{P}{LA}$

a) E is related to V and J is related to I, so use $P = VI \Rightarrow p = \frac{VI}{LA}$

$\frac{V}{L} = E$ and $\frac{I}{A} = J \Rightarrow p = EJ$

b) J is related to I and ρ is related to R, so use $P = I^2R \Rightarrow p = \frac{I^2R}{LA}$

$I = JA$ and $R = \frac{\rho L}{A} \Rightarrow p = \frac{J^2A^2\rho L}{LA^2} = \rho J^2$

26-31 (cont)

c) E is related to V and ρ is related to R, so use $P=\frac{V^2}{R} \Rightarrow p=\frac{V^2}{RLA}$

$V=EL$ and $R=\frac{\rho L}{A}$ $\Rightarrow p=\frac{E^2L^2}{LA}\left(\frac{A}{\rho L}\right)=\frac{E^2}{\rho}$

26-37

a) $P=VI=(12V)(50A)=600W$

The battery can provide this for 1.0 h, so the energy the battery has stored is $U=Pt=(600W)(3600s)=\underline{2.16\times10^6 J}$

b) For gasoline the heat of combustion is $L_c=46\times10^6$ J/kg.
The mass of gasoline that supplies 2.16×10^6 J is $m=\frac{2.16\times10^6 J}{46\times10^6 J/kg}=0.0470kg$

The volume of this mass of gasoline is
$V=\frac{m}{\rho}=\frac{0.0470 kg}{900 kg/m^3}=5.22\times10^{-5} m^3\left(\frac{1000L}{1 m^3}\right)=\underline{0.0522 L}$

c) $U=Pt$, $t=\frac{U}{P}=\frac{2.16\times10^6 J}{400 W}=5400 s=90 \text{ min}=\underline{1.5h}$

26-39

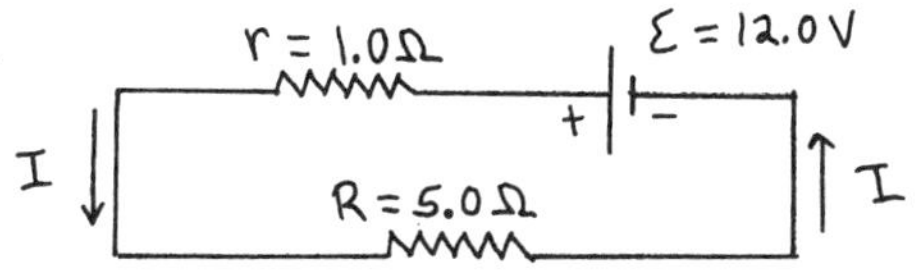

Compute I:
$\mathcal{E}-Ir-IR=0$
$I=\frac{\mathcal{E}}{r+R}=\frac{12.0 V}{1.0\Omega+5.0\Omega}=2.00 A$

a) The rate of conversion of chemical energy to electrical energy in the emf of the battery is $P=\mathcal{E}I=(12.0V)(2.00A)=\underline{24.0 W}$.

b) The rate of dissipation of electrical energy in the internal resistance of the battery is $P=I^2r=(2.00A)^2(1.0\Omega)=\underline{4.0W}$.

c) The rate of dissipation of electrical energy in the external resistor R is $P=I^2R=(2.00A)^2(5.0\Omega)=\underline{20.0W}$.

Note: The rate of production of electrical energy in the circuit is 24.0 W. The total rate of consumption of electrical energy in the circuit is 4.0W + 20.0W = 24.0W. Equal rates of production and consumption of electrical energy are required by energy conservation.

26-41

Table 26-1 gives $\rho=1.47\times10^{-8}\Omega\cdot m$ for silver at room temperature.

Eq. (26-24): $\rho=\frac{m}{ne^2\tau} \Rightarrow \tau=\frac{m}{ne^2\rho}=\frac{9.109\times10^{-31}kg}{(5.80\times10^{28}m^{-3})(1.602\times10^{-19}C)^2(1.47\times10^{-8}\Omega\cdot m)}$

$\tau=\underline{4.16\times10^{-14} s}$

26-43

a) $I = \frac{V}{R}$

The total resistance R along the current path is the $R_{ps} = 2000\,\Omega$ of the power supply plus the $R_b = 10\times10^3\,\Omega$ body resistance; $R = R_{ps} + R_b = 2000\,\Omega + 10\times10^3\,\Omega$.

$$I = \frac{20\times10^3\text{ V}}{2000\,\Omega + 10\times10^3\,\Omega} = \underline{1.7\text{ A}}$$

b) $P_b = I^2 R_b = (1.7\text{ A})^2(10\times10^3\,\Omega) = \underline{2.9\times10^4\text{ W}}$

c) $I = 1.00\text{ mA} \Rightarrow R = \frac{V}{I} = \frac{20\times10^3\text{ V}}{1.00\times10^{-3}\text{ A}} = 2.0\times10^7\,\Omega$

$R = R_b + R_{ps} \Rightarrow R_{ps} = R - R_b = 2.0\times10^7\,\Omega - 10\times10^3\,\Omega = \underline{2.0\times10^7\,\Omega}$

Problems

26-45

a) $R = \frac{\rho L}{A} \Rightarrow \rho = \frac{RA}{L} = \frac{(0.0625\,\Omega)(2.00\times10^{-3}\text{ m})^2}{12.0\text{ m}} = \underline{2.08\times10^{-8}\,\Omega\cdot\text{m}}$

b) Use $V = IR \Rightarrow I = \frac{V}{R}$

$V = EL = (1.28\text{ V/m})(12.0\text{ m}) = 15.4\text{ V}$

$I = \frac{V}{R} = \frac{15.4\text{ V}}{0.0625\,\Omega} = \underline{246\text{ A}}$

or

$E = \rho J \Rightarrow J = \frac{E}{\rho} = \frac{1.28\text{ V/m}}{2.08\times10^{-8}\,\Omega\cdot\text{m}} = 6.15\times10^7\text{ A/m}^2$

$I = JA = (6.15\times10^7\text{ A/m}^2)(2.00\times10^{-3}\text{ m})^2 = 246\text{ A}$, which checks

c) $J = n|q|v_d = nev_d$

$v_d = \frac{J}{ne} = \frac{6.15\times10^7\text{ A/m}^2}{(8.5\times10^{28}\text{ m}^{-3})(1.602\times10^{-19}\text{ C})} = 4.52\times10^{-3}\text{ m/s} = \underline{4.52\text{ mm/s}}$

26-49

a)

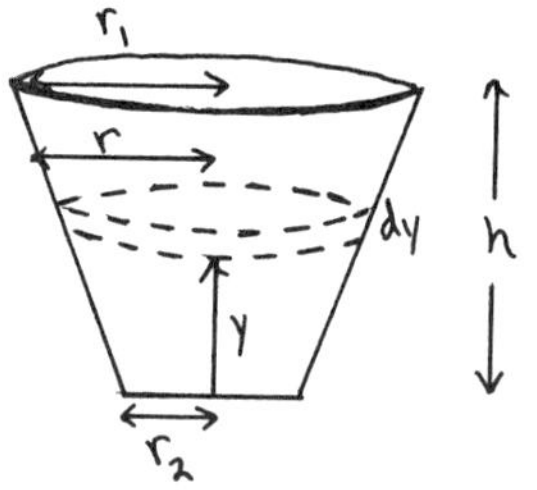

The radius of the truncated cone a distance y above the bottom is given by

$r = r_2 + \frac{y}{h}(r_1 - r_2) = r_2 + y\beta$ with $\beta = \frac{r_1 - r_2}{h}$

Consider a thin slice a distance y above the bottom. The slice has thickness dy and radius r. The resistance of the slice is

$$dR = \frac{\rho\,dy}{A} = \frac{\rho\,dy}{\pi r^2} = \frac{\rho\,dy}{\pi(r_2 + \beta y)^2}$$

The total resistance of the cone is obtained by integrating over these thin slices

$$R = \int dR = \frac{\rho}{\pi}\int_0^h \frac{dy}{(r_2+\beta y)^2} = \frac{\rho}{\pi}\left(-\frac{1}{\beta}(r_2 + y\beta)^{-1}\Big|_0^h\right) = -\frac{\rho}{\pi\beta}\left[\frac{1}{r_2 + h\beta} - \frac{1}{r_2}\right]$$

But $r_2 + h\beta = r_1$

$$\Rightarrow R = \frac{\rho}{\pi\beta}\left[\frac{1}{r_2} - \frac{1}{r_1}\right] = \frac{\rho}{\pi}\left(\frac{h}{r_1 - r_2}\right)\left(\frac{r_1 - r_2}{r_1 r_2}\right) = \frac{\rho h}{\pi r_1 r_2}$$

26-49 (cont)

b) Let $r_1 = r_2 = r$. Then $R = \frac{\rho h}{\pi r^2} = \frac{\rho L}{A}$ where $A = \pi r^2$ and $L = h$.
This agrees with Eq. (26-10).

26-51

Use $E = \rho J$ to calculate the current density between the plates. Let A be the area of each plate; then $I = JA$.

$$J = \frac{E}{\rho}$$

$$E = \frac{\sigma}{K\epsilon_0} = \frac{Q}{KA\epsilon_0}$$

$$J = \frac{E}{\rho} = \frac{Q}{KA\epsilon_0 \rho} \; ; \; I = JA = \frac{Q}{K\epsilon_0 \rho}$$, as was to be shown.

26-55

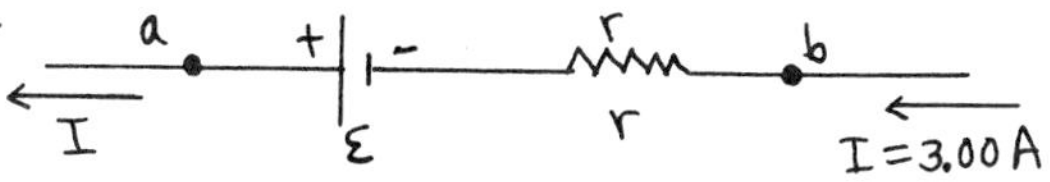

$V_{ab} = 8.6\text{ V}$

$V_{ab} = \mathcal{E} - Ir$

$\boxed{\mathcal{E} - (3.00\text{A})r = 8.6\text{ V}}$

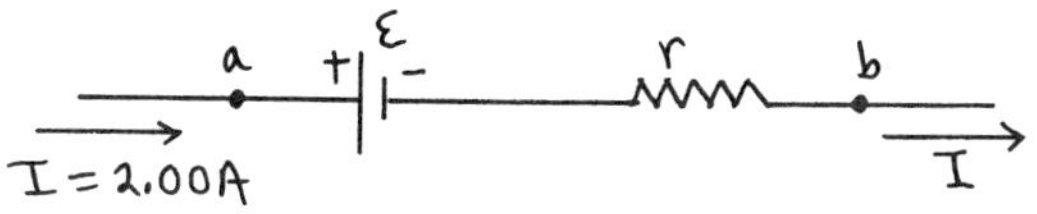

$V_{ab} = 10.1\text{ V}$

$V_{ab} = \mathcal{E} + Ir$

$\boxed{\mathcal{E} + (2.00\text{A})r = 10.1\text{V}}$

a) Solve the first equation for $\mathcal{E}$ and use that result in the second equation:

$$\mathcal{E} = 8.6\text{ V} + (3.00\text{ A})r$$

$$8.6\text{ V} + (3.00\text{ A})r + (2.00\text{ A})r = 10.1\text{ V}$$

$$(5.00\text{A})\,r = 1.5\text{ V}$$

$$r = \frac{1.5\text{ V}}{5.00\text{ A}} = \underline{0.30\,\Omega}$$

b) Then $\mathcal{E} = 8.6\text{ V} + (3.00\text{ A})r = 8.6\text{V} + (3.00\text{A})(0.30\Omega) = \underline{9.5\text{ V}}$

26-57

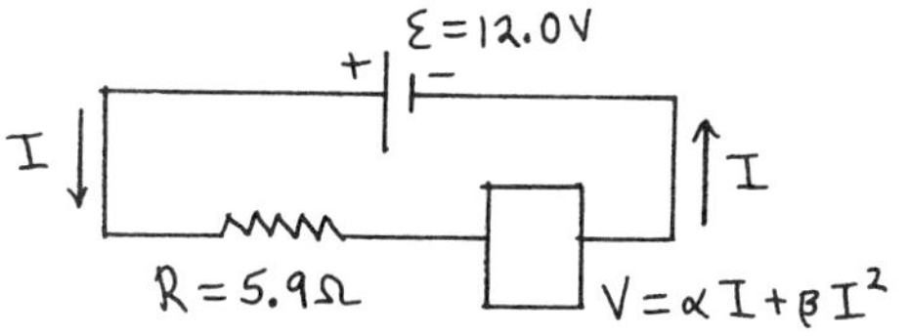

$$\mathcal{E} - IR - V = 0$$

$$\mathcal{E} - IR - \alpha I - \beta I^2 = 0$$

$$\beta I^2 + (R+\alpha)I - \mathcal{E} = 0$$

The quadratic formula gives

$$I = \frac{1}{2\beta}\left[-(R+\alpha) \pm \sqrt{(R+\alpha)^2 + 4\beta\mathcal{E}}\,\right]$$

I must be positive, so take the + sign

$$I = \frac{1}{2\beta}\left[-(R+\alpha) + \sqrt{(R+\alpha)^2 + 4\beta\mathcal{E}}\,\right] = \frac{1}{2(1.3\,\Omega/\text{A})}\left[-9.7\Omega + \sqrt{(9.7\Omega)^2 + 4(1.3\Omega/\text{A})(12.0\text{V})}\,\right]$$

$$I = -3.73\text{ A} + 4.81\text{ A} = \underline{1.08\text{ A}}$$

26-59

a) With the ammeter in the circuit:

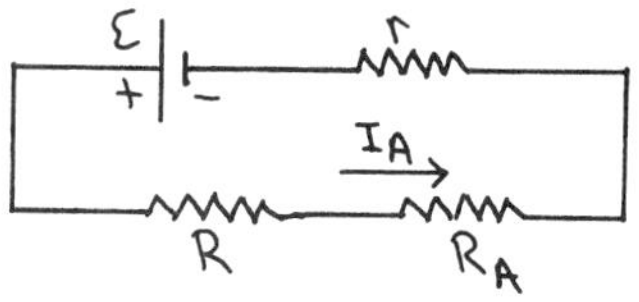

$$I_A = \frac{\varepsilon}{r+R+R_A} \Rightarrow \varepsilon = I_A(r+R+R_A)$$

With the ammeter removed:

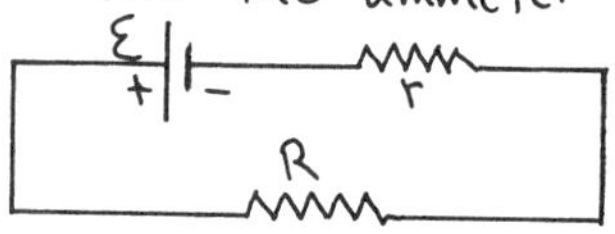

$$I = \frac{\varepsilon}{R+r} = \left(\frac{1}{R+r}\right) I_A(r+R+R_A) = I_A\left(1+\frac{R_A}{r+R}\right)$$

b) Want $I_A = 0.990\,I$. Use this in the result for part (a)

$$I = 0.990\,I\left(1+\frac{R_A}{r+R}\right)$$

$$0.010 = 0.990\left(\frac{R_A}{r+R}\right)$$

$$R_A = (r+R)\left(\frac{0.010}{0.990}\right) = (0.75\,\Omega + 6.50\,\Omega)\frac{0.010}{0.990} = \underline{0.0732\,\Omega}$$

c) $$I - I_A = \frac{\varepsilon}{r+R} - \frac{\varepsilon}{r+R+R_A} = \varepsilon\,\frac{r+R+R_A-r-R}{(r+R)(r+R+R_A)} = \frac{\varepsilon R_A}{(r+R)(r+R+R_A)}$$

The difference between I and I_A increases as R_A increases. If R_A is larger than the value calculated in part (b) then I_A differs from I by more than 1.0%.

26-63

a)

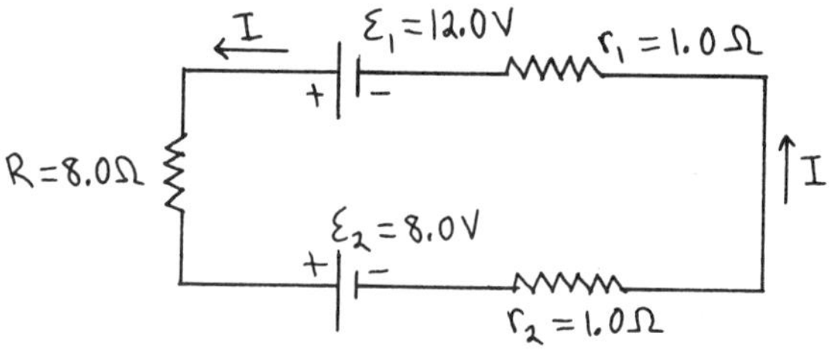

$$\varepsilon_1 - \varepsilon_2 - I(r_1+r_2+R) = 0$$

$$I = \frac{\varepsilon_1-\varepsilon_2}{r_1+r_2+R} = \frac{12.0\,V - 8.0\,V}{1.0\,\Omega+1.0\,\Omega+8.0\,\Omega} = \underline{0.40\,A}$$

b) $$P = I^2R + I^2r_1 + I^2r_2 = I^2(R+r_1+r_2) = (0.40\,A)^2(8.0\,\Omega+1.0\,\Omega+1.0\,\Omega) = \underline{1.6\,W}$$

c) Chemical → electrical energy in a battery when the current goes through the battery from the − to + terminal, so the electrical energy of the charges increases as the current passes through. This happens in the 12.0 V battery, and the rate of production of electrical energy is $P = \varepsilon_1 I = (12.0\,V)(0.40\,A) = \underline{4.8\,W}$.

d) Electrical → chemical energy in a battery when the current goes through the battery from the + to the − terminal, so the electrical energy of the charges decreases as the current passes through. This happens in the 8.0 V battery and the rate of consumption of electrical energy is $P = \varepsilon_2 I = (8.0\,V)(0.40\,A) = \underline{3.2\,W}$.

e) Total rate of production of electrical energy = 4.8 W.
Total rate of consumption of electrical energy = 1.6 W + 3.2 W = 4.8 W, which checks.

26-67

a) $\ominus \rightarrow |q|E$ (with $\vec{a}$ pointing right)

$$\sum \vec{F} = m\vec{a}$$

$$|q|E = ma$$

$$\frac{|q|}{m} = \frac{a}{E}$$

b) $V_{bc} = EL \Rightarrow E = \frac{V_{bc}}{L}$

$$\frac{|q|}{m} = \frac{aL}{V_{bc}}$$

c) $\ominus \rightarrow F_E$; $\leftarrow E$

If the acceleration is to the right the force on the free charges must be to the right. For a negative charge the force and the electric field are in opposite directions, so $\vec{E}$ is to the left.

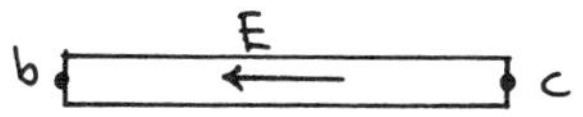

$\vec{E}$ points from high potential to low potential so point c is at higher potential.

d) From part (b), $a = \frac{V_{bc}}{L}\frac{|q|}{m} = \left(\frac{1.0\times10^{-3}\text{ V}}{0.50\text{ m}}\right)\left(\frac{1.60\times10^{-19}\text{ C}}{9.11\times10^{-31}\text{ kg}}\right) = \underline{3.5\times10^{8}\text{ m/s}^2}$

e) The rotating spool with many turns allows for a very large length L of wire. From $V_{bc} = aL\left(\frac{m}{|q|}\right)$ this increases V_{bc} and makes it easier to measure.

CHAPTER 27

Exercises 3, 5, 7, 9, 11, 17, 19, 21, 23, 25, 33, 37, 39, 41

Problems 43, 47, 49, 51, 53, 55, 57, 63, 65, 67

Exercises

27-3

a)

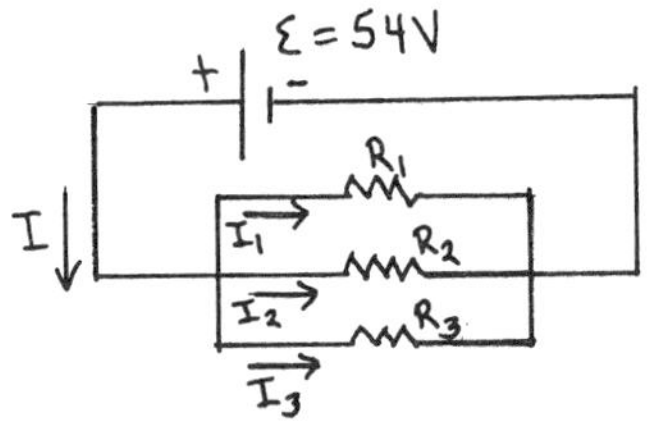

parallel $\Rightarrow \frac{1}{R_{eq}} = \frac{1}{R_1} + \frac{1}{R_2} + \frac{1}{R_3}$

$\frac{1}{R_{eq}} = \frac{1}{2.4\Omega} + \frac{1}{3.6\Omega} + \frac{1}{4.8\Omega} = 0.9028\,\Omega^{-1}$

$R_{eq} = \underline{1.1\Omega}$

b) For resistors in parallel the voltage is the same across each and equal to the applied voltage $\Rightarrow V_1 = V_2 = V_3 = \mathcal{E} = 54\text{ V}$

$V = IR \Rightarrow I_1 = \frac{V_1}{R_1} = \frac{54\text{ V}}{2.4\Omega} = 23\text{ A}$

$I_2 = \frac{V_2}{R_2} = \frac{54\text{V}}{3.6\Omega} = 15\text{A}$

$I_3 = \frac{V_3}{R_3} = \frac{54\text{ V}}{4.8\Omega} = 11\text{ A}$

c) The currents through the resistors add to give the current through the battery:
$I = I_1 + I_2 + I_3 = 23\text{A} + 15\text{A} + 11\text{A} = 49\text{A}$

Alternatively, we can use the equivalent resistance R_{eq}:

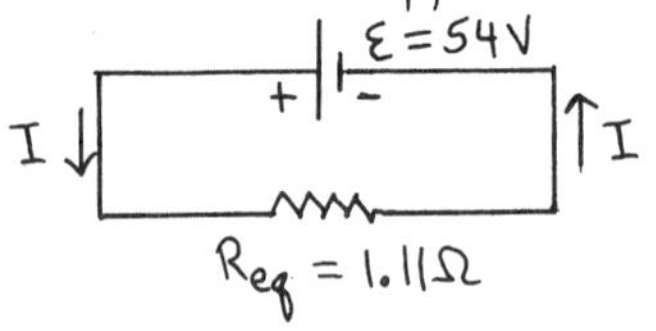

$\mathcal{E} - IR_{eq} = 0$

$I = \frac{\mathcal{E}}{R_{eq}} = \frac{54\text{V}}{1.11\Omega} = 49\text{A}$, which checks

d) As shown in part (b), the voltage across each resistor is 54 V.

e) We can use any of the three expressions for P: $P = VI = I^2R = \frac{V^2}{R}$. They will all give the same results, if we keep enough significant figures in intermediate calculations. Using $P = \frac{V^2}{R}$

$P_1 = \frac{V_1^2}{R_1} = \frac{(54\text{V})^2}{2.4\Omega} = 1220\text{ W}$

$P_2 = \frac{V_2^2}{R_2} = \frac{(54\text{V})^2}{3.6\Omega} = 810\text{W}$

$P_3 = \frac{V_3^2}{R_3} = \frac{(54\text{V})^2}{4.8\Omega} = 610\text{W}$

Note that the total power dissipated is $P_{out} = P_1 + P_2 + P_3 = 2640\text{ W}$. This is the same as the power $P_{in} = \mathcal{E}I$ delivered by the battery.

27-5

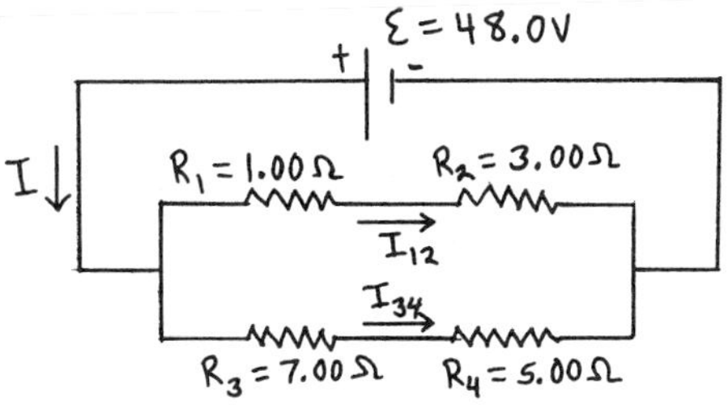

R_1 and R_2 in series have an equivalent resistance of $R_{12} = R_1 + R_2 = 4.00\,\Omega$

R_3 and R_4 in series have an equivalent resistance of $R_{34} = R_3 + R_4 = 12.0\,\Omega$

The circuit is equivalent to

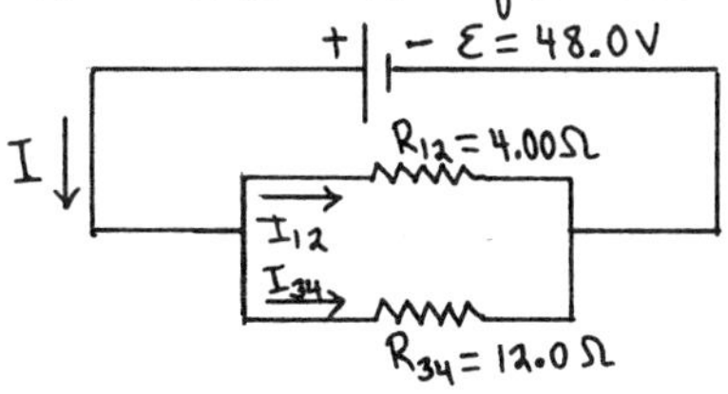

R_{12} and R_{34} in parallel are equivalent to R_{eq} given by

$$\frac{1}{R_{eq}} = \frac{1}{R_{12}} + \frac{1}{R_{34}} = \frac{R_{12} + R_{34}}{R_{12} R_{34}}$$

$$R_{eq} = \frac{R_{12} R_{34}}{R_{12} + R_{34}} = \frac{(4.00\,\Omega)(12.0\,\Omega)}{4.00\,\Omega + 12.0\,\Omega} = \underline{3.00\,\Omega}$$

The voltage across each branch of the parallel combination is $\mathcal{E}$, so $\mathcal{E} - I_{12} R_{12} = 0$.

$$I_{12} = \frac{\mathcal{E}}{R_{12}} = \frac{48.0\text{ V}}{4.00\,\Omega} = 12.0\text{ A}$$

$$\mathcal{E} - I_{34} R_{34} = 0$$

$$I_{34} = \frac{\mathcal{E}}{R_{34}} = \frac{48.0\text{ V}}{12.0\,\Omega} = 4.0\text{ A}$$

The current is $\underline{12.0\text{ A}}$ through the $1.00\,\Omega$ and $3.00\,\Omega$ resistors, and it is $\underline{4.0\text{ A}}$ through the $7.00\,\Omega$ and $5.00\,\Omega$ resistors.

Note: The current through the battery is $I = I_{12} + I_{34} = 12.0\text{ A} + 4.0\text{ A} = 16.0\text{ A}$, and this is equal to $\frac{\mathcal{E}}{R_{eq}} = \frac{48.0\text{ V}}{3.00\,\Omega} = 16.0\text{ A}$.

27-7

a) $P = \frac{V_{ab}^2}{R}$ (Eq. 26-18) $\Rightarrow V_{ab} = \sqrt{PR} = \sqrt{(5.0\text{ W})(10\times10^3\,\Omega)} = \underline{224\text{ V}}$

b) $P = \frac{V_{ab}^2}{R} = \frac{(220\text{ V})^2}{15\times10^3\,\Omega} = \underline{3.2\text{ W}}$

27-9

a)

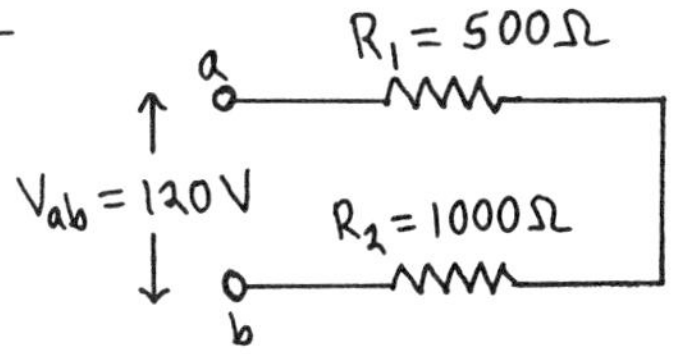

For resistors in series the current is the same through each. $R_{eq} = R_1 + R_2 = 1500\,\Omega$.

$$I = \frac{V}{R_{eq}} = \frac{120\text{ V}}{1500\,\Omega} = \underline{0.080\text{ A}}$$

b) $P_1 = I_1^2 R_1 = (0.080\text{ A})^2 (500\,\Omega) = 3.2\text{ W}$

$P_2 = I_2^2 R_2 = (0.080\text{ A})^2 (1000\,\Omega) = 6.4\text{ W}$

$P_{out} = P_1 + P_2 = 9.6\text{ W}$, the total power dissipated in both bulbs.

Note that $P_{in} = V_{ab} I = (120\text{ V})(0.080\text{ A}) = 9.6\text{ W}$, the power delivered by the potential source, equals P_{out}.

27-9 (cont)

c) 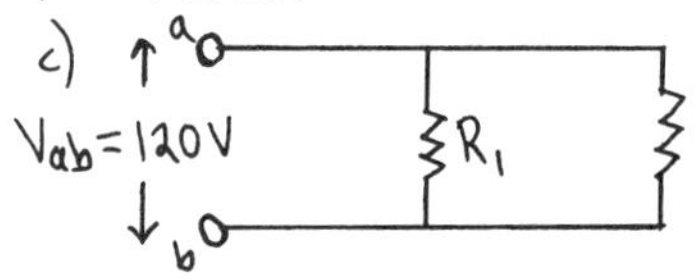

For resistors in parallel the voltage across each resistor is the same.

$$I_1 = \frac{V_1}{R_1} = \frac{120\ V}{500\ \Omega} = 0.24\ A, \quad I_2 = \frac{V_2}{R_2} = \frac{120\ V}{1000\ \Omega} = 0.12\ A$$

Note that each current is larger than the current when the resistors are connected in series.

d) $P_1 = I_1^2 R_1 = (0.24\ A)^2 (500\ \Omega) = 28.8\ W$

$P_2 = I_2^2 R_2 = (0.12\ A)^2 (1000\ \Omega) = 14.4\ W$

$P_{out} = P_1 + P_2 = 43.2\ W$

Note that the total current drawn from the line is $I = I_1 + I_2 = 0.36\ A$. The power input from the line is $P_{in} = V_{ab} I = (120\ V)(0.36\ A) = 43.2\ W$, which equals the total power dissipated by the bulbs.

e) The bulb that is dissipating the most power glows most brightly. For the series connection the currents are the same and by $P = I^2 R$ the bulb with the larger R has the larger P; the $1000\ \Omega$ bulb glows more brightly.

For the parallel combination the voltages are the same and by $P = \frac{V^2}{R}$ the bulb with the smaller R has the larger P; the $500\ \Omega$ bulb glows more brightly.

The total power output P_{out} equals $P_{in} = V_{ab} I$, so P_{out} is larger for the parallel connection where the current drawn from the line is larger (because the equivalent resistance is smaller).

27-11

a) 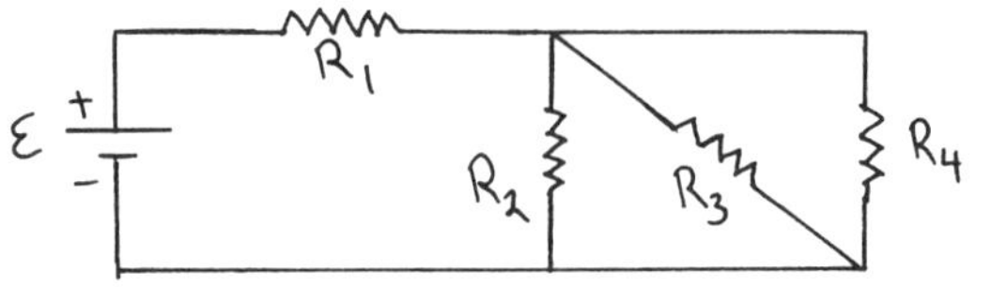

R_2, R_3, and R_4 are in parallel, so their equivalent resistance R_{eq} is given by

$$\frac{1}{R_{eq}} = \frac{1}{R_2} + \frac{1}{R_3} + \frac{1}{R_4} = \frac{3}{6.00\ \Omega} \text{ and } R_{eq} = 2.00\ \Omega$$

The equivalent circuit is

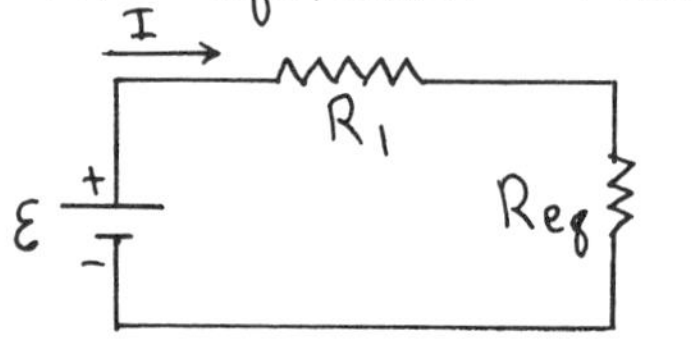

$$\mathcal{E} - I(R_1 + R_{eq}) = 0$$

$$I = \frac{\mathcal{E}}{R_1 + R_{eq}} = \frac{3.00\ V}{6.00\ \Omega + 2.00\ \Omega} = 0.375\ A \Rightarrow \underline{I_1 = 0.375\ A}$$

$$V_1 = I_1 R_1 = (0.375\ A)(6.00\ \Omega) = 2.25\ V$$

$I_{eq} = 0.375\ A$, $V_{eq} = I_{eq} R_{eq} = (0.375\ A)(2.00\ \Omega) = 0.75\ V$

For resistors in parallel the voltages are equal and are the same as the voltage across the equivalent resistor, so $V_2 = V_3 = V_4 = 0.75\ V$.

$$I_2 = \frac{V_2}{R_2} = \frac{0.75\ V}{6.00\ \Omega} = \underline{0.125\ A}, \quad I_3 = \frac{V_3}{R_3} = \underline{0.125\ A}, \quad I_4 = \frac{V_4}{R_4} = \underline{0.125\ A}$$

27-11 (cont)

Note that $I_2 + I_3 + I_4 = 0.375\,A$, which is I_{eq}. For resistors in parallel the currents add and their sum is the current through the equivalent resistor.

b) $P = I^2 R$

$P_1 = (0.375\,A)^2 (6.00\,\Omega) = \underline{0.844\,W}$

$P_2 = P_3 = P_4 = (0.125\,A)^2 (6.00\,\Omega) = \underline{0.0938\,W}$

Note that $P_2 + P_3 + P_4 = 0.281\,W$. This equals $P_{eq} = I_{eq}^2 R_{eq} = (0.375\,A)^2 (2.00\,\Omega) = 0.281\,W$, the power dissipated in the equivalent resistor.

c) With R_4 removed the circuit becomes

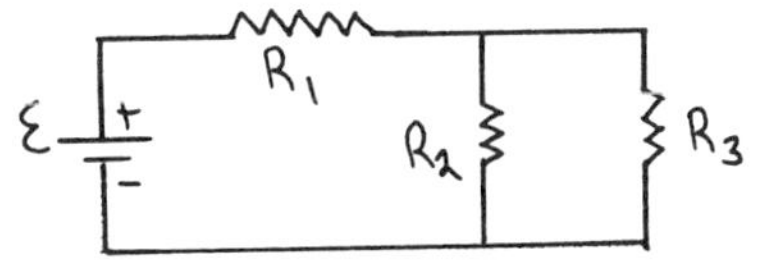

R_2 and R_3 are in parallel and their equivalent resistance R_{eq} is given by $\frac{1}{R_{eq}} = \frac{1}{R_2} + \frac{1}{R_3} = \frac{2}{6.00\,\Omega}$ and $R_{eq} = 3.00\,\Omega$

The equivalent circuit is

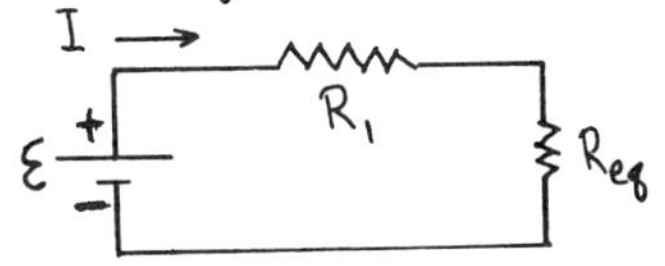

$\mathcal{E} - I\,(R_1 + R_{eq}) = 0$

$I = \frac{\mathcal{E}}{R_1 + R_{eq}} = \frac{3.00\,V}{6.00\,\Omega + 3.00\,\Omega} = 0.333\,A$

$I_1 = \underline{0.333\,A}$, $V_1 = I_1 R_1 = (0.333\,A)(6.00\,\Omega) = 2.00\,V$

$I_{eq} = 0.333\,A$, $V_{eq} = I_{eq} R_{eq} = (0.333\,A)(3.00\,\Omega) = 1.00\,V$ and $V_2 = V_3 = 1.00\,V$.

$I_2 = \frac{V_2}{R_2} = \frac{1.00\,V}{6.00\,\Omega} = \underline{0.167\,A}$, $I_3 = \frac{V_3}{R_3} = \underline{0.167\,A}$

d) $P = I^2 R$

$P_1 = (0.333\,A)^2 (6.00\,\Omega) = \underline{0.667\,W}$

$P_2 = P_3 = (0.167\,A)^2 (6.00\,\Omega) = \underline{0.167\,W}$

e) When R_4 is removed, P_1 decreases and P_2 and P_3 increase. Bulb R_1 glows less brightly and bulbs R_2 and R_3 glow more brightly. When R_4 is removed the equivalent resistance of the circuit increases and the current through R_1 decreases. But in the parallel combination this current divides into two equal currents rather than three so the currents through R_2 and R_3 increase. Can also see this by noting that with R_4 removed and less current through R_1 the voltage drop across R_1 is less so the voltage drop across R_2 and across R_3 must become larger.

27-17

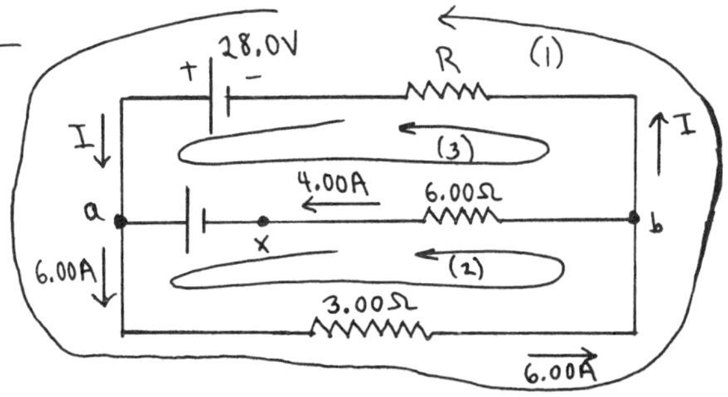

a) Apply Kirchhoff's point rule to point a:

$\sum I = 0 \Rightarrow I + 4.00\text{A} - 6.00\text{A} = 0$

$I = \underline{2.00\text{A}}$ (in the direction shown in the diagram)

b) Apply Kirchhoff's loop rule to loop (1):

$-(6.00\text{A})(3.00\Omega) - (2.00\text{A})R + 28.0\text{V} = 0$

$-18.0\text{V} - (2.00\text{A})R + 28.0\text{V} = 0$

$R = \dfrac{28.0\text{V} - 18.0\text{V}}{2.00\text{A}} = \underline{5.00\Omega}$

c) Apply Kirchhoff's loop rule to loop (2):

$-(6.00\text{A})(3.00\Omega) - (4.00\text{A})(6.00\Omega) + \mathcal{E} = 0$

$\mathcal{E} = 18.0\text{V} + 24.0\text{V} = \underline{42.0\text{V}}$

Note: Can check that the loop rule is satisfied for loop (3), as a check of our work:

$28.0\text{V} - \mathcal{E} + (4.00\text{A})(6.00\Omega) - (2.00\text{A})R = 0$

$28.0\text{V} - 42.0\text{V} + 24.0\text{V} - (2.00\text{A})(5.00\Omega) = 0$

$52.0\text{V} = 42.0\text{V} + 10.0\text{V}$

$52.0\text{V} = 52.0\text{V}$ ✓

d) If the circuit is broken at point X there can be no current in the 6.00Ω resistor. There is now only a single current path:

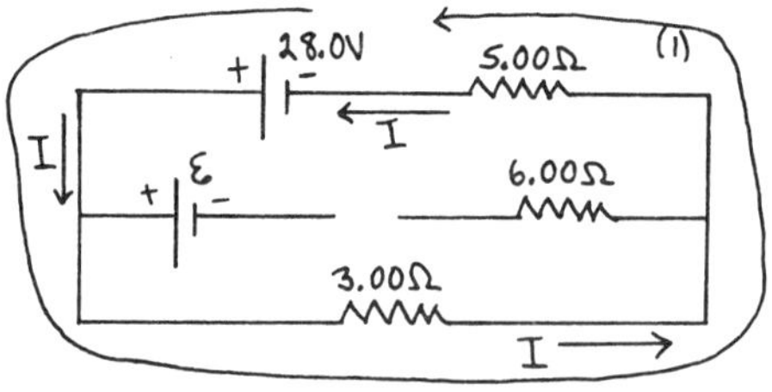

Apply the loop rule:

$+28.0\text{V} - (3.00\Omega)I - (5.00\Omega)I = 0$

$I = \dfrac{28.0\text{V}}{8.00\Omega} = \underline{3.50\text{A}}$

27-19

a)

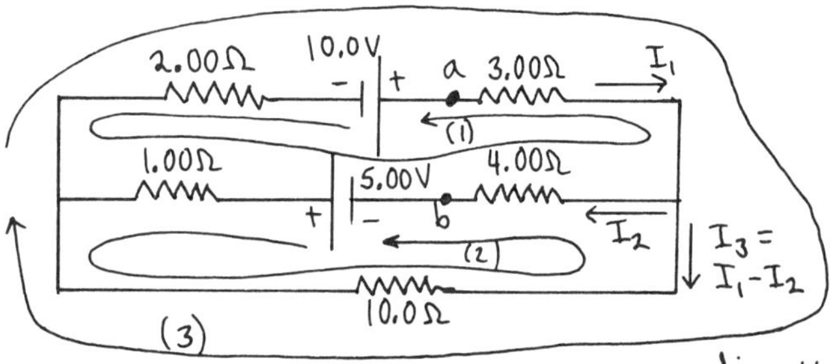

Let I_1 be the current in the 3.00Ω resistor and I_2 be the current in the 4.00Ω resistor and assume that these currents are in the directions shown. Then the current in the 10.0Ω resistor is $I_3 = I_1 - I_2$, in the direction shown, where we have used Kirchhoff's point rule to relate I_3 to I_1 and I_2. If we get a negative answer for any of these currents we know that the current is actually in the opposite direction to what we have assumed.

Three loops and directions to travel around the loops are shown in the circuit diagram. Apply Kirchhoff's loop rule to each loop.

27-19 (cont)

Loop (1)

$+10.00\text{ V} - I_1(3.00\,\Omega) - I_2(4.00\,\Omega) + 5.00\text{ V} - I_2(1.00\,\Omega) - I_1(2.00\,\Omega) = 0$

$15.00\text{ V} - (5.00\,\Omega)I_1 - (5.00\,\Omega)I_2 = 0$

$\boxed{3.00\text{ A} - I_1 - I_2 = 0}$

Loop (2)

$+5.00\text{ V} - I_2(1.00\,\Omega) + (I_1 - I_2)\,10.0\,\Omega - I_2(4.00\,\Omega) = 0$

$5.00\text{ V} + (10.0\,\Omega)I_1 - (15.0\,\Omega)I_2 = 0$

$\boxed{1.00\text{ A} + 2.00\,I_1 - 3.00\,I_2 = 0}$

The first equation says $I_2 = 3.00\text{ A} - I_1$.

Use this in the second equation: $1.00\text{ A} + 2.00 I_1 - 9.00\text{ A} + 3.00 I_1 = 0$

$5.00\,I_1 = 8.00\text{ A}$, $I_1 = \underline{1.60\text{ A}}$

$I_2 = 3.00\text{ A} - I_1 = 3.00\text{ A} - 1.60\text{ A} = \underline{1.40\text{ A}}$

$I_3 = I_1 - I_2 = 1.60\text{ A} - 1.40\text{ A} = \underline{0.20\text{ A}}$

loop (3) can be used as a check:

$+10.0\text{ V} - (1.60\text{ A})(3.00\,\Omega) - (0.20\text{ A})(10.00\,\Omega) - (1.60\text{ A})(2.00\,\Omega) = 0$

$10.0\text{ V} = 4.8\text{ V} + 2.0\text{ V} + 3.2\text{ V}$

$10.0\text{ V} = 10.0\text{ V}$ ✓

We find that with our calculated currents the loop rule is satisfied for loop (3). Also, all the currents came out to be positive, so the current directions in the circuit diagram are correct.

b) To find $V_{ab} = V_a - V_b$ start at point b and travel to point a. Many different routes can be taken from b to a and all must yield the same result for V_{ab}.

Travel through the 4.00Ω resistor and then through the 3.00Ω resistor:

$V_b + I_2(4.00\,\Omega) + I_1(3.00\,\Omega) = V_a$

$V_a - V_b = (1.40\text{ A})(4.00\,\Omega) + (1.60\text{ A})(3.00\,\Omega) = 5.60\text{ V} + 4.80\text{ V} = 10.4\text{ V}$ (point a is at higher potential than b)

Or, travel through the 5.00 V emf, the 1.00Ω resistor, the 2.00Ω resistor, and the 10.0 V emf:

$V_b + 5.00\text{ V} - I_2(1.00\,\Omega) - I_1(2.00\,\Omega) + 10.0\text{ V} = V_a$

$V_a - V_b = 15.0\text{ V} - (1.40\text{ A})(1.00\,\Omega) - (1.60\text{ A})(2.00\,\Omega) = 15.0\text{ V} - 1.40\text{ V} - 3.20\text{ V} = 10.4\text{ V}$ ✓

27-21

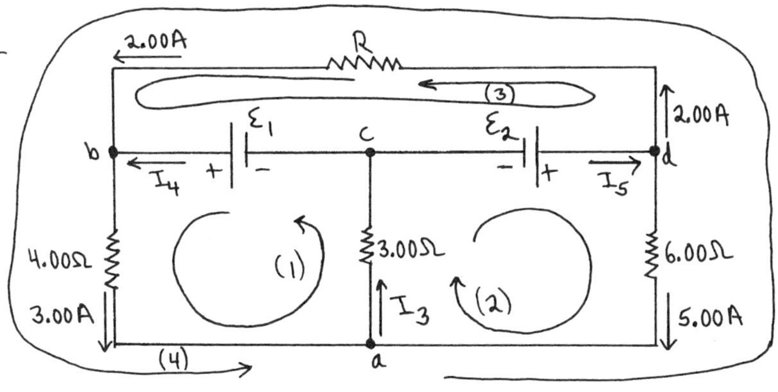

a) Apply the junction rule to point a:

$3.00\text{ A} + 5.00\text{ A} - I_3 = 0$

$I_3 = \underline{8.00\text{ A}}$

Apply the junction rule to point b:

$2.00\text{ A} + I_4 - 3.00\text{ A} = 0$

$I_4 = \underline{1.00\text{ A}}$

27-21 (cont)

Apply the junction rule to point c:

$$I_3 - I_4 - I_5 = 0$$

$$I_5 = I_3 - I_4 = 8.00\,A - 1.00\,A = \underline{7.00\,A}$$

As a check, apply the junction rule to point d:

$$I_5 - 2.00\,A - 5.00\,A = 0$$

$$I_5 = 7.00\,A \checkmark$$

b) Apply the loop rule to loop (1):

$$\mathcal{E}_1 - (3.00\,A)(4.00\,\Omega) - I_3(3.00\,\Omega) = 0$$

$$\mathcal{E}_1 = 12.0\,V + (8.00\,A)(3.00\,\Omega) = \underline{36.0\,V}$$

Apply the loop rule to loop (2):

$$\mathcal{E}_2 - (5.00\,A)(6.00\,\Omega) - I_3(3.00\,\Omega) = 0$$

$$\mathcal{E}_2 = 30.0\,V + (8.00\,A)(3.00\,\Omega) = \underline{54.0\,V}$$

c) Apply the loop rule to loop (3):

$$-(2.00\,A)R - \mathcal{E}_1 + \mathcal{E}_2 = 0$$

$$R = \frac{\mathcal{E}_2 - \mathcal{E}_1}{2.00\,A} = \frac{54.0\,V - 36.0\,V}{2.00\,A} = \underline{9.00\,\Omega}$$

Note: Apply the loop rule to loop (4), as a check of our calculations:

$$-(2.00\,A)R - (3.00\,A)(4.00\,\Omega) + (5.00\,A)(6.00\,\Omega) = 0$$

$$-(2.00\,A)(9.00\,\Omega) - 12.0\,V + 30.0\,V = 0$$

$$-18.0\,V + 18.0\,V = 0 \checkmark$$

27-23

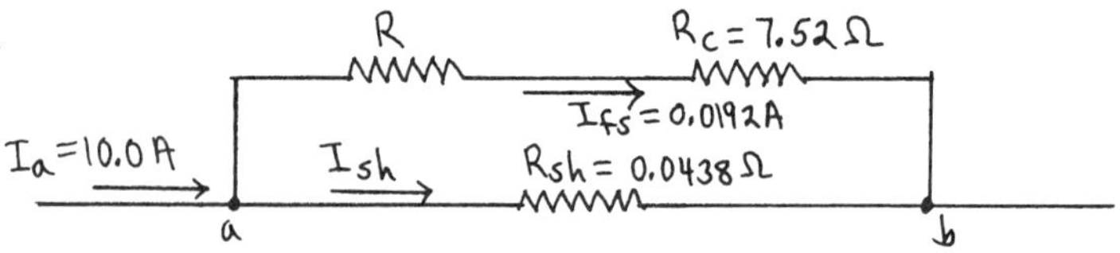

We want that $I_a = 10.0\,A$ in the external circuit produces $I_{fs} = 0.0192\,A$ through the galvanometer coil. Applying the junction rule to point a gives

$$I_a - I_{fs} - I_{sh} = 0$$

$$I_{sh} = I_a - I_{fs} = 10.0\,A - 0.0192\,A = 9.98\,A$$

The potential difference V_{ab} between points a and b must be the same for both paths between these two points

$$\Rightarrow I_{fs}(R + R_c) = I_{sh}R_{sh}$$

$$R = \frac{I_{sh}R_{sh}}{I_{fs}} - R_c = \frac{(9.98\,A)(0.0438\,\Omega)}{0.0192\,A} - 7.52\,\Omega = 22.77\,\Omega - 7.52\,\Omega = \underline{15.2\,\Omega}$$

27-25

For each range setting the circuit has the form

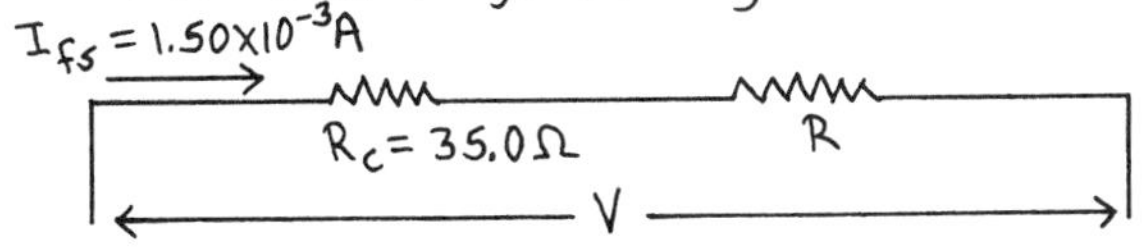

3.00V

For $V = 3.00$V, $R = R_1$ and the total meter resistance R_m is $R_m = R_c + R_1$.

$$V = I_{fs}R_m \Rightarrow R_m = \frac{V}{I_{fs}} = \frac{3.00\text{ V}}{1.50\times10^{-3}\text{ A}} = \underline{2.00\times10^{3}\,\Omega}$$

$$R_m = R_c + R_1 \Rightarrow R_1 = R_m - R_c = 2.00\times10^{3}\,\Omega - 35.0\,\Omega = \underline{1965\,\Omega}$$

15.0 V

For $V = 15.0$ V, $R = R_1 + R_2$ and the total meter resistance is $R_m = R_c + R_1 + R_2$.

$$V = I_{fs}R_m \Rightarrow R_m = \frac{V}{I_{fs}} = \frac{15.0\text{ V}}{1.50\times10^{-3}\text{ A}} = \underline{1.00\times10^{4}\,\Omega}$$

$$R_2 = R_m - R_c - R_1 = 1.00\times10^{4}\,\Omega - 35.0\,\Omega - 1965\,\Omega = \underline{8.00\times10^{3}\,\Omega}$$

150 V

For $V = 150$ V, $R = R_1 + R_2 + R_3$ and the total meter resistance R_m is $R_m = R_c + R_1 + R_2 + R_3$

$$V = I_{fs}R_m \Rightarrow R_m = \frac{V}{I_{fs}} = \frac{150\text{ V}}{1.50\times10^{-3}\text{ A}} = \underline{1.00\times10^{5}\,\Omega}$$

$$R_3 = R_m - R_c - R_1 - R_2 = 1.00\times10^{5}\,\Omega - 35.0\,\Omega - 1965\,\Omega - 8.00\times10^{3}\,\Omega = \underline{9.00\times10^{4}\,\Omega}$$

27-33

a)

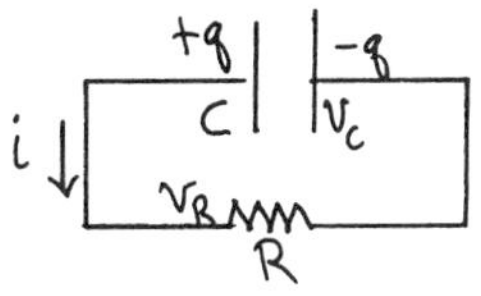

By Kirchhoff's loop rule,

$$V_c - V_R = 0$$

$$\frac{q}{C} - iR = 0 \Rightarrow i = \frac{q}{RC}$$

Just after the connection is made the charge q on the capacitor hasn't had any time to decrease from its initial value of 7.38×10^{-8} C, so

$$i = \frac{q}{RC} = \frac{7.83\times10^{-8}\text{ C}}{(5.30\times10^{5}\,\Omega)(3.43\times10^{-10}\text{ F})} = \underline{4.31\times10^{-4}\text{ A}}$$

b) Eq. (27-14): $\tau = RC = (5.30\times10^{5}\,\Omega)(3.43\times10^{-10}\text{ F}) = \underline{1.82\times10^{-4}\text{ s}}$

27-37

a) With the switch in position 2 the circuit is the charging circuit.

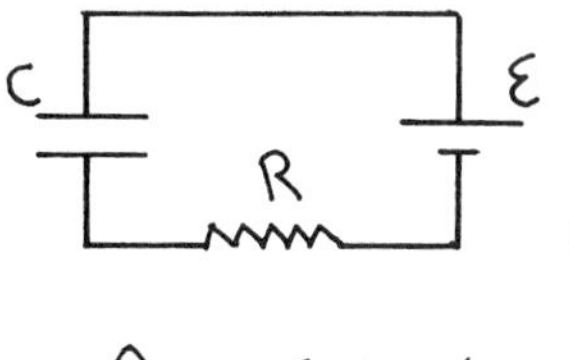

At $t = 0$, $q = 0$.

The charge q on the capacitor is given as a function of time by Eq. (27-12): $q = C\mathcal{E}(1 - e^{-t/RC})$

$$Q_f = C\mathcal{E} = (4.00\times10^{-5}\text{ F})(24.0\text{ V}) = 9.6\times10^{-4}\text{ C}$$

27-37 (cont)

$RC = (950\,\Omega)(4.00\times10^{-5}\,F) = 0.0380\,s$

At $t = 0.050\,s$, $q = 9.6\times10^{-4}\,C\,(1 - e^{-0.050\,s/0.0380\,s}) = \underline{702\,\mu C}$.

b) $V_C = \frac{q}{C} = \frac{702\,\mu C}{4.00\times10^{-5}\,F} = \underline{17.6\,V}$

$\mathcal{E} - V_C - V_R = 0$

$V_R = \mathcal{E} - V_C = 24.0\,V - 17.6\,V = \underline{6.4\,V}$

c) Throwing the switch back to position 1 produces the discharging circuit

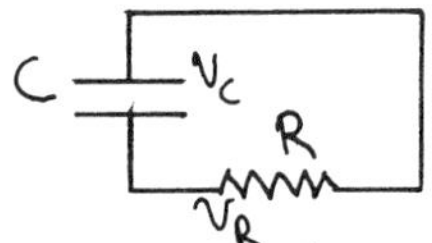

The initial charge Q_0 is the charge calculated in part (a), $Q_0 = 702\,\mu C$. $V_C = \frac{q}{C} = \frac{702\,\mu C}{4.00\times10^{-5}\,F} = \underline{17.6\,V}$, the same as just before the switch is thrown. But now $V_C - V_R = 0$, so $V_R = V_C = \underline{17.6\,V}$.

d) In the discharging circuit the charge on the capacitor as a function of time is given by Eq. (27-16): $q = Q_0 e^{-t/RC}$. $RC = 0.0380\,s$, the same as in part (a). Thus at $t = 0.050\,s$, $q = (702\,\mu C)\,e^{-(0.050\,s/0.0380\,s)} = \underline{188\,\mu C}$

27-39

Find the current through the heater: $P = VI \Rightarrow I = \frac{P}{V} = \frac{1350\,W}{120\,V} = 11.2\,A$.
The maximum total current allowed is 20 A, so the current through the dryer must be less than $20\,A - 11.2\,A = 8.8\,A$. The power dissipated by the dryer if the current has this value is $P = VI = (120\,V)(8.8\,A) = 1050\,W$. For P larger than this the dryer draws more current than 8.8 A and the circuit breaker trips. The 900 W setting is ok but at the 1200 W setting the circuit breaker trips.

27-41

a) When the heater element is first turned on it is at room temperature and has resistance $R = 24\,\Omega$.

$I = \frac{V}{R} = \frac{120\,V}{24\,\Omega} = \underline{5.0\,A}$

$P = \frac{V^2}{R} = \frac{(120\,V)^2}{24\,\Omega} = \underline{600\,W}$

b) Find the resistance R_T of the element at the operating temperature of 280°C:
Eq. (26-12): $R(T) = R_0(1 + \alpha(T - T_0)) = 24\,\Omega(1 + (2.8\times10^{-3}\,(C°)^{-1}))(280°C - 23°C) = 41.27\,\Omega$.

$I = \frac{V}{R} = \frac{120\,V}{41.27\,\Omega} = \underline{2.9\,A}$

$P = \frac{V^2}{R} = \frac{(120\,V)^2}{41.27\,\Omega} = \underline{350\,W}$

Problems

27-43

a) Two of the resistors in series would each dissipate one-half the total, or 1.6 W, which is ok. But the series combination would have an equivalent resistance of 1600 Ω, not the 800 Ω that is required. Resistors in parallel have an equivalent resistance less than that of the individual resistors, so a solution is two in series in parallel with another two in series.

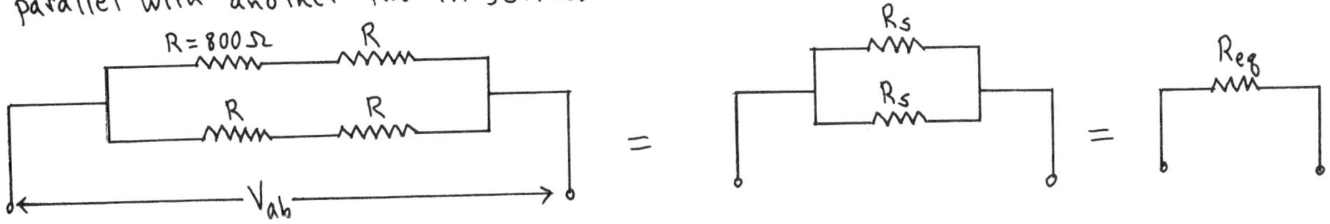

R_s is the resistance equivalent to two of the 800 Ω resistors in series:

$R_s = R + R = 1600\ \Omega$.

R_{eq} is the resistance equivalent to the two $R_s = 1600\ \Omega$ resistors in parallel:

$\frac{1}{R_{eq}} = \frac{1}{R_s} + \frac{1}{R_s} = \frac{2}{R_s} \Rightarrow R_{eq} = \frac{R_s}{2} = \frac{1600\ \Omega}{2} = 800\ \Omega$

This combination does have the required 800 Ω equivalent resistance. It will be shown in part (b) that a total of 3.2 W can be dissipated without exceeding the power rating of each individual resistor.

Another solution is two resistors in parallel in series with two more in parallel:

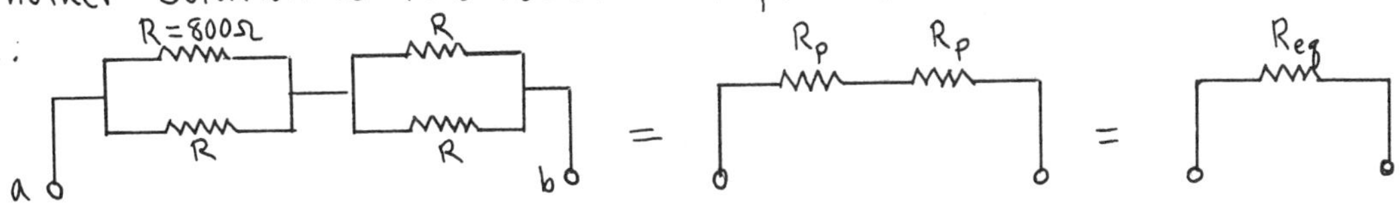

$\frac{1}{R_p} = \frac{1}{R} + \frac{1}{R} = \frac{2}{800\ \Omega}$; $R_p = 400\ \Omega$

$R_{eq} = R_p + R_p = 800\ \Omega$

This combination has the required 800 Ω equivalent resistance. It will be shown in part (b) that a total of 3.2 W can be dissipated without exceeding the power rating of each individual resistor.

b) For a combination with equivalent resistance $R_{eq} = 800\ \Omega$ to dissipate 3.2 W the voltage V_{ab} applied to the network must be given by

$P = \frac{V_{ab}^2}{R_{eq}} \Rightarrow V_{ab} = \sqrt{P R_{eq}} = \sqrt{(3.2\text{ W})(800\ \Omega)} = 50.6\text{ V}$ and the current through the equivalent resistance is $I = \frac{V_{ab}}{R} = \frac{50.6\text{ V}}{800\ \Omega} = 0.0632\text{ A}$.

For the first combination this means 50.6 V across each parallel branch and $\frac{1}{2}(50.6\text{ V}) = 25.3\text{ V}$ across each 800 Ω resistor. The power dissipated by each individual resistor is then $P = \frac{V^2}{R} = \frac{(25.3\text{ V})^2}{800\ \Omega} = \underline{0.80\text{ W}}$, which is less than the maximum allowed value of 1.6 W.

27-43 (cont)

For the second combination this means a voltage of $IR_p = (0.0632\text{ A})(400\ \Omega) = 25.3\text{ V}$ across each parallel combination and hence across each separate resistor. The power dissipated by each resistor is again $P = \frac{V^2}{R} = \frac{(25.3\text{ V})^2}{800\ \Omega} = 0.80\text{ W}$, which is less than the maximum allowed value of 1.6 W.

27-47

a) Break in the circuit between points a and b ⇒ no current in the middle branch that contains the 3.00 Ω resistor and the 10.0 V battery. The circuit therefore has a single current path. Find the current, so that potential drops across the resistor can be calculated.

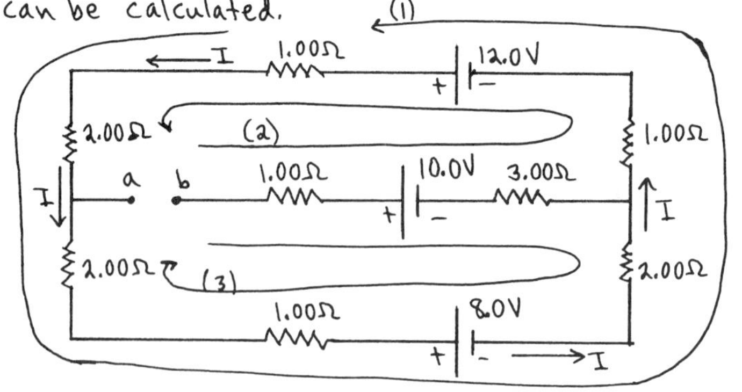

Apply the loop rule to loop (1):

$+12.0\text{ V} - I(1.00\ \Omega + 2.00\ \Omega + 2.00\ \Omega + 1.00\ \Omega) - 8.0\text{ V} - I(2.00\ \Omega + 1.00\ \Omega) = 0$

$I = \frac{12.0\text{ V} - 8.0\text{ V}}{9.00\ \Omega} = 0.4444\text{ A}$

To find V_{ab} start at point b and travel to a, adding up the potential rises and drops. Travel on path (2) shown on the diagram. The 1.00 Ω and 3.00 Ω resistors in the middle branch have no current through them and hence no voltage across them. Therefore,

$V_b - 10.0\text{ V} - I(1.00\ \Omega) + 12.0\text{ V} - I(1.00\ \Omega + 2.00\ \Omega) = V_a$

$V_{ab} = V_a - V_b = 12.0\text{ V} - 10.0\text{ V} - I(4.00\ \Omega) = 2.0\text{ V} - (0.4444\text{ A})(4.00\ \Omega) = \underline{+0.22\text{ V}}$

(point a is at higher potential)

As a check on this calculation we can also compute V_{ab} by traveling from b to a on path (3):

$V_b - 10.0\text{ V} + 8.0\text{ V} + I(2.00\ \Omega + 1.00\ \Omega + 2.00\ \Omega) = V_a$

$V_{ab} = -2.00\text{ V} + (0.4444\text{ A})(5.00\ \Omega) = +0.22\text{ V}$, which checks.

b) With points a and b connected by a wire there are three current branches:

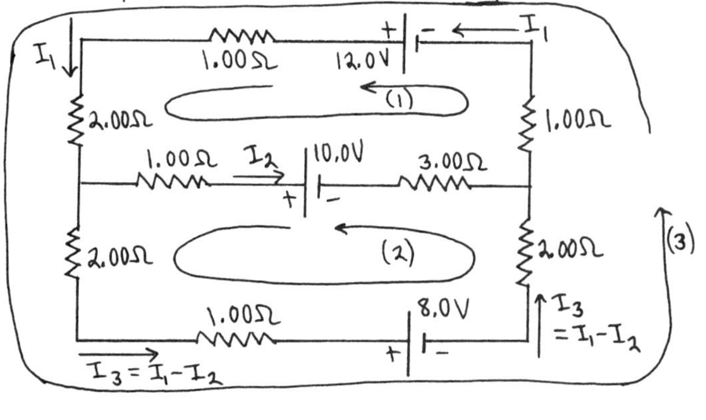

The junction rule has been used to write the third current (in the 8.0 V battery) in terms of the other two.

27-47 (cont)

Apply the loop rule to loop (1):

$$12.0\text{ V} - I_1(1.00\,\Omega) - I_1(2.00\,\Omega) - I_2(1.00\,\Omega) - 10.0\text{ V} - I_2(3.00\,\Omega) - I_1(1.00\,\Omega) = 0$$

$$2.0\text{ V} - I_1(4.00\,\Omega) - I_2(4.00\,\Omega) = 0$$

$$\boxed{(2.00\,\Omega)I_1 + (2.00\,\Omega)I_2 = 1.0\text{ V}} \quad \text{equation (1)}$$

Apply the loop rule to loop (2):

$$-(I_1 - I_2)(2.00\,\Omega) - (I_1 - I_2)(1.00\,\Omega) - 8.0\text{ V} - (I_1 - I_2)(2.00\,\Omega) + I_2(3.00\,\Omega) + 10.0\text{ V} + I_2(1.00\,\Omega) = 0$$

$$\boxed{2.0\text{ V} - (5.00\,\Omega)I_1 + (9.00\,\Omega)I_2 = 0} \quad \text{equation (2)}$$

Solve eq. (1) for I_2 and use this to replace I_2 in eq. (2):

$$I_2 = 0.50\text{ A} - I_1$$

$$2.0\text{ V} - (5.00\,\Omega)I_1 + (9.00\,\Omega)(0.50\text{ A} - I_1) = 0$$

$$(14.0\,\Omega)I_1 = 6.50\text{ V} \Rightarrow I_1 = \frac{6.50\text{ V}}{14.0\,\Omega} = \underline{0.464\text{ A}}$$

$$I_2 = 0.500\text{ A} - 0.464\text{ A} = 0.036\text{ A}.$$

The current in the 12.0 V battery is $I_1 = 0.464$ A.

27-49

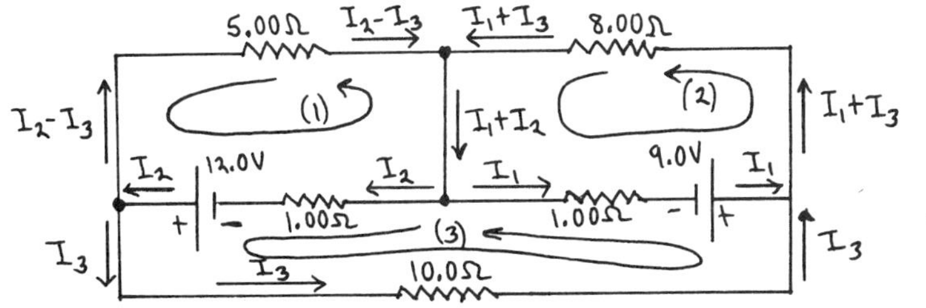

The current in each branch has been written in terms of I_1, I_2, and I_3 such that the junction rule is satisfied at each junction point.

Apply the loop rule to loop (1):

$$-12.0\text{ V} + I_2(1.00\,\Omega) + (I_2 - I_3)\,5.00\,\Omega = 0$$

$$\boxed{I_2(6.00\,\Omega) - I_3(5.00\,\Omega) = 12.0\text{ V}} \quad \text{eq. (1)}$$

Apply the loop rule to loop (2):

$$-I_1(1.00\,\Omega) + 9.00\text{ V} - (I_1 + I_3)(8.00\,\Omega) = 0$$

$$\boxed{I_1(9.00\,\Omega) + I_3(8.00\,\Omega) = 9.00\text{ V}} \quad \text{eq. (2)}$$

Apply the loop rule to loop (3):

$$-I_3(10.0\,\Omega) - 9.00\text{ V} + I_1(1.00\,\Omega) - I_2(1.00\,\Omega) + 12.0\text{ V} = 0$$

$$\boxed{-I_1(1.00\,\Omega) + I_2(1.00\,\Omega) + I_3(10.0\,\Omega) = 3.00\text{ V}} \quad \text{eq. (3)}$$

eq. (1) $\Rightarrow I_2 = 2.00\text{ A} + \frac{5}{6}I_3$; eq. (2) $\Rightarrow I_1 = 1.00\text{ A} - \frac{8}{9}I_3$

Use these in eq. (3) $\Rightarrow -(1.00\text{ A} - \frac{8}{9}I_3)(1.00\,\Omega) + (2.00\text{ A} + \frac{5}{6}I_3)(1.00\,\Omega) + I_3(10.0\,\Omega) = 3.00\text{ V}$

$$\left(\frac{16+15+180}{18}\right)I_3 = 2.00\text{ A} \;;\; I_3 = \frac{18}{211}(2.00\text{ A}) = \underline{0.171\text{ A}}$$

Then $I_2 = 2.00\text{ A} + \frac{5}{6}I_3 = 2.00\text{ A} + \frac{5}{6}(0.171\text{ A}) = \underline{2.14\text{ A}}$

$I_1 = 1.00\text{ A} - \frac{8}{9}I_3 = 1.00\text{ A} - \frac{8}{9}(0.171\text{ A}) = \underline{0.848\text{ A}}$

27-51

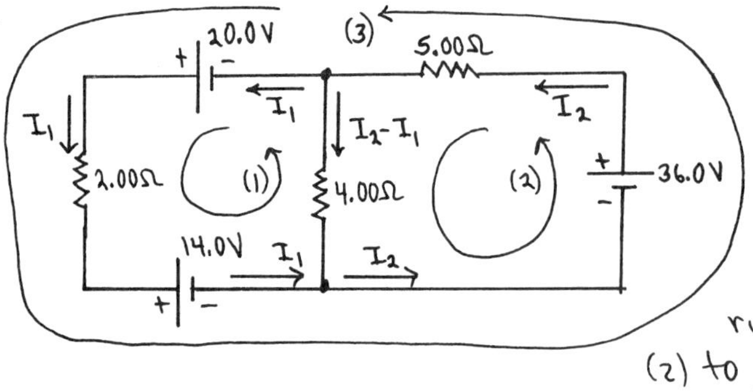

Two unknown currents I_1 (through the 2.00Ω resistor) and I_2 (through the 5.00Ω resistor) are labeled on the circuit diagram. The current through the 4.00Ω resistor has been written as $I_2 - I_1$ by using the junction rule. Apply the loop rule to loops (1) and (2) to get two equations for the unknown currents I_1 and I_2. Loop (3) can then be used to check the results.

loop (1):

$$+20.0\text{ V} - I_1(2.00\ \Omega) - 14.0\text{ V} + (I_2 - I_1)(4.00\ \Omega) = 0$$

$$6.00\, I_1 - 4.00\, I_2 = 6.00\text{ A}$$

$$\boxed{3.00\, I_1 - 2.00\, I_2 = 3.00\text{ A}} \qquad \text{eq. (1)}$$

loop (2):

$$+36.0\text{ V} - I_2(5.00\ \Omega) - (I_2 - I_1)(4.00\ \Omega) = 0$$

$$\boxed{-4.00\, I_1 + 9.00\, I_2 = 36.0\text{ A}} \qquad \text{eq. (2)}$$

Solve eq. (1) for $I_1 \Rightarrow I_1 = 1.00\text{ A} + \frac{2}{3} I_2$

Use in eq. (2) $\Rightarrow -4.00(1.00\text{ A} + \frac{2}{3} I_2) + 9.00\, I_2 = 36.0\text{ A}$

$(-\frac{8}{3} + 9.00) I_2 = 40.0\text{ A} \Rightarrow I_2 = \underline{6.32\text{ A}}$

Then $I_1 = 1.00\text{ A} + \frac{2}{3} I_2 = 1.00\text{ A} + \frac{2}{3}(6.32\text{ A}) = \underline{5.21\text{ A}}$

Current through the 2.00Ω resistor: $I_1 = \underline{5.21\text{ A}}$.

Current through the 5.00Ω resistor: $I_2 = \underline{6.32\text{ A}}$.

Current through the 4.00Ω resistor: $I_2 - I_1 = 6.32\text{ A} - 5.21\text{ A} = \underline{1.11\text{ A}}$.

Loop (3) to check: $+20.0\text{ V} - I_1(2.00\ \Omega) - 14.0\text{ V} + 36.0\text{ V} - I_2(5.00\ \Omega) = 0$

$$(5.21\text{ A})(2.00\ \Omega) + (6.32\text{ A})(5.00\ \Omega) = 42.0\text{ V}$$

$$10.4\text{ V} + 31.6\text{ V} = 42.0\text{ V} \checkmark$$

27-53

a)

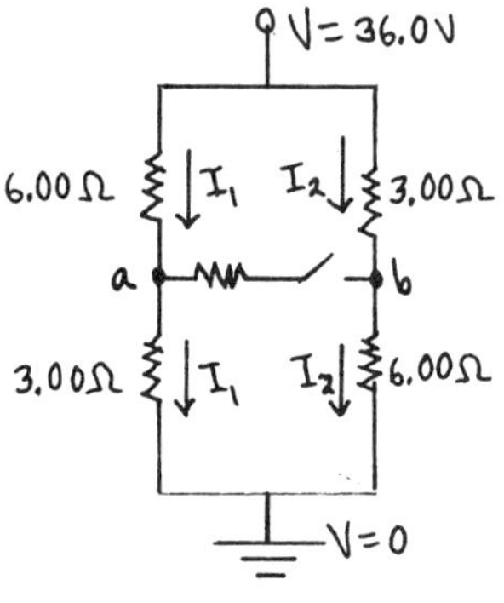

With the switch open there is no current through it and there are only the two currents I_1 and I_2 indicated in the sketch.

The potential drop across each parallel branch is 36.0 V:

$$-I_1(6.00\ \Omega + 3.00\ \Omega) + 36.0\text{ V} = 0$$

$$I_1 = \frac{36.0\text{ V}}{6.00\ \Omega + 3.00\ \Omega} = 4.00\text{ A}$$

$$-I_2(3.00\ \Omega + 6.00\ \Omega) + 36.0\text{ V} = 0$$

$$I_2 = \frac{36.0\text{ V}}{3.00\ \Omega + 6.00\ \Omega} = 4.00\text{ A}$$

27-53 (cont)

To calculate $V_{ab} = V_a - V_b$ start at point b and travel to point a, adding up all the potential rises and drops along the way. We can do this by going from b up through the 3.00 Ω resistor:

$$V_b + I_2(3.00\,\Omega) - I_1(6.00\,\Omega) = V_a$$

$$V_a - V_b = (4.00\text{ A})(3.00\,\Omega) - (4.00\text{ A})(6.00\,\Omega) = 12.0\text{ V} - 24.0\text{ V} = -12.0\text{ V}$$

$$V_{ab} = \underline{-12.0\text{ V}}$$ (point a is 12.0 V lower in potential than point b)

Alternatively, we can go from point b down through the 6.00 Ω resistor:

$$V_b - I_2(6.00\,\Omega) + I_1(3.00\,\Omega) = V_a$$

$$V_a - V_b = -(4.00\text{ A})(6.00\,\Omega) + (4.00\text{ A})(3.00\,\Omega) = -24.0\text{ V} + 12.0\text{ V} = -12.0\text{ V}\ ✓$$

b)

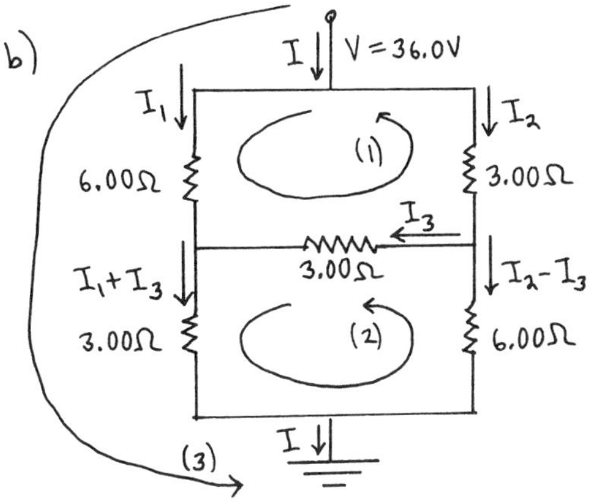

The three unknown currents I_1, I_2, and I_3 are labeled on the sketch. Apply the loop rule to loops (1), (2), and (3).

Loop (1):

$$-I_1(6.00\,\Omega) + I_3(3.00\,\Omega) + I_2(3.00\,\Omega) = 0$$

$$\boxed{I_2 = 2I_1 - I_3} \quad \text{eq. (1)}$$

loop (2):

$$-(I_1+I_3)(3.00\,\Omega) + (I_2-I_3)(6.00\,\Omega) - I_3(3.00\,\Omega) = 0$$

$$6I_2 - 12I_3 - 3I_1 = 0$$

$$2I_2 - 4I_3 - I_1 = 0$$

Use eq. (1) to replace I_2:

$$4I_1 - 2I_3 - 4I_3 - I_1 = 0$$

$$3I_1 = 6I_3 \Rightarrow \boxed{I_1 = 2I_3} \quad \text{eq. (2)}$$

Loop (3) (The loop is completed through the battery [not shown], in the direction from the − to the + terminal.):

$$-I_1(6.00\,\Omega) - (I_1+I_3)(3.00\,\Omega) + 36.0\text{ V} = 0$$

$$9I_1 + 3I_3 = 36.0\text{ A}$$

$$\boxed{3I_1 + I_3 = 12.0\text{ A}} \quad \text{eq. (3)}$$

Use eq. (2) in eq. (3) to replace I_1:

$$3(2I_3) + I_3 = 12.0\text{ A}$$

$$I_3 = \frac{12.0\text{ A}}{7} = 1.71\text{ A}$$

$$I_1 = 2I_3 = 3.42\text{ A}$$

$$I_2 = 2I_1 - I_3 = 2(3.42\text{ A}) - 1.71\text{ A} = 5.13\text{ A}$$

The current through the switch is $I_3 = \underline{1.71\text{ A}}$.

c) From the results in part (a) the current through the battery is $I = I_1 + I_2 = 3.42\text{ A} + 5.13\text{ A} = 8.55\text{ A}$. The equivalent circuit is a single resistor that produces the same current through the 36.0 V battery.

27-53 (cont)

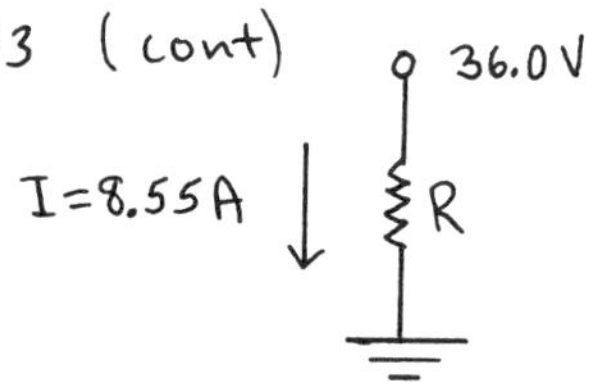

$$-IR + 36.0\text{ V} = 0$$

$$R = \frac{36.0\text{ V}}{I} = \frac{36.0\text{ V}}{8.55\text{ A}} = \underline{4.21\ \Omega}$$

27-55

a)

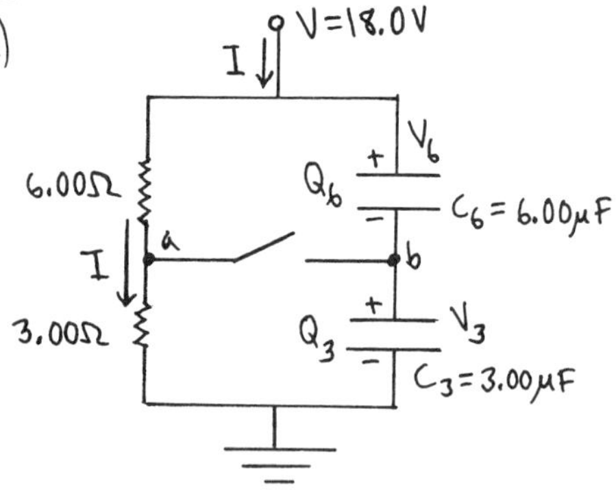

After the capacitors have been charged to their final charge there is no current through them. There is only one current path.

$$-I(6.00\ \Omega + 3.00\ \Omega) + 18.0\text{ V} = 0$$

$$I = \frac{18.0\text{ V}}{9.00\ \Omega} = 2.00\text{ A}$$

There is also a potential difference of 18.0 V applied across the two capacitors in series. The capacitors have the same charge: $Q_3 = Q_6 = Q$.

$$V_1 + V_3 = 18.0\text{ V}$$

$$Q\left(\frac{1}{C_6} + \frac{1}{C_3}\right) = 18.0\text{ V}$$

$$Q\left(\frac{C_3 + C_6}{C_3 C_6}\right) = 18.0\text{ V}$$

$$Q = \left(\frac{C_3 C_6}{C_3 + C_6}\right)(18.0\text{ V}) = \left[\frac{(3.00\times10^{-6}\text{ F})(6.00\times10^{-6}\text{ F})}{3.00\times10^{-6}\text{ F} + 6.00\times10^{-6}\text{ F}}\right] 18.0\text{ V} = 36.0\ \mu\text{C}$$

And then $V_6 = \frac{Q}{C_6} = \frac{36.0\ \mu C}{6.00\ \mu F} = 6.00\text{V}$, $V_3 = \frac{Q}{C_3} = \frac{36.0\ \mu C}{3.00\ \mu F} = 12.0\text{V}$

$$V_b + V_6 - I(6.00\ \Omega) = V_a$$

$$V_a - V_b = 6.00\text{ V} - (2.00\text{ A})(6.00\ \Omega) = \underline{-6.00\text{ V}}$$

or

$$V_b - V_3 + I(3.00\ \Omega) = V_a$$

$$V_a - V_b = (2.00\text{ A})(3.00\ \Omega) - 12.0\text{ V} = -6.00\text{ V} \checkmark$$

b) $V_{ab} < 0$; point b is at higher potential

c)

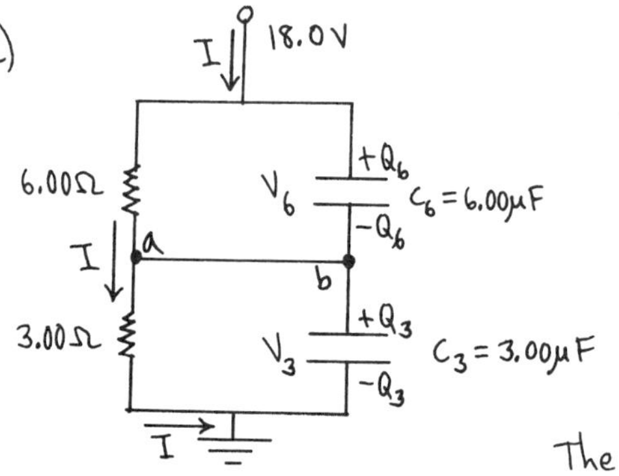

The only current path is still the one through both resistors, so $I = 2.00\text{ A}$ as calculated in part (a).
The effect of closing the switch is to put points a and b at the same potential, so that $I(6.00\ \Omega) = 12.0\text{ V}$ is the potential V_6 across the $6.00\ \mu F$ capacitor and $I(3.00\ \Omega) = 6.00\text{ V}$ is the potential V_3 across the $3.00\ \mu F$ capacitor.

The potential of point b above ground is

$$V_b = I(3.00\ \Omega) = V_3 = \underline{6.00\text{ V}}$$

d) The charges on the capacitors after the switch is closed are

$$Q_6 = C_6 V_6 = (6.00\ \mu F)(12.0\text{ V}) = 72.0\ \mu C$$

27-55 (cont)

$Q_3 = C_3 V_3 = (3.00\,\mu F)(6.00\,V) = 18.0\,\mu C$

As calculated in part (a), with the switch open the charge on each capacitor was $36.0\,\mu C$.

switch open

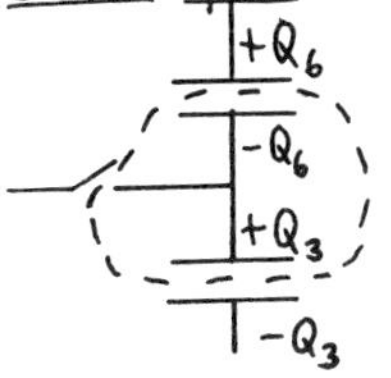

The net charge on the conductor enclosed by the dashed line is zero.

switch closed

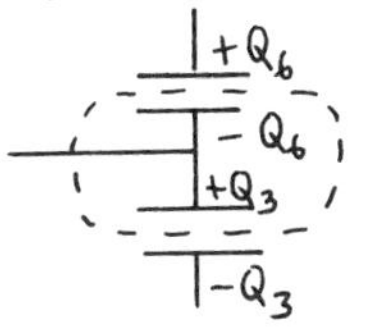

The net charge on the conductor enclosed by the dashed line is now $-Q_6 + Q_3 = -72.0\,\mu C + 18.0\,\mu C = \underline{-54.0\,\mu C}$

This charge must have flowed through the switch when it was closed.

27-57

a)

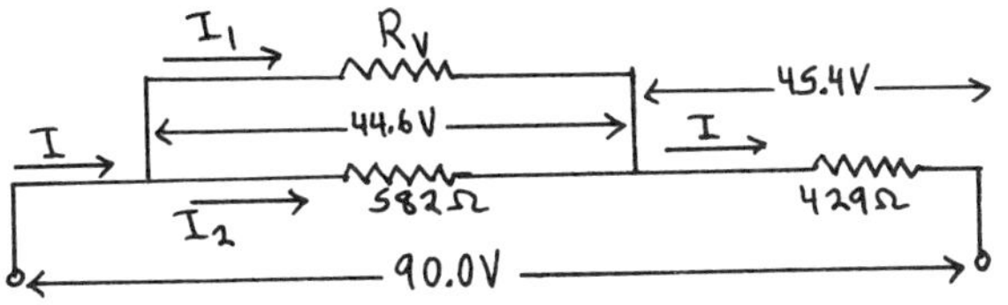

The voltmeter reads the potential difference across its terminals, which is 44.6 V. If we can find the current I_1 through the voltmeter then we can use Ohm's law to find its resistance.

The voltage drop across the 429Ω resistor is $90.0\,V - 44.6\,V = 45.4\,V$, so

$I = \frac{V}{R} = \frac{45.4\,V}{429\,\Omega} = 0.1058\,A$

The voltage drop across the 582Ω resistor is 44.6 V, so $I_2 = \frac{V}{R} = \frac{44.6\,V}{582\,\Omega} = 0.07663\,A$.

Then $I = I_1 + I_2 \Rightarrow I_1 = I - I_2 = 0.1058\,A - 0.07663\,A = 0.02917\,A$.

$R_V = \frac{V}{I_1} = \frac{44.6\,V}{0.02917\,A} = \underline{1530\,\Omega}$

b)

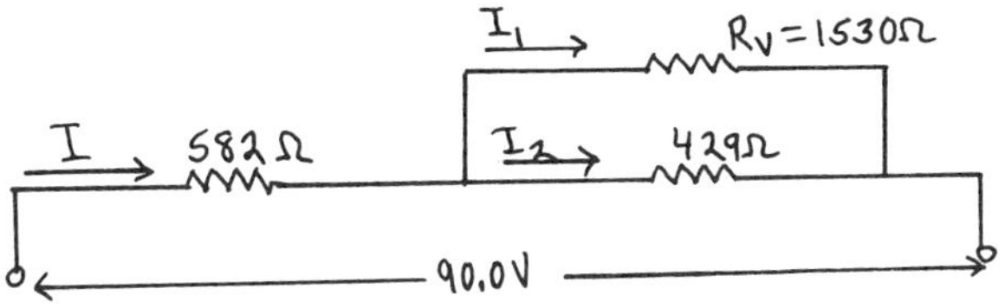

Replace the two resistors in parallel by their equivalent:

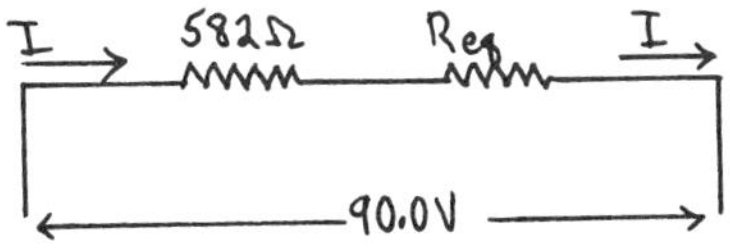

$\frac{1}{R_{eq}} = \frac{1}{1530\,\Omega} + \frac{1}{429\,\Omega} = \frac{1530\,\Omega + 429\,\Omega}{(1530\,\Omega)(429\,\Omega)}$; $R_{eq} = 335.1\,\Omega$

$I = \frac{90.0\,V}{582\,\Omega + 335.1\,\Omega} = 0.09814\,A$

The potential drop across the 582Ω resistor then is $IR = (0.09814\,A)(582\,\Omega) = 57.12\,V$, so the potential drop across the 429Ω resistor and across the voltmeter (what the voltmeter reads) is $90.0\,V - 57.12\,V = \underline{32.9\,V}$.

27-63

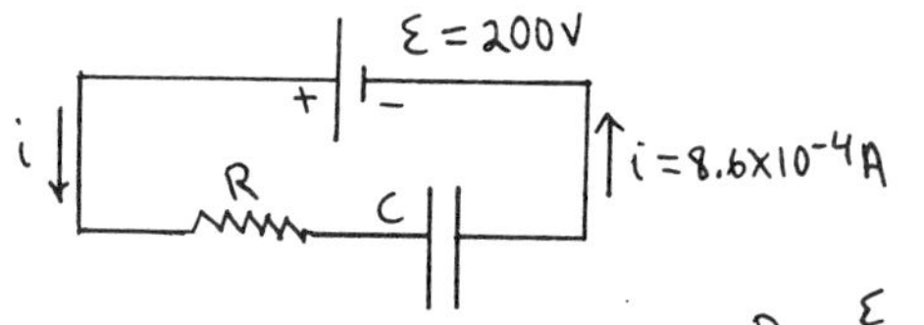

Initially, the charge on the capacitor is zero, so by $v = \frac{q}{C}$ the voltage across the capacitor is zero. The loop rule therefore gives $\mathcal{E} - iR = 0$

$$R = \frac{\mathcal{E}}{i} = \frac{200\ \text{V}}{8.6\times10^{-4}\ \text{A}} = \underline{2.3\times10^{5}\ \Omega}$$

The time constant is given by $\tau = RC$ (Eq. 27-14), so $C = \frac{\tau}{R} = \frac{5.7\ \text{s}}{2.3\times10^{5}\ \Omega} = \underline{25\ \mu\text{F}}$

27-65

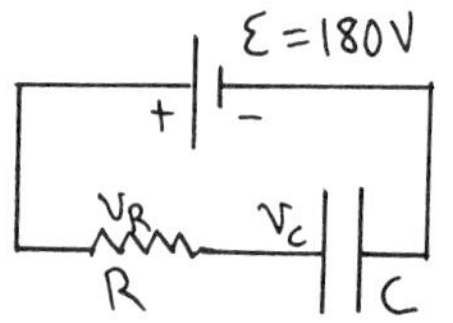

At any instant of time Kirchhoff's loop rule is satisfied:

$$\mathcal{E} - v_R - v_C = 0$$

$i = 0.0185\ \text{A} \Rightarrow v_R = iR = (0.0185\ \text{A})(7.25\times10^{3}\ \Omega) = 134\ \text{V}$

$$v_C = \mathcal{E} - v_R = 180\ \text{V} - 134\ \text{V} = 46\ \text{V}$$

Then $q = v_C C = (46\ \text{V})(3.40\times10^{-6}\ \text{F}) = \underline{156\ \mu\text{C}}$.

27-67

Eq. (27-13) $i = \frac{\mathcal{E}}{R} e^{-t/RC}$

a) $P = \mathcal{E} i$

The total energy supplied by the battery is $\int_0^\infty P\,dt = \int_0^\infty \mathcal{E} i\,dt = \frac{\mathcal{E}^2}{R}\int_0^\infty e^{-t/RC}\,dt$
$= \frac{\mathcal{E}^2}{R}\left(-RC e^{-t/RC}\Big|_0^\infty\right) = \underline{C\mathcal{E}^2}$

b) $P = i^2 R$

The total energy dissipated in the resistor is $\int_0^\infty P\,dt = \int_0^\infty i^2 R\,dt = \frac{\mathcal{E}^2}{R}\int_0^\infty e^{-2t/RC}\,dt$
$= \frac{\mathcal{E}^2}{R}\left(-\frac{RC}{2} e^{-2t/RC}\Big|_0^\infty\right) = \underline{\tfrac{1}{2}C\mathcal{E}^2}$.

c) The final charge on the capacitor is $Q = C\mathcal{E}$. The energy stored is $U = \frac{Q^2}{2C} = \underline{\tfrac{1}{2}C\mathcal{E}^2}$.

Final energy stored in capacitor $(\tfrac{1}{2}C\mathcal{E}^2)$ = total energy supplied by battery $(C\mathcal{E}^2)$ − energy dissipated in the resistor $(\tfrac{1}{2}C\mathcal{E}^2)$.

d) $\tfrac{1}{2}$ of the energy supplied by the battery is stored in the capacitor. This fraction is independent of R.

CHAPTER 28

Exercises 1, 3, 9, 11, 15, 17, 21, 23, 27, 31, 33, 35, 37

Problems 39, 41, 45, 49, 53, 55, 57, 61, 63, 65

Exercises

28-1

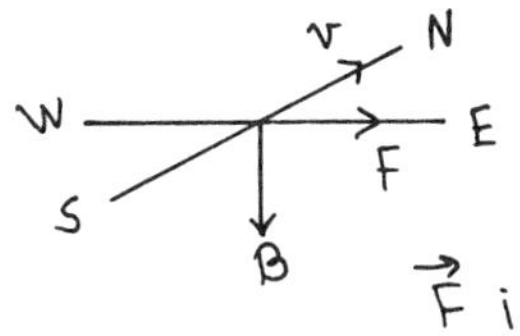

Or, view from above

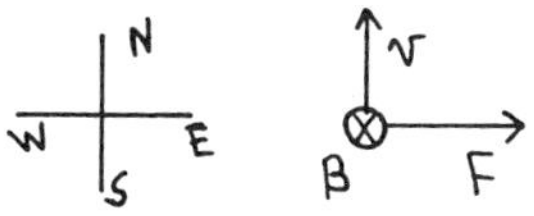

The force must be in the direction the particle is deflected.

$\vec{F}$ is opposite to the direction of $\vec{v}\times\vec{B}$ as given by the right-hand rule, so q is <u>negative</u>.

28-3

$$\vec{v} = (-3.85\times10^4\,\text{m/s})\hat{\imath} + (4.19\times10^4\,\text{m/s})\hat{\jmath}$$

a) $\vec{B} = (1.40\,\text{T})\hat{\imath}$

$\vec{F} = q\,\vec{v}\times\vec{B} = (-2.48\times10^{-8}\,\text{C})(1.40\,\text{T})[(-3.85\times10^4\,\text{m/s})\hat{\imath}\times\hat{\imath} + (4.19\times10^4\,\text{m/s})\hat{\jmath}\times\hat{\imath}]$

$\hat{\imath}\times\hat{\imath} = 0, \quad \hat{\jmath}\times\hat{\imath} = -\hat{k}$

$\vec{F} = (-2.48\times10^{-8}\,\text{C})(1.40\,\text{T})(4.19\times10^4\,\text{m/s})(-\hat{k}) = +(1.45\times10^{-3}\,\text{N})\hat{k}$

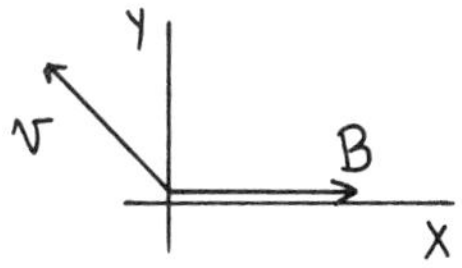

The right-hand rule gives that $\vec{v}\times\vec{B}$ is directed into the paper (−z-direction). The charge is negative so $\vec{F}$ is opposite to $\vec{v}\times\vec{B}$; $\vec{F}$ is in the +z-direction. This agrees with the direction calculated with unit vectors.

b) $\vec{B} = (1.40\,\text{T})\hat{k}$

$\vec{F} = q\,\vec{v}\times\vec{B} = (-2.48\times10^{-8}\,\text{C})(1.40\,\text{T})[(-3.85\times10^4\,\text{m/s})\hat{\imath}\times\hat{k} + (4.19\times10^4\,\text{m/s})\hat{\jmath}\times\hat{k}]$

$\hat{\imath}\times\hat{k} = -\hat{\jmath}, \quad \hat{\jmath}\times\hat{k} = \hat{\imath}$

$\vec{F} = (+1.34\times10^{-3}\,\text{N})(-\hat{\jmath}) - (1.45\times10^{-3}\,\text{N})\hat{\imath} = -[(1.45\times10^{-3}\,\text{N})\hat{\imath} + (1.34\times10^{-3}\,\text{N})\hat{\jmath}]$

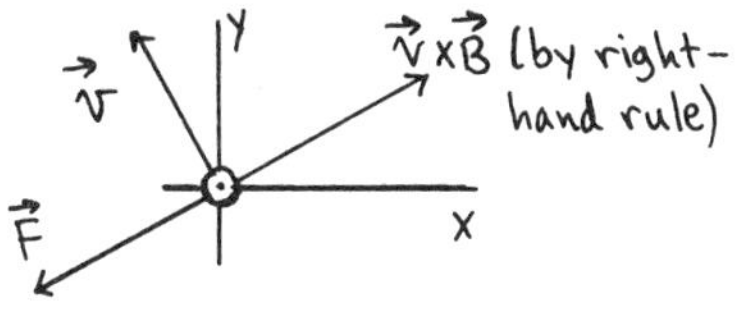

The direction of $\vec{F}$ is opposite to $\vec{v}\times\vec{B}$ since q is negative. The direction of $\vec{F}$ computed from the right-hand rule agrees qualitatively with the direction calculated with unit vectors.

28-9

a) $\vec{B} = \beta y^2\hat{\jmath}$ with $\beta = 5.00\,\text{T/m}^2$

<u>surface abcd</u>

$d\vec{A} = dA(-\hat{\imath})$ $\qquad \vec{B}\cdot d\vec{A} = -\beta y^2\,dA\,(\hat{\jmath}\cdot\hat{\imath}) = 0$, so $\Phi_B = 0$

28-9 (cont)

<u>surface befc</u>

$d\vec{A} = dA(-\hat{k})$ $\quad \vec{B}\cdot d\vec{A} = -\beta y^2 dA\,(\hat{j}\cdot\hat{k}) = 0$, so $\Phi_B = 0$

<u>surface aefd</u>

$d\vec{A} = dA(\hat{i}+\hat{k})/\sqrt{2}$ $\quad \vec{B}\cdot d\vec{A} = +(\beta y^2 dA/\sqrt{2})(\hat{i}+\hat{k})\cdot\hat{j} = 0$, so $\Phi_B = 0$

<u>surface abc</u>

$d\vec{A} = dA\,\hat{j}$ $\quad \vec{B}\cdot d\vec{A} = \beta y^2 dA\,\hat{j}\cdot\hat{j} = \beta y^2 dA$

$\Phi_B = \int \vec{B}\cdot d\vec{A} = \int \beta y^2 dA = \beta y^2 \int dA = \beta y^2 A$ (since y is constant on the surface)

$y = 0.300$ m, $\quad A = \frac{1}{2}(0.400\text{ m})(0.300\text{ m}) = 0.0600\text{ m}^2$

$\Phi_B = +(5.00\text{ T/m}^2)(0.300\text{m})^2(0.0600\text{ m}^2) = 0.0270$ Wb

<u>surface cfd</u>

On this surface $y=0$ so $B=0$ and $\Phi_B = 0$.

The net flux is the sum of the fluxes through each surface. Only for surface abc is the flux nonzero, so the net flux is $\Phi_B = $ <u>0.0270 Wb</u>.

b) Gauss's law for magnetism (Eq. (28-8)) says $\oint \vec{B}\cdot d\vec{A} = 0$. But in part (a) we found $\oint \vec{B}\cdot d\vec{A} = $ net $\Phi_B = 0.0270$ Wb. The magnetic field produces a flux that violates Gauss's law for magnetism so is not possible.

<u>28-11</u>

$\Phi_B = \int \vec{B}\cdot d\vec{A}$

Circular area in the xy-plane $\Rightarrow A = \pi r^2 = \pi(0.374\text{m})^2 = 0.4394\text{ m}^2$ and $d\vec{A}$ is in the z-direction.

a) $\vec{B} = (1.16\text{ T})\hat{k}$; $\vec{B}$ and $d\vec{A}$ are parallel ($\phi = 0°$) so $\vec{B}\cdot d\vec{A} = B\,dA$

B is constant over the circular area so

$\Phi_B = \int \vec{B}\cdot d\vec{A} = \int B\,dA = B\int dA = BA = (1.16\text{T})(0.4394\text{ m}^2) = $ <u>0.510 Wb</u>

b) 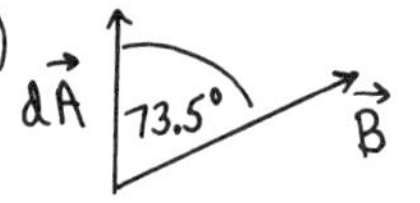

$\vec{B}\cdot d\vec{A} = B\cos\phi\, dA$ with $\phi = 73.5°$

B and ϕ are constant over the circular area so

$\Phi_B = \int \vec{B}\cdot d\vec{A} = \int B\cos\phi\, dA = B\cos\phi\int dA = B\cos\phi\, A$

$\Phi_B = (1.16\text{ T})(\cos 73.5°)(0.4394\text{ m}^2) = $ <u>0.145 Wb</u>

c) 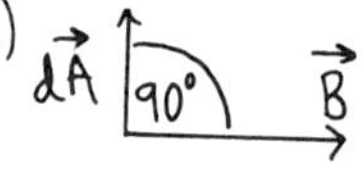

$\vec{B}\cdot d\vec{A} = 0$ since $d\vec{A}$ and $\vec{B}$ are perpendicular ($\phi = 90°$)

$\Phi_B = \int \vec{B}\cdot d\vec{A} = 0$

<u>28-15</u>

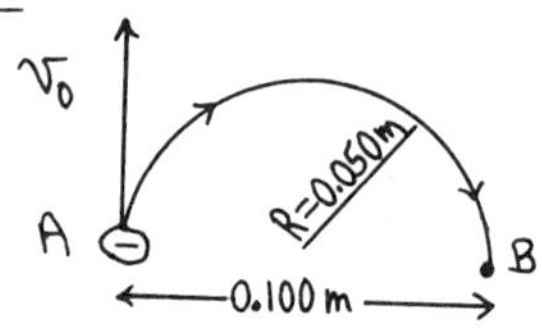

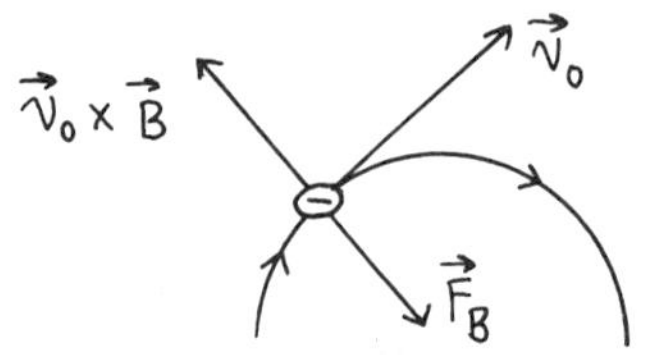

As the electron moves in the semicircle, its velocity is tangent to the circular path.

28-15 (cont)

Circular motion ⇒ the acceleration of the electron $\vec{a}_{rad}$ is directed in toward the center of the circle. Thus the force $\vec{F}_B$ exerted by the magnetic field, since it is the only force on the electron, must be radially inward. Since q is negative, $\vec{F}_B$ is opposite to the direction given by the right-hand rule for $\vec{v}_0 \times \vec{B}$. Thus $\vec{B}$ is <u>into the page</u>.

Apply Newton's 2nd law to calculate the magnitude of $\vec{B}$:

$$\sum \vec{F} = m\vec{a} \Rightarrow \sum F_{rad} = ma_{rad}$$

$$F_B = m\frac{v^2}{R}$$

$$F_B = |q| vB \sin\phi = |q| vB \Rightarrow |q| vB = m\frac{v^2}{R}$$

$$B = \frac{mv}{|q| R} = \frac{(9.109\times10^{-31}\text{ kg})(2.94\times10^6\text{ m/s})}{(1.602\times10^{-19}\text{ C})(0.050\text{ m})} = \underline{3.34\times10^{-4}\text{ T}}$$

b) The speed of the electron as it moves along the path is constant. ($\vec{F}_B$ changes the direction of $\vec{v}$ but not its magnitude.) The time is given by the distance divided by v_0. The distance along the semicircular path is πR, so

$$t = \frac{\pi R}{v_0} = \frac{\pi(0.050\text{ m})}{2.94\times10^6\text{ m/s}} = \underline{5.34\times10^{-8}\text{ s}}$$

<u>28-17</u>

a) $\sum \vec{F} = m\vec{a} \Rightarrow |q| vB = m\frac{v^2}{R}$

$$v = \frac{|q|BR}{m} = \frac{(1.602\times10^{-19}\text{ C})(1.50\text{ T})(0.0348\text{ m})}{3.34\times10^{-27}\text{ kg}} = \underline{2.50\times10^6\text{ m/s}}$$

b) $t = \frac{\pi R}{v} = \frac{\pi(0.0348\text{ m})}{2.50\times10^6\text{ m/s}} = \underline{4.37\times10^{-8}\text{ s}}$

c) Kinetic energy gained = electric potential energy lost

$$\tfrac{1}{2}mv^2 = |q| V$$

$$V = \frac{mv^2}{2|q|} = \frac{(3.34\times10^{-27}\text{ kg})(2.50\times10^6\text{ m/s})^2}{2(1.602\times10^{-19}\text{ C})} = 6.52\times10^4\text{ V} = \underline{65.2\text{ kV}}$$

<u>28-21</u>

a) In the plane perpendicular to $\vec{B}$ (the yz-plane) the motion is circular. But there is a velocity component in the direction of $\vec{B}$, so the motion is a helix. The electric field in the $-\hat{i}$ direction exerts a force in the $-\hat{i}$ direction. This force produces an acceleration in the $-\hat{i}$ direction and this causes the pitch of the helix to vary. The force does not affect the circular motion in the yz-plane, so the electric field does not affect the radius of the helix.

b) Calculate the period T:

$$\omega = |q|B/m$$

$$T = \frac{2\pi}{\omega} = \frac{2\pi m}{|q| B} = \frac{2\pi(1.67\times10^{-27}\text{ kg})}{(1.60\times10^{-19}\text{ C})(0.500\text{ T})} = 1.312\times10^{-7}\text{ s}$$

28-21 (cont)

Then $t = \frac{T}{2} = 6.558\times10^{-8}\,s.$

$$x - x_0 = v_{0x}t + \tfrac{1}{2}a_x t^2$$

$v_{0x} = 1.50\times10^5\,m/s$

$$a_x = \frac{F_x}{m} = \frac{|q|E_x}{m} = \frac{(1.60\times10^{-19}\,C)(-3.00\times10^4\,V/m)}{1.67\times10^{-27}\,kg} = -2.874\times10^{12}\,m/s^2$$

Then $x - x_0 = (1.50\times10^5\,m/s)(6.558\times10^{-8}\,s) + \tfrac{1}{2}(-2.874\times10^{12}\,m/s^2)(6.558\times10^{-8}\,s)^2 = \underline{3.66\,mm}$

28-23

In the velocity selector $|q|E = |q|vB$

$$v = \frac{E}{B} = \frac{1.20\times10^6\,V/m}{0.600\,T} = 2.00\times10^6\,m/s$$

In the region of the circular path $\Sigma\vec{F} = m\vec{a}$ gives

$$|q|vB = m\frac{v^2}{R} \Rightarrow m = \frac{|q|RB}{v}$$

Singly charged ion $\Rightarrow |q| = +e = 1.602\times10^{-19}\,C$

$$m = \frac{(1.602\times10^{-19}\,C)(0.899\,m)(0.600\,T)}{2.00\times10^6\,m/s} = \underline{4.32\times10^{-26}\,kg}$$

Mass number = number of protons and neutrons. The mass of one proton or neutron is 1.67×10^{-27} kg, so the mass number for this isotope is $\frac{4.32\times10^{-26}\,kg}{1.67\times10^{-27}\,kg} = \underline{26}.$

28-27

$$\vec{F} = I\vec{l}\times\vec{B}$$

$I = 7.00\,A$, $\vec{l} = (0.0100\,m)\,\hat{\imath}$ (since the current is in the +x-direction)

a) $\vec{B} = -(0.65\,T)\,\hat{\jmath}$

$\vec{F} = (7.00\,A)(0.0100\,m)(-0.65\,T)\,\hat{\imath}\times\hat{\jmath}$

$\hat{\imath}\times\hat{\jmath} = \hat{k}$, so $\vec{F} = -(0.046\,N)\,\hat{k}$

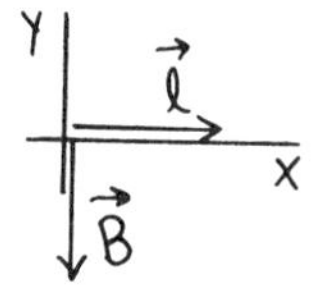

$\vec{l}\times\vec{B}$ ⊗ (−z-direction)

b) $\vec{B} = +(0.56\,T)\,\hat{k}$

$\vec{F} = (7.00\,A)(0.0100\,m)(0.56\,T)\,\hat{\imath}\times\hat{k}$

$\hat{\imath}\times\hat{k} = -\hat{\jmath}$, so $\vec{F} = -(0.039\,N)\,\hat{\jmath}$

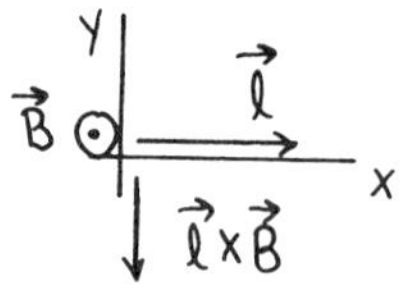

c) $\vec{B} = (-0.31\,T)\,\hat{\imath}$

$\vec{F} = (7.00\,A)(0.0100\,m)(-0.31\,T)\,\hat{\imath}\times\hat{\imath}$

$\hat{\imath}\times\hat{\imath} = 0$, so $F = 0$

d) $\vec{B} = +(0.33\,T)\,\hat{\imath} - (0.28\,T)\,\hat{k}$

$\vec{F} = (7.00\,A)(0.0100\,m)\,[(0.33\,T)\,\hat{\imath}\times\hat{\imath} - (0.28\,T)\,\hat{\imath}\times\hat{k}]$

$\hat{\imath}\times\hat{\imath} = 0$, $\hat{\imath}\times\hat{k} = -\hat{\jmath}$ so $\vec{F} = (7.00\,A)(0.0100\,m)(-0.28\,T)(-\hat{\jmath}) = +(0.020\,N)\,\hat{\jmath}$

e) $\vec{B} = +(0.74\,T)\,\hat{\jmath} - (0.36\,T)\,\hat{k}$

$\vec{F} = (7.00\,A)(0.0100\,m)\,[+(0.74\,T)(\hat{\imath}\times\hat{\jmath}) - (0.36\,T)(\hat{\imath}\times\hat{k})]$

28-27 (cont)

$\hat{i}\times\hat{j} = \hat{k}$, $\hat{i}\times\hat{k} = -\hat{j}$

$$\vec{F} = (0.052\,\text{N})\,\hat{k} - (0.025\,\text{N})(-\hat{j}) = +(0.025\,\text{N})\,\hat{j} + (0.052\,\text{N})\,\hat{k}$$

28-31

$$U = -\vec{\mu}\cdot\vec{B}$$

$\vec{\mu}$ and $\vec{B}$ parallel $\Rightarrow \phi = 0°$ and $\vec{\mu}\cdot\vec{B} = \mu B\cos\phi = \mu B$

$\vec{\mu}$ and $\vec{B}$ antiparallel $\Rightarrow \phi = 180°$ and $\vec{\mu}\cdot\vec{B} = \mu B\cos\phi = -\mu B$

$$U_1 = -\mu B,\quad U_2 = +\mu B$$

$$\Delta U = U_2 - U_1 = 2\mu B = 2(1.20\,\text{A}\cdot\text{m}^2)(0.728\,\text{T}) = \underline{+1.75\,\text{J}}$$

28-33

$$\vec{\tau} = \vec{\mu}\times\vec{B},\quad U = -\vec{\mu}\cdot\vec{B},\quad \vec{B} = B\hat{j}$$

a)

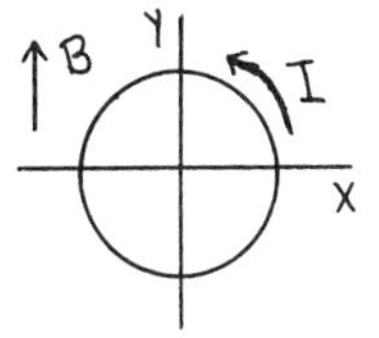

$\vec{\mu}\odot \Rightarrow \vec{\mu} = \mu\hat{k} = NIA\hat{k}$

$\vec{\tau} = \vec{\mu}\times\vec{B} = NIAB\,\hat{k}\times\hat{j} = -NIAB\,\hat{i}$

$U = -\vec{\mu}\cdot\vec{B} = -NIAB\,\hat{k}\cdot\hat{j} = 0$

b)

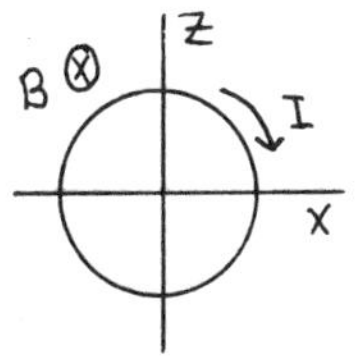

$\vec{\mu}\otimes \Rightarrow \vec{\mu} = \mu\hat{j} = NIA\hat{j}$

$\vec{\tau} = \vec{\mu}\times\vec{B} = NIAB\,\hat{j}\times\hat{j} = 0$

$U = -\vec{\mu}\cdot\vec{B} = -NIAB\,\hat{j}\cdot\hat{j} = -NIAB$

c)

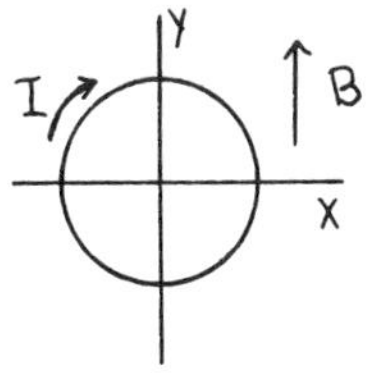

$\vec{\mu}\otimes \Rightarrow \vec{\mu} = -\mu\hat{k} = -NIA\hat{k}$

$\vec{\tau} = \vec{\mu}\times\vec{B} = -NIAB\,\hat{k}\times\hat{j} = -NIAB(-\hat{i}) = +NIAB\,\hat{i}$

$U = -\vec{\mu}\cdot\vec{B} = -(-NIAB)\,\hat{k}\cdot\hat{j} = 0$

d)

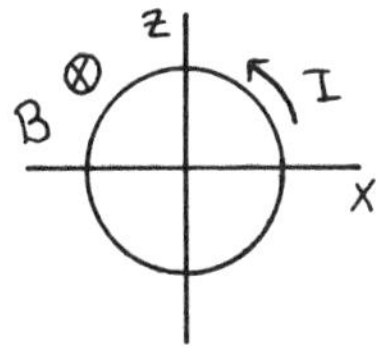

$\vec{\mu}\odot \Rightarrow \vec{\mu} = -\mu\hat{j} = -NIA\hat{j}$

$\vec{\tau} = \vec{\mu}\times\vec{B} = -NIAB\,(\hat{j}\times\hat{j}) = 0$

$U = -\vec{\mu}\cdot\vec{B} = -(-NIAB)\,\hat{j}\cdot\hat{j} = +NIAB$

28-35

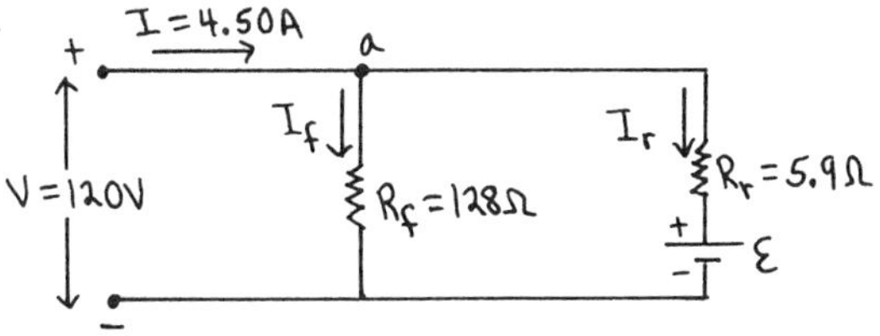

$\mathcal{E}$ is the induced emf developed by the motor. It is directed so as to oppose the current through the rotor.

28-35 (cont)

a) The field coils and the rotor are in parallel with the applied potential difference V, so $V = I_f R_f$. $I_f = \frac{V}{R_f} = \frac{120\,V}{128\,\Omega} = \underline{0.94\,A}$

b) Apply the junction rule to point a in the circuit diagram $\Rightarrow I - I_f - I_r = 0$.
$I_r = I - I_f = 4.87\,A - 0.94\,A = \underline{3.93\,A}$

c) The potential drop across the rotor, $I_r R_r + \mathcal{E}$, must equal the applied potential difference V: $V = I_r R_r + \mathcal{E}$
$\mathcal{E} = V - I_r R_r = 120\,V - (3.93\,A)(5.9\,\Omega) = \underline{97\,V}$

d) The mechanical power output is the electrical power input minus the rate of dissipation of electrical energy in the resistance of the motor:
electrical power input to the motor $P_{in} = IV = (4.87\,A)(120\,V) = 584\,W$
electrical power loss in the two resistances
$P_{loss} = I_f^2 R_f + I_r^2 R_r = (0.94\,A)^2(128\,\Omega) + (3.93\,A)^2(5.9\,\Omega) = 203\,W$
mechanical power output
$P_{out} = P_{in} - P_{loss} = 584\,W - 203\,W = \underline{381\,W}$

Note: The mechanical power output is the power associated with the induced emf $\mathcal{E}$
$P_{out} = P_{\mathcal{E}} = \mathcal{E} I_r = (97\,V)(3.93\,A) = 381\,W$, which agrees with the above calculation.

28-37

a)

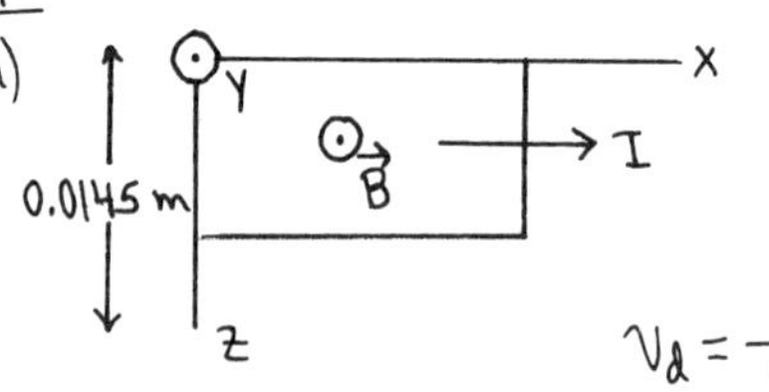

$J_x = n|q|v_d \Rightarrow v_d = \frac{J_x}{n|q|}$

$J_x = \frac{I}{A} = \frac{I}{y_1 z_1} = \frac{150\,A}{(0.29\times10^{-3}\,m)(0.0145\,m)} = 3.57\times10^{7}\,A/m^2$

$v_d = \frac{J_x}{n|q|} = \frac{3.57\times10^{7}\,A/m^2}{(5.85\times10^{28}/m^3)(1.602\times10^{-19}\,C)} = 3.81\times10^{-3}\,m/s = 3.81\,mm/s$

b) magnitude of $\vec{E}$
$|q| E_z = |q| v_d B_y$
$E_z = v_d B_y = (3.81\times10^{-3}\,m/s)(1.5\,T) = \underline{5.7\times10^{-3}\,V/m}$

direction of $\vec{E}$
The drift velocity of the electrons is in the opposite direction to the current.

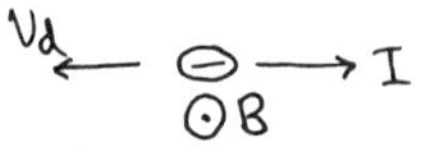

$\vec{v}\times\vec{B}\uparrow$ $\vec{F}_B = q\,\vec{v}\times\vec{B} = -e\,\vec{v}\times\vec{B}\downarrow$

F_E must oppose $\vec{F}_B \Rightarrow F_E$ is in the $-z$-direction
$\vec{F}_E = q\vec{E} = -e\vec{E} \Rightarrow \vec{E}$ is opposite to the direction of $\vec{F}_E$
$\Rightarrow \vec{E}$ is in the $\underline{+z\text{-direction}}$

c) The Hall emf is the potential difference between the two edges of the strip (at $z = 0$ and $z = z_1$) that results from the electric field calculated in part (b).
$\mathcal{E}_{Hall} = E z_1 = (5.7\times10^{-3}\,V/m)(0.0145\,m) = \underline{8.3\times10^{-5}\,V}$

Problems

28-39

$\vec{v}_0 = (3.29\times10^3\,\text{m/s})\,\hat{\imath}\,, \quad \vec{B} = -(0.515\,\text{T})\,\hat{\jmath}$

a) $q = +0.292\times10^{-8}\,\text{C}$

$\vec{F}_B = q\,\vec{v}_0\times\vec{B} = -(qv_0B)(\hat{\imath}\times\hat{\jmath}) = -qv_0B\,\hat{k}$

$\vec{F}_E = q\vec{E}$

undeflected $\Rightarrow \Sigma\vec{F} = 0 \Rightarrow \vec{F}_E = -\vec{F}_B$

$\cancel{q}\vec{E} = \cancel{q}\,v_0 B\,\hat{k}$

$\vec{E} = +v_0 B\,\hat{k}$

$\vec{E}$ is in the +z-direction and has magnitude $E = v_0B = (3.29\times10^3\,\text{m/s})(0.515\,\text{T}) = \underline{1.69\times10^3\,\text{V/m}}$

b) $q = -0.292\times10^{-8}\,\text{C}$

$\vec{E} = +v_0B\,\hat{k}$, independent of q

$\vec{E}$ is the same as calculated in part (a): $E = 1.69\times10^3\,\text{V/m}$ and is in the +z-direction.

28-41

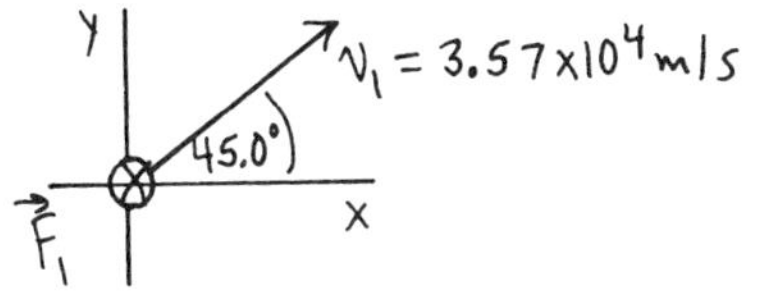

$\vec{F} = q\,\vec{v}\times\vec{B} \Rightarrow \vec{F}$ is perpendicular to $\vec{v}$ and $\vec{B}$.
The information given here means that $\vec{B}$ can have no z-component.

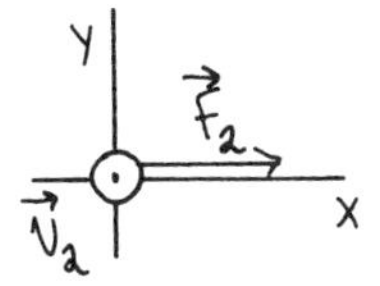

$\vec{F}$ is perpendicular to $\vec{v}$ and $\vec{B} \Rightarrow \vec{B}$ can have no x-component.

Both pieces of information taken together say that $\vec{B}$ is in the y-direction;
$\vec{B} = B_y\hat{\jmath}$.

Use the information given about $\vec{F}_2$ to calculate B_y:

$\vec{F}_2 = F_2\hat{\imath}\,, \quad \vec{v}_2 = v_2\hat{k}\,, \quad \vec{B} = B_y\hat{\jmath}$

$\vec{F}_2 = q\,\vec{v}_2\times\vec{B} \Rightarrow F_2\hat{\imath} = qv_2B_y\,\hat{k}\times\hat{\jmath} = qv_2B_y(-\hat{\imath}) \Rightarrow F_2 = -qv_2B_y$

$$B_y = -\frac{F_2}{qv_2} = -\frac{4.00\times10^{-5}\,\text{N}}{(4.97\times10^{-9}\,\text{C})(1.62\times10^4\,\text{m/s})} = -0.497\,\text{T}$$

$\vec{B}$ has magnitude $\underline{0.497\,\text{T}}$ and is in the $\underline{-y\text{-direction}}$.

28-45

a) The maximum radius of the orbit determines the maximum speed v of the protons.

$\Sigma\vec{F} = m\vec{a} \Rightarrow |q|vB = m\dfrac{v^2}{R}$

$$v = \frac{|q|BR}{m} = \frac{(1.60\times10^{-19}\,\text{C})(1.5\,\text{T})(0.50\,\text{m})}{1.67\times10^{-27}\,\text{kg}} = 7.19\times10^7\,\text{m/s}$$

28-45 (cont)

The kinetic energy of a proton moving with this speed is

$K = \frac{1}{2}mv^2 = \frac{1}{2}(1.67\times10^{-27}\,\text{kg})(7.19\times10^{7}\,\text{m/s})^2 = \underline{4.3\times10^{-12}\,\text{J}} = \underline{27\ \text{MeV}}$.

b) The time for one revolution is the period $T = \frac{2\pi R}{v} = \frac{2\pi(0.50\,\text{m})}{7.19\times10^{7}\,\text{m/s}} = \underline{4.4\times10^{-8}\,\text{s}}$

c) $K = \frac{1}{2}mv^2 = \frac{1}{2}m\left(\frac{|q|BR}{m}\right)^2 = \frac{1}{2}\frac{|q|^2B^2R^2}{m}$. Or, $B = \frac{\sqrt{2Km}}{|q|R}$. B is proportional to $\sqrt{K}$, so if K is increased by a factor of 2 then B must be increased by a factor of $\sqrt{2}$.

$B = \sqrt{2}\,(1.5\,\text{T}) = \underline{2.1\,\text{T}}$.

d) $v = \frac{|q|BR}{m} = \frac{(3.20\times10^{-19}\,\text{C})(1.5\,\text{T})(0.50\,\text{m})}{6.65\times10^{-27}\,\text{kg}} = 3.61\times10^{7}\,\text{m/s}$

$K = \frac{1}{2}mv^2 = \frac{1}{2}(6.65\times10^{-27}\,\text{kg})(3.61\times10^{7}\,\text{m/s})^2 = \underline{4.3\times10^{-12}\,\text{J}} = \underline{27\ \text{MeV}}$, the same as the maximum energy for protons. We can see that the maximum energy must be approximately the same as follows:

From part (c) $K = \frac{1}{2}\frac{|q|^2B^2R^2}{m}$. For alpha particles $|q|$ is larger by a factor of 2 and m is larger by a factor of 4 (approximately). Thus $\frac{|q|^2}{m}$ is unchanged and K is the same.

28-49

$\vec{B} = -(0.220\,\text{T})\hat{k}$, $\vec{v} = (1.22\times10^{6}\,\text{m/s})(4\hat{i} - 3\hat{j} - 12\hat{k})$, $F = 1.75\,\text{N}$

a) $\vec{F} = q\,\vec{v}\times\vec{B}$

$\vec{F} = q(-0.220\,\text{T})(1.22\times10^{6}\,\text{m/s})(4\,\hat{i}\times\hat{k} - 3\,\hat{j}\times\hat{k} - 12\,\hat{k}\times\hat{k})$

$\hat{i}\times\hat{k} = -\hat{j}$, $\hat{j}\times\hat{k} = \hat{i}$, $\hat{k}\times\hat{k} = 0$

$\vec{F} = -q(2.68\times10^{5}\,\text{N/C})(-4\hat{j} - 3\hat{i}) = +q(2.68\times10^{5}\,\text{N/C})(3\hat{i} + 4\hat{j})$

The magnitude of the vector $3\hat{i} + 4\hat{j}$ is $\sqrt{3^2+4^2} = 5$.

Thus $F = q(2.68\times10^{5}\,\text{N/C})(5)$

$q = \frac{F}{5(2.68\times10^{5}\,\text{N/C})} = \frac{1.75\,\text{N}}{5(2.68\times10^{5}\,\text{N/C})} = \underline{1.31\times10^{-6}\,\text{C}}$

b) $\sum\vec{F} = m\vec{a} \Rightarrow \vec{a} = \frac{\vec{F}}{m}$

$\vec{F} = q(2.68\times10^{5}\,\text{N/C})(3\hat{i}+4\hat{j}) = (1.31\times10^{-6}\,\text{C})(2.68\times10^{5}\,\text{N/C})(3\hat{i}+4\hat{j}) = 0.3511\,\text{N}(3\hat{i}+4\hat{j})$

$\vec{a} = \frac{\vec{F}}{m} = \frac{0.3511\,\text{N}(3\hat{i}+4\hat{j})}{1.53\times10^{-15}\,\text{kg}} = (2.29\times10^{14}\,\text{m/s}^2)(3\hat{i}+4\hat{j})$

c) $\vec{F}$ is in the xy-plane, so in the z-direction the particle moves with constant speed 14.6×10^{6} m/s.

In the xy-plane the force $\vec{F}$ causes the particle to move in a circle, with $\vec{F}$ directed in toward the center of the circle.

$\vec{F} = m\vec{a} \Rightarrow F = m\frac{v^2}{R}$ and $R = \frac{mv^2}{F}$

$v^2 = v_x^2 + v_y^2 = (4.88\times10^{6}\,\text{m/s})^2 + (-3.66\times10^{6}\,\text{m/s})^2 = 3.72\times10^{13}\,\text{m}^2/\text{s}^2$

$F = \sqrt{F_x^2+F_y^2} = 0.3511\,\text{N}\sqrt{3^2+4^2} = 1.756\,\text{N}$

$R = \frac{mv^2}{F} = \frac{(1.53\times10^{-15}\,\text{kg})(3.72\times10^{13}\,\text{m}^2/\text{s}^2)}{1.756\,\text{N}} = \underline{0.0324\,\text{m}}$

28-49 (cont)

d) Eq. (28-12): cyclotron frequency $f = \frac{\omega}{2\pi} = \frac{v}{2\pi R}$.
The circular motion is in the xy-plane, so $v = \sqrt{v_x^2 + v_y^2} = 6.10 \times 10^6$ m/s.

$$f = \frac{v}{2\pi R} = \frac{6.10 \times 10^6 \text{ m/s}}{2\pi (0.0324 \text{ m})} = \underline{3.00 \times 10^7 \text{ Hz}}$$

e) The period of the motion in the xy-plane is given by

$$T = \frac{1}{f} = \frac{1}{3.00 \times 10^7 \text{ Hz}} = 3.34 \times 10^{-8} \text{ s}$$

In $t = 2T$ the particle has returned to the same x and y coordinates. The z-component of the motion is motion with a constant velocity of $v_z = -14.6 \times 10^6$ m/s

$\Rightarrow z = z_0 + v_z t = 0 + (-14.64 \times 10^6 \text{ m/s})(2)(3.34 \times 10^{-8} \text{ s}) = -0.978$ m.

The coordinates at $t = 2T$ are $x = R$, $y = 0$, $z = -0.98$ m.

28-53

$$\vec{F} = I\vec{\ell} \times \vec{B}, \quad \vec{B} = (0.435\text{T})\hat{\imath}, \quad I = 4.92 \text{ A}$$

a) segment ab

$\vec{\ell} = (0.500 \text{ m})\hat{\jmath}$

$\vec{F}_{ab} = I\vec{\ell} \times \vec{B} = (4.92\text{A})(0.500 \text{ m})(0.435\text{T})\, \hat{\jmath} \times \hat{\imath} = (1.07\text{N})(-\hat{k}) = -(1.07\text{N})\hat{k}$

The force has magnitude $\underline{1.07 \text{ N}}$ and is in the $\underline{-z\text{-direction}}$.

segment bc

$\vec{\ell} = (0.500\text{m})(\hat{\imath} - \hat{k})$

$\vec{F}_{bc} = I\vec{\ell} \times \vec{B} = (4.92 \text{ A})(0.500\text{m})(0.435\text{T})(\hat{\imath} \times \hat{\imath} - \hat{k} \times \hat{\imath})$

$\hat{\imath} \times \hat{\imath} = 0, \quad \hat{k} \times \hat{\imath} = \hat{\jmath}$

$\vec{F}_{bc} = -(1.07\text{N})\hat{\jmath}$

The force has magnitude $\underline{1.07 \text{ N}}$ and is in the $\underline{-y\text{-direction}}$.

segment cd

$\vec{\ell} = (0.500 \text{ m})(-\hat{\jmath} + \hat{k})$

$\vec{F}_{cd} = I\vec{\ell} \times \vec{B} = (4.92\text{A})(0.500 \text{ m})(0.435\text{T})(-\hat{\jmath} \times \hat{\imath} + \hat{k} \times \hat{\imath})$

$\hat{\jmath} \times \hat{\imath} = -\hat{k}, \quad \hat{k} \times \hat{\imath} = \hat{\jmath}$

$\vec{F}_{cd} = (1.07\text{N})(\hat{\jmath} + \hat{k})$, so $F_{cd} = 1.07\text{N}\sqrt{1^2 + 1^2} = 1.51$ N

The force has magnitude 1.51 N and is directed midway between the +y-axis and the +z-axis.

Segment de

$\vec{\ell} = -(0.500 \text{ m})\hat{k}$

$\vec{F}_{de} = I\vec{\ell} \times \vec{B} = -(4.92\text{ A})(0.500 \text{ m})(0.435\text{ T})\, \hat{k} \times \hat{\imath} = -(1.07\text{N})\hat{\jmath}$

The force has magnitude 1.07 N and is in the -y-direction.

Segment ef

$\vec{\ell} = -(0.500 \text{ m})\hat{\imath}$

28-53 (cont)

$\vec{F}_{ef} = I\vec{l}\times\vec{B} = -(4.92\,A)(0.500\,m)(0.435\,T)(\hat{i}\times\hat{i}) = 0$

b) $\vec{F}_{tot} = \vec{F}_{ab} + \vec{F}_{bc} + \vec{F}_{cd} + \vec{F}_{de} + \vec{F}_{ef}$

$\vec{F}_{tot} = (1.07N)(-\hat{k} - \hat{j} + \hat{j} + \hat{k} - \hat{j}) = -(1.07N)\,\hat{j}$

The total force has magnitude <u>1.07N</u> and is in the <u>-y-direction</u>.

<u>28-55</u>

For the wire to remain at rest the force exerted on it by the magnetic field must have a component directed up the incline. To produce a force in this direction, the current in the wire must be directed <u>from right to left</u> in Fig. 28-47.

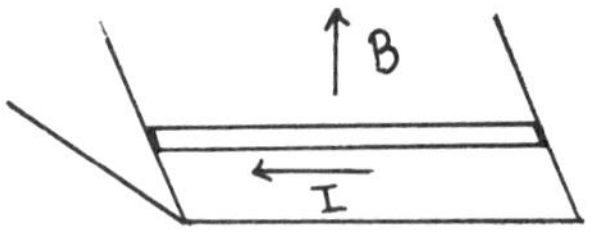

or, viewing the wire from its left-hand end

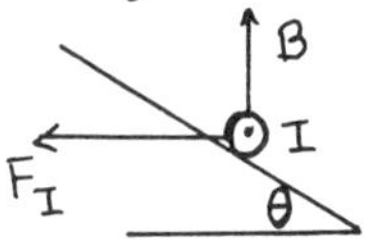

Free-body diagram for the wire:

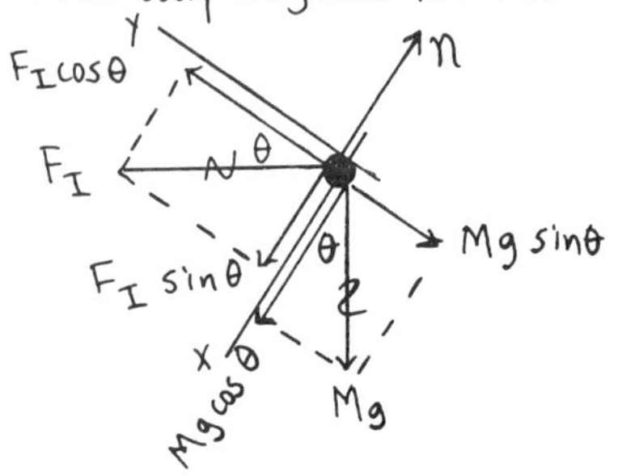

$\sum F_y = 0$

$F_I \cos\theta - Mg\sin\theta = 0$

$F_I = ILB\sin\phi$; $\phi = 90°$ since $\vec{B}$ is perpendicular to the current direction

$(ILB)\cos\theta - Mg\sin\theta = 0$

$I = \dfrac{Mg\tan\theta}{LB}$

<u>28-57</u>

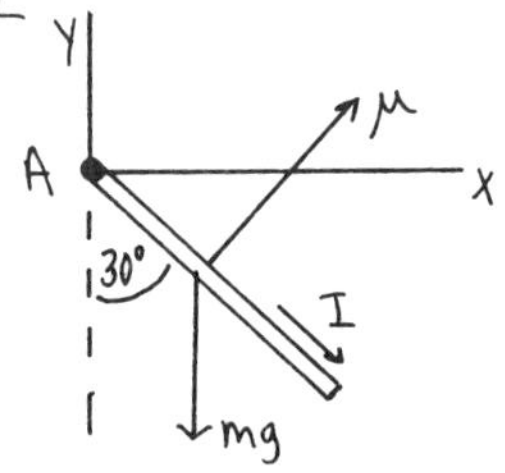

Use $\sum \tau_A = 0$, where point A is at the origin.

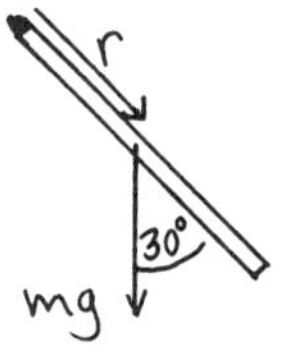

$\tau_{mg} = mgr\sin\phi = mg(0.0400m)\sin 30.0°$

The torque is clockwise; $\vec{\tau}_{mg}$ is directed into the paper.

For the loop to be in equilibrium the torque due to $\vec{B}$ must be counterclockwise (opposite to $\vec{\tau}_{mg}$) and $\tau_B = \tau_{mg}$. $\vec{\tau}_B = \vec{\mu}\times\vec{B}$. For this torque to be counterclockwise ($\vec{\tau}_B$ directed out of the paper) $\vec{B}$ must be in the <u>+y-direction</u>.

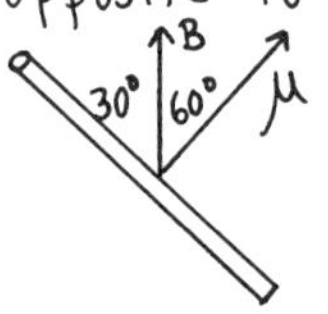

$\tau_B = \mu B\sin\phi = IAB\sin 60.0°$

$\tau_B = \tau_{mg} \Rightarrow IAB\sin 60° = mg\,(0.0400m)\sin 30.0°$

$m = (0.19\,g/cm)\,2\,(8.00\,cm + 6.00\,cm) = 5.32\,g = 5.32\times10^{-3}\,kg$

$A = (0.0800\,m)(0.0600\,m) = 4.80\times10^{-3}\,m^2$

28-57 (cont)

$$B = \frac{mg(0.0400\text{ m})(\sin 30.0^\circ)}{IA\sin 60.0^\circ} = \frac{(5.32\times10^{-3}\text{ kg})(9.80\text{ m/s}^2)(0.0400\text{ m})\sin 30.0^\circ}{(6.8\text{ A})(4.80\times10^{-3}\text{ m}^2)\sin 60.0^\circ} = \underline{0.037\text{ T}}$$

28-61

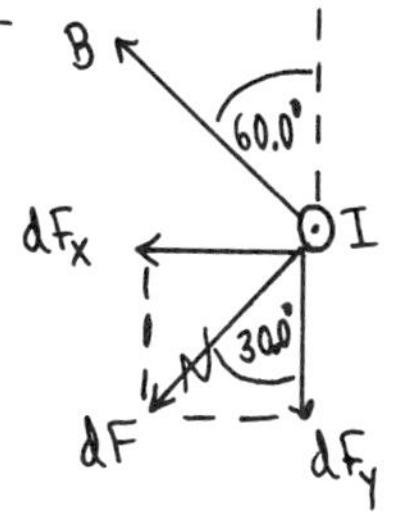

Consider the force $d\vec{F}$ on a short segment dl at the left-hand side of the coil, as viewed in Fig. 28-52. The current at this point is directed out of the page. $d\vec{F}$ is perpendicular both to $\vec{B}$ and to the direction of I.

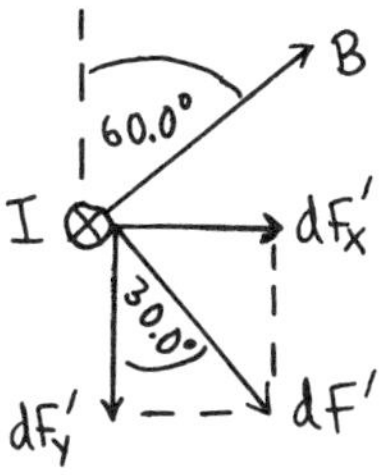

Consider also the force $d\vec{F}'$ on a short segment on the opposite side of the coil, at the right-hand side of the coil in Fig. 28-52. The current at this point is directed into the page.

The two sketches show that the x-components cancel and the y-components add. This is true for all pairs of short segments on opposite sides of the coil. The net magnetic force on the coil is in the -y-direction and its magnitude is given by $F = \int dF_y$.

$dF = I\,dl\,B\sin\phi$. But $\vec{B}$ is perpendicular to the current direction so $\phi = 90°$.

$dF_y = dF\cos 30.0^\circ = IB\cos 30.0^\circ\,dl$

$F = \int dF_y = IB\cos 30.0^\circ \int dl$

But $\int dl = N(2\pi r)$, the total length of wire in the coil.

$F = IB\cos 30.0^\circ N(2\pi r) = (0.810\text{ A})(0.170\text{ T})(\cos 30.0^\circ)(40)\pi(0.0130\text{ m}) = 0.195\text{ N}$

$\underline{\vec{F} = -(0.195\text{ N})\hat{j}}$

28-63

a) After the wire leaves the mercury its acceleration is g, downward. The wire travels upward a total distance of 0.700 m from its initial position. Its ends lose contact with the mercury after the wire has traveled 0.050 m, so the wire travels upward 0.650 m after it leaves the mercury. Consider the motion of the wire after it leaves the mercury. Take +y to be upward and take the origin at the position of the wire as it leaves the mercury.

$v_{0y} = ?$

$a_y = -9.80\text{ m/s}^2$

$y - y_0 = +0.650\text{ m}$

$v_y = 0$ (at maximum height)

$v_y^2 = v_{0y}^2 + 2a_y(y-y_0)$, with $v_y^2 = 0$

$v_{0y} = \sqrt{-2a_y(y-y_0)} = \sqrt{-2(-9.80\text{ m/s}^2)(+0.650\text{ m})} = \underline{3.57\text{ m/s}}$

b) Now consider the motion of the wire while it is in contact with the mercury. Take +y to be upward and the origin at the initial position of the wire.

28-63 (cont)

Calculate the acceleration:

$y - y_0 = +0.050\,m$

$v_{0y} = 0$ (starts from rest)

$v_y = +3.57\,m/s$ (from part (a))

$a_y = ?$

$$v_y^2 = v_{0y}^2 + 2a_y(y-y_0) \quad (v_{0y}^2 = 0)$$

$$a_y = \frac{v_y^2}{2(y-y_0)} = \frac{(3.57\,m/s)^2}{2(0.050\,m)} = 127\,m/s^2$$

Free-body diagram for the wire:

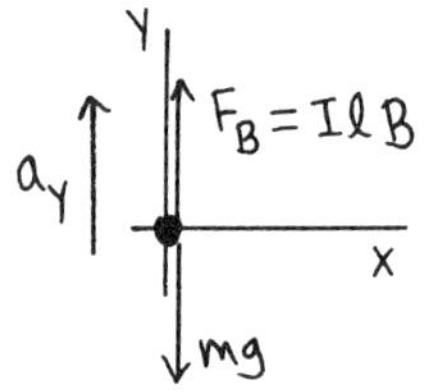

$$\Sigma F_y = ma_y$$
$$F_B - mg = ma_y$$
$$I\ell B = m(g + a_y)$$
$$I = \frac{m(g+a_y)}{\ell B}$$

ℓ is the length of the horizontal section of the wire; $\ell = 0.250\,m$

$$I = \frac{(9.09\times10^{-5}\,kg)(9.80\,m/s^2 + 127\,m/s^2)}{(0.250\,m)\ (0.0127\,T)} = \underline{3.92\,A}$$

c) $V = IR \Rightarrow R = \frac{V}{I} = \frac{1.50\,V}{3.92\,A} = \underline{0.383\,\Omega}$

28-65

a)

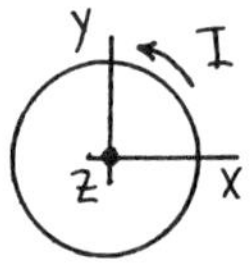

$\vec{\mu}$ is out of the plane of the paper, in the +z-direction

$\vec{\mu} = \mu\hat{k}$

$$\mu = IA = (15.0A)(5.28\times10^{-4}\,m^2) = \underline{0.00792\,A\cdot m^2}$$

b) $\vec{\tau} = (1.00\times10^{-3}\,N\cdot m)(-6\hat{i} + 8\hat{j})$

$\vec{\tau} = \vec{\mu}\times\vec{B}$, $\vec{\mu} = \mu\hat{k}$, $\vec{B} = B_x\hat{i} + B_y\hat{j} + B_z\hat{k}$

$\vec{\tau} = \vec{\mu}\times\vec{B} = \mu(B_x\,\hat{k}\times\hat{i} + B_y\,\hat{k}\times\hat{j} + B_z\,\hat{k}\times\hat{k}) = -(\mu B_y)\hat{i} + \mu B_x\hat{j}$

Compare this to the expression given for $\vec{\tau}$

$$\Rightarrow B_y = \frac{6.00\times10^{-3}\,N\cdot m}{0.00792\,A\cdot m^2} = \underline{+0.76T}, \quad B_x = \frac{8.00\times10^{-3}\,N\cdot m}{0.00792\,A\cdot m^2} = \underline{+1.01T}$$

B_z doesn't contribute to the torque since $\vec{\mu}$ is along the z-direction.

But $B = 2.60T$ and $B_x^2 + B_y^2 + B_z^2 = B^2$

$$\Rightarrow B_z = \pm\sqrt{B^2 - B_x^2 - B_y^2} = \pm\sqrt{(2.60T)^2 - (0.76T)^2 - (1.01T)^2} = \pm 2.27T$$

$U = -\vec{\mu}\cdot\vec{B}$ is negative determines the sign of B_z:

$$U = -\vec{\mu}\cdot\vec{B} = -(\mu\hat{k})\cdot(B_x\hat{i} + B_y\hat{j} + B_z\hat{k}) = -\mu B_z$$

U negative $\Rightarrow B_z$ positive, so $B_z = \underline{+2.27T}$

CHAPTER 29

Exercises 1, 3, 5, 11, 13, 17, 19, 21, 25, 29, 33, 37, 39

Problems 41, 43, 47, 49, 55, 59, 61, 63, 67, 69

Exercises

29-1

$$\vec{B} = \frac{\mu_0}{4\pi}\frac{q\,\vec{v}\times\hat{r}}{r^2} = \frac{\mu_0}{4\pi}\frac{q\,\vec{v}\times\vec{r}}{r^3}, \quad \text{since } \hat{r} = \frac{\vec{r}}{r}.$$

$\vec{v} = (4.00\times10^{6}\ \text{m/s})\hat{i}$, $\vec{r}$ is the vector from the charge to the point where the field is calculated.

a) $\vec{r} = (0.500\ \text{m})\hat{i}$, $r = 0.500\ \text{m}$

$\vec{v}\times\vec{r} = (4.00\times10^{6}\ \text{m/s})(0.500\ \text{m})\ \hat{i}\times\hat{i} = 0$, $\underline{B = 0}$

b) $\vec{r} = -(0.500\ \text{m})\hat{j}$, $r = 0.500\ \text{m}$

$\vec{v}\times\vec{r} = -(4.00\times10^{6}\ \text{m/s})(0.500\ \text{m})\ \hat{i}\times\hat{j} = -(2.00\times10^{6}\ \text{m}^2/\text{s})\hat{k}$

$$\vec{B} = (1\times10^{-7}\ \text{T}\cdot\text{m/A})\frac{(3.00\times10^{-6}\ \text{C})(2.00\times10^{6}\ \text{m}^2/\text{s})}{(0.500\ \text{m})^3}(-\hat{k}) = \underline{-(4.80\times10^{-6}\ \text{T})\hat{k}}$$

c) $\vec{r} = +(0.500\ \text{m})\hat{k}$, $r = 0.500\ \text{m}$

$\vec{v}\times\vec{r} = (4.00\times10^{6}\ \text{m/s})(0.500\ \text{m})\ \hat{i}\times\hat{k} = -(2.00\times10^{6}\ \text{m}^2/\text{s})\hat{j}$

$$\vec{B} = (1\times10^{-7}\ \text{T}\cdot\text{m/A})\frac{(3.00\times10^{-6}\ \text{C})(2.00\times10^{6}\ \text{m}^2/\text{s})}{(0.500\ \text{m})^3}(-\hat{j}) = -(4.80\times10^{-6}\ \text{T})\hat{j}$$

d) $\vec{r} = -(0.500\ \text{m})\hat{j} + (0.500\ \text{m})\hat{k}$, $r = \sqrt{(0.500\ \text{m})^2 + (0.500\ \text{m})^2} = 0.7071\ \text{m}$

$\vec{v}\times\vec{r} = (4.00\times10^{6}\ \text{m/s})(0.500\ \text{m})(-\hat{i}\times\hat{j} + \hat{i}\times\hat{k}) = (2.00\times10^{6}\ \text{m}^2/\text{s})(-\hat{j}-\hat{k})$

$$\vec{B} = -(1\times10^{-7}\ \text{T}\cdot\text{m/A})\frac{(3.00\times10^{-6}\ \text{C})(2.00\times10^{6}\ \text{m}^2/\text{s})}{(0.7071\ \text{m})^3}(\hat{j}+\hat{k}) = \underline{-(1.70\times10^{-6}\ \text{T})(\hat{j}+\hat{k})}$$

29-3

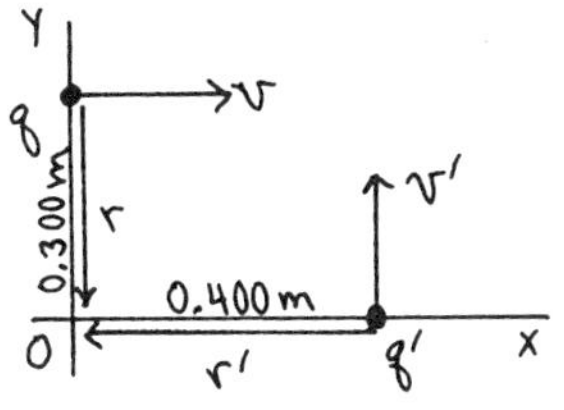

$$\vec{B}_{tot} = \vec{B}_q + \vec{B}_{q'}$$

$v = 6.00\times10^{5}\ \text{m/s}$, $v' = 8.00\times10^{5}\ \text{m/s}$

$q = +5.00\times10^{-6}\ \text{C}$, $q' = -3.00\times10^{-6}\ \text{C}$

$$\vec{B} = \left(\frac{\mu_0}{4\pi}\right)\frac{q\,\vec{v}\times\vec{r}}{r^3}$$

Field $\vec{B}_q$ due to q:

$\vec{v} = (6.00\times10^{5}\ \text{m/s})\hat{i}$, $\vec{r} = -(0.300\ \text{m})\hat{j}$

$\vec{v}\times\vec{r} = -(6.00\times10^{5}\ \text{m/s})(0.300\ \text{m})(\hat{i}\times\hat{j}) = -(1.80\times10^{5}\ \text{m}^2/\text{s})\hat{k}$

$$\vec{B}_q = (1\times10^{-7}\ \text{T}\cdot\text{m/A})\frac{(5.00\times10^{-6}\ \text{C})(1.80\times10^{5}\ \text{m}^2/\text{s})}{(0.300\ \text{m})^3}\hat{k} = -(3.333\times10^{-6}\ \text{T})\hat{k}$$

29-3 (cont)

Field $\vec{B}_{q'}$ due to q':

$\vec{v} = (8.00\times10^5\ \mathrm{m/s})\,\hat{j}$, $\vec{r} = -(0.400\ \mathrm{m})\,\hat{i}$

$\vec{v}\times\vec{r} = -(8.00\times10^5\ \mathrm{m/s})(0.400\ \mathrm{m})\,(\hat{j}\times\hat{i}) = +(3.20\times10^5\ \mathrm{m^2/s})\,\hat{k}$

$$B_{q'} = (1\times10^{-7}\ \mathrm{T\cdot m/A})\,\frac{(-3.00\times10^{-6}\ \mathrm{C})(3.20\times10^5\ \mathrm{m^2/s})}{(0.400\ \mathrm{m})^3}\,\hat{k} = -(1.50\times10^{-6}\ \mathrm{T})\,\hat{k}$$

$$\vec{B}_{tot} = \vec{B}_q + \vec{B}_{q'} = -(3.333\times10^{-6}\ \mathrm{T} + 1.50\times10^{-6}\ \mathrm{T})\,\hat{k} = -(4.83\times10^{-6}\ \mathrm{T})\,\hat{k}$$

The resultant field at the origin has magnitude 4.83×10^{-6} T and is in the $-z$-direction.

29-5

$$d\vec{B} = \frac{\mu_0}{4\pi}\,\frac{I\,d\vec{l}\times\hat{r}}{r^2}$$

As in Example 29-2 use this equation for the finite 2.00-mm segment of wire since the $\Delta l = 2.00$ mm length is much smaller than the distances to the field points.

$$\vec{B} = \frac{\mu_0}{4\pi}\,\frac{I\,\Delta\vec{l}\times\hat{r}}{r^2} = \frac{\mu_0}{4\pi}\,\frac{I\,\Delta\vec{l}\times\vec{r}}{r^3}$$

I is in the $+x$-direction, so $\Delta\vec{l} = (2.00\times10^{-3}\ \mathrm{m})\,\hat{i}$

a) Field point is at $x = 3.00$ m, $y=0$, $z=0$ so the vector $\vec{r}$ from the source point (at the origin) to the field point is $\vec{r} = (3.00\ \mathrm{m})\,\hat{i}$.

$\Delta\vec{l}\times\vec{r} = (2.00\times10^{-3}\ \mathrm{m})(3.00\ \mathrm{m})\,\hat{i}\times\hat{i} = 0$, so $\underline{\vec{B} = 0}$.

b) $\vec{r} = (3.00\ \mathrm{m})\,\hat{j}$, $r = 3.00$ m

$\Delta\vec{l}\times\vec{r} = (2.00\times10^{-3}\ \mathrm{m})(3.00\ \mathrm{m})\,\hat{i}\times\hat{j} = (6.00\times10^{-3}\ \mathrm{m^2})\,\hat{k}$

$$\vec{B} = (1\times10^{-7}\ \mathrm{T\cdot m/A})\,\frac{(8.00\ \mathrm{A})(6.00\times10^{-3}\ \mathrm{m^2})\,\hat{k}}{(3.00\ \mathrm{m})^3} = \underline{(1.78\times10^{-10}\ \mathrm{T})\,\hat{k}}$$

c) $\vec{r} = (3.00\mathrm{m})\,(\hat{i}+\hat{j})$, $r = \sqrt{2}\,(3.00\ \mathrm{m})$

$\Delta\vec{l}\times\vec{r} = (2.00\times10^{-3}\ \mathrm{m})(3.00\ \mathrm{m})\,\hat{i}\times(\hat{i}+\hat{j}) = (6.00\times10^{-3}\ \mathrm{m^2})\,\hat{k}$

$$\vec{B} = (1\times10^{-7}\ \mathrm{T\cdot m/A})\,\frac{(8.00\mathrm{A})(6.00\times10^{-3}\mathrm{m^2})\,\hat{k}}{[\sqrt{2}\,(3.00\mathrm{m})]^3} = \underline{(6.28\times10^{-11}\ \mathrm{T})\,\hat{k}}$$

d) $\vec{r} = (3.00\ \mathrm{m})\,\hat{k}$, $r = 3.00$ m

$\Delta\vec{l}\times\vec{r} = (2.00\times10^{-3}\ \mathrm{m})(3.00\ \mathrm{m})\,\hat{i}\times\hat{k} = -(6.00\times10^{-3}\ \mathrm{m^2})\,\hat{j}$

$$\vec{B} = (1\times10^{-7}\ \mathrm{T\cdot m/A})\,\frac{(8.00\mathrm{A})(-6.00\times10^{-3}\mathrm{m^2})\,\hat{j}}{(3.00\ \mathrm{m})^3} = \underline{-(1.78\times10^{-10}\ \mathrm{T})\,\hat{j}}$$

29-11

a) Viewed from above

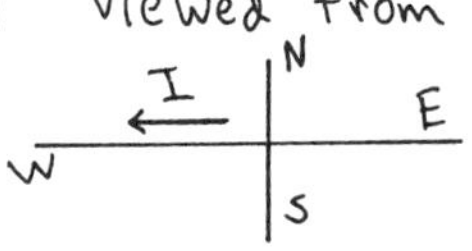

Directly below the wire the direction of the magnetic field due to the current in the wire is south.

$$B = \frac{\mu_0 I}{2\pi r} = (2\times10^{-7}\ \mathrm{T\cdot m/A})\,\frac{900\ \mathrm{A}}{5.00\ \mathrm{m}} = \underline{3.60\times10^{-5}\ \mathrm{T}}$$

29-11 (cont)

b) B from the current is nearly equal to the earth's field so, yes, the current really is a problem.

<u>29-13</u>

For each wire $B = \frac{\mu_0 I}{2\pi r}$ (Eq. 29-9), and the direction of $\vec{B}$ is given by the right-hand rule (Fig. 29-6).

a) 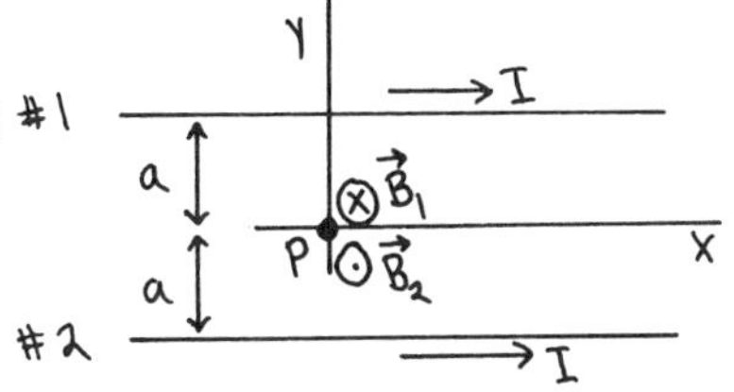

At point P midway between the two wires the fields $\vec{B}_1$ and $\vec{B}_2$ due to the two currents are in opposite directions, so $B = B_2 - B_1$.

But $B_1 = B_2 = \frac{\mu_0 I}{2\pi a}$, so $\underline{B = 0}$.

b) 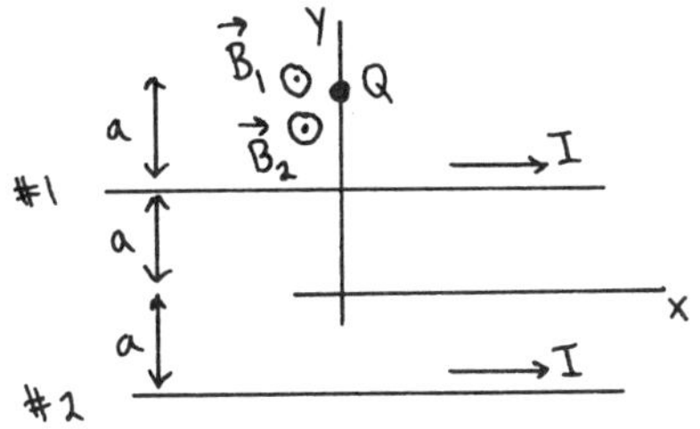

At point Q above the upper wire $\vec{B}_1$ and $\vec{B}_2$ are both directed out of the page (+z-direction), so $B = B_1 + B_2$

$B_1 = \frac{\mu_0 I}{2\pi a}$, $B_2 = \frac{\mu_0 I}{2\pi (3a)}$

$B = \frac{\mu_0 I}{2\pi a}\left(1 + \frac{1}{3}\right) = \frac{2\mu_0 I}{3\pi a}$; $\vec{B} = \frac{2\mu_0 I}{3\pi a}\hat{k}$

c) 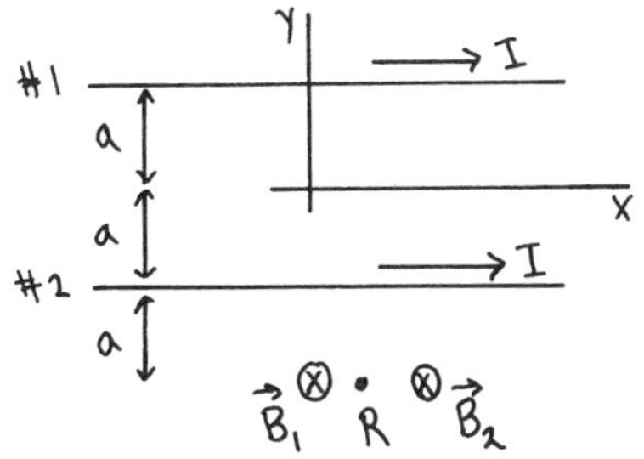

At point R below the lower wire $\vec{B}_1$ and $\vec{B}_2$ are both directed into the page (−z-direction), so $B = B_1 + B_2$.

$B_1 = \frac{\mu_0 I}{2\pi (3a)}$, $B_2 = \frac{\mu_0 I}{2\pi a}$

$B = B_1 + B_2 = \frac{\mu_0 I}{2\pi a}\left(1 + \frac{1}{3}\right) = \frac{2\mu_0 I}{3\pi a}$; $\vec{B} = -\frac{2\mu_0 I}{3\pi a}\hat{k}$

<u>29-17</u>

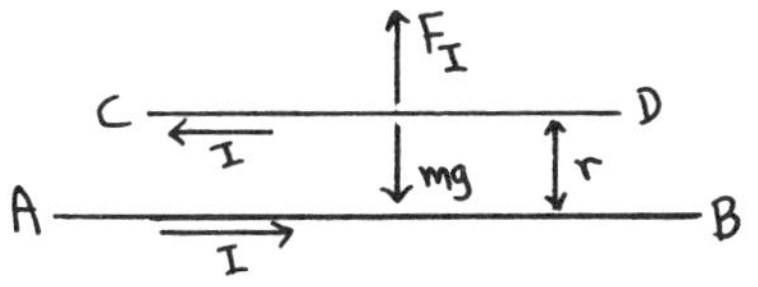

The wire CD rises until the upward force F_I due to the currents balances the downward force of gravity. Currents in opposite directions ⇒ the force is repulsive ⇒ F_I is upward, as shown.

Eq. (29-11) ⇒ $F_I = \frac{\mu_0 I^2}{2\pi h} L$ where L is the length of wire CD and h is the distance between the wires.

$$mg = (5.00\times10^{-3}\ \text{kg/m})\,Lg$$

$$F_I - mg = 0 \qquad \frac{\mu_0 I^2 L}{2\pi h} = (5.00\times10^{-3}\ \text{kg/m})\,Lg$$

$$h = \frac{\mu_0 I^2}{2\pi g (5.00\times10^{-3}\ \text{kg/m})} = \frac{(2\times10^{-7}\ \text{T}\cdot\text{m/A})(40.0\ \text{A})^2}{(9.80\ \text{m/s}^2)(5.00\times10^{-3}\ \text{kg/m})} = 6.53\times10^{-3}\ \text{m} = \underline{6.53\ \text{mm}}$$

29-19

The magnetic field at the center of N circular loops is given by Eq. (29-17): $B = \frac{\mu_0 N I}{2a}$.

Thus $N = \frac{2aB}{\mu_0 I} = \frac{2(9.00 \times 10^{-2}\ \text{m})(8.38 \times 10^{-4}\ \text{T})}{(4\pi \times 10^{-7}\ \text{T}\cdot\text{m/A})(2.50\ \text{A})} = \underline{48}$.

29-21

First consider the field at P produced by the current I_1 in the upper semicircle of wire.

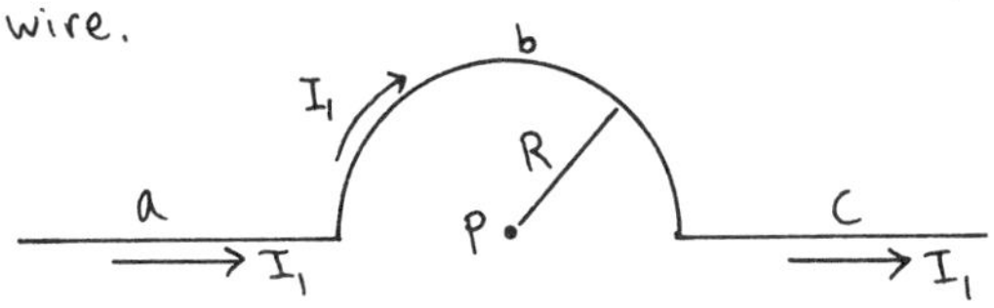

Consider the three parts of this wire
a: long, straight section, b: semicircle,
c: long, straight section

Apply the Biot-Savart law $d\vec{B} = \frac{\mu_0}{4\pi}\frac{I\, d\vec{l} \times \hat{r}}{r^2} = \frac{\mu_0}{4\pi}\frac{I\, d\vec{l} \times \vec{r}}{r^3}$ to each piece.

part a

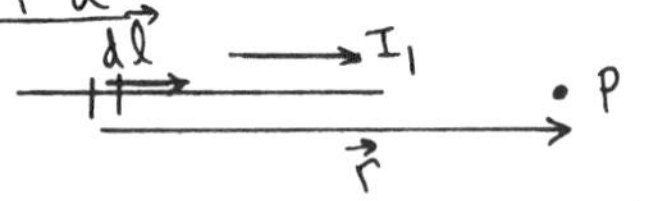

$d\vec{l} \times \vec{r} = 0$, so $dB = 0$
Same is true for all the infinitesimal segments that make up this wire, so $B = 0$.

part c

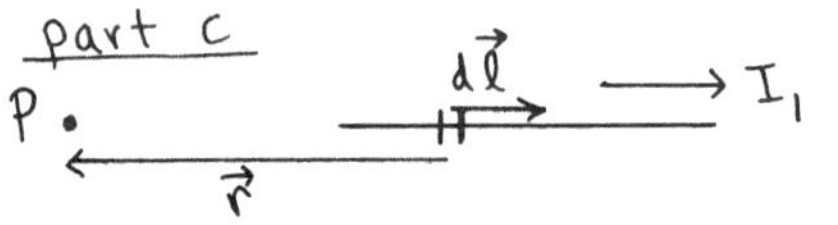

$d\vec{l} \times \vec{r} = 0$, so $dB = 0$
$B = 0$

part b

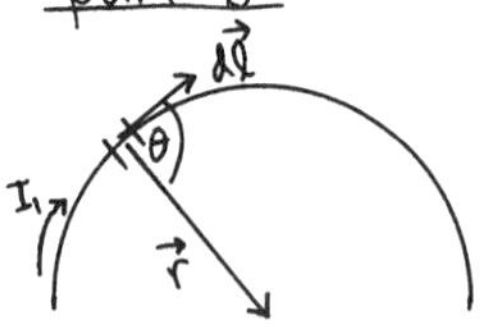

$d\vec{l} \times \vec{r}$ is directed into the paper for all infinitesimal segments that make up this semicircular piece, so $\vec{B}$ is directed into the paper and $B = \int dB$ (the vector sum of the $d\vec{B}$ is obtained by adding the magnitudes dB since all the $d\vec{B}$ are in the same direction).

$|d\vec{l} \times \vec{r}| = r\, dl \sin\theta$. The angle θ between $d\vec{l}$ and $\vec{r}$ is 90° and $r = R$, the radius of the semicircle. $|d\vec{l} \times \vec{r}| = R\, dl$

$$dB = \frac{\mu_0}{4\pi}\frac{I_1 |d\vec{l} \times \vec{r}|}{r^3} = \frac{\mu_0 I_1}{4\pi}\frac{R}{R^3}\, dl = \frac{\mu_0 I_1}{4\pi R^2}\, dl$$

$$B = \int dB = \frac{\mu_0 I_1}{4\pi R^2}\int dl = \frac{\mu_0 I_1}{4\pi R^2}(\pi R) = \frac{\mu_0 I_1}{2R}$$

($\int dl$ is equal to πR, the length of wire in the semicircle)

The two straight sections make zero contribution to $\vec{B}$, so $\vec{B}_1 = \frac{\mu_0 I_1}{2R}$ ⊗

For

a similar analysis gives $\vec{B}_2 = \frac{\mu_0 I_2}{2R}$ ⊙

$\vec{B}_1$ and $\vec{B}_2$ are in opposite directions, so the magnitude of the net field at P is $B = |B_1 - B_2| = \frac{\mu_0 |I_1 - I_2|}{2R}$. When $I_1 = I_2$, $B = 0$.

29-25

a) The magnetic field near the center of a long solenoid is given by Eq.(29-23): $B=\mu_0 n I$.

turns per unit length $n = \frac{B}{\mu_0 I} = \frac{0.190\ T}{(4\pi \times 10^{-7}\ T\cdot m/A)(10.0\ A)} = \underline{1.51\times 10^4\ turns/m}$

b) $N = nL = (1.51\times 10^4\ turns/m)(0.800\ m) = 1.21\times 10^4$ turns.

Each turn of radius R has a length $2\pi R$ of wire. The total length of wire required is $N(2\pi R) = (1.21\times 10^4)(2\pi)(3.00\times 10^{-2}\ m) = \underline{2.28\times 10^3\ m}$.

29-29

a) $\underline{a<r<b}$

end view

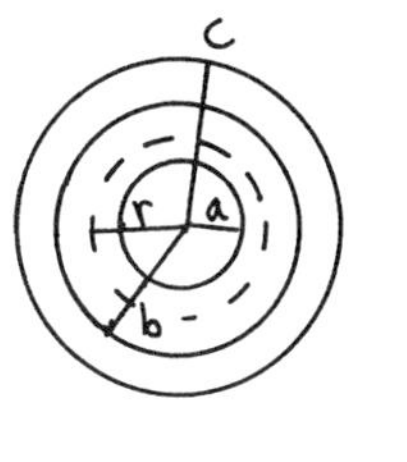

Apply Ampere's law to a circle of radius r, where $a<r<b$.
Take the currents I_1 and I_2 to be directed into the page
Take this current direction to be positive, so go around the integration path in the clockwise direction.

$$\oint \vec{B}\cdot d\vec{l} = \mu_0 I_{encl}$$

$\oint \vec{B}\cdot d\vec{l} = B(2\pi r)$, $\quad I_{encl} = I_1$

$\Rightarrow B(2\pi r) = \mu_0 I_1$

$$B = \frac{\mu_0 I_1}{2\pi r}$$

b) $\underline{r>c}$

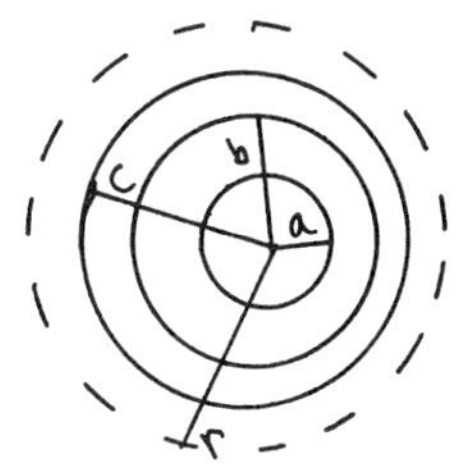

Apply Ampere's law to a circle of radius r, where $r>c$.

$$\oint \vec{B}\cdot d\vec{l} = \mu_0 I_{encl}$$

$\oint \vec{B}\cdot d\vec{l} = B(2\pi r)$, $\quad I_{encl} = I_1 + I_2$ (both currents are in the positive direction)

$$B(2\pi r) = \mu_0 (I_1 + I_2)$$

$$B = \frac{\mu_0 (I_1+I_2)}{2\pi r}$$

29-33

$B = K_m \frac{\mu_0 N I}{2\pi r}$ (Eq. 29-24, with μ_0 replaced by $K_m \mu_0$)

a) $K_m = 1400$

$I = \frac{2\pi r B}{\mu_0 K_m N} = \frac{(3.80\times 10^{-2} m)(0.250 T)}{(2\times 10^{-7}\ T\cdot m/A)(1400)(400)} = \underline{0.0848\ A}$

b) $K_m = 5200$

$I = \frac{2\pi r B}{\mu_0 K_m N} = \frac{(3.80\times 10^{-2} m)(0.250\ T)}{(2\times 10^{-7}\ T\cdot m/A)(5200)(400)} = \underline{0.0228\ A}$

29-37

a) $i_c = 1.20\times10^{-3}\ A$

$q = 0$ at $t = 0$

The amount of charge brought to the plates by the charging current in time t is

$q = i_c t = (1.20\times10^{-3}\ A)(5.00\times10^{-6}\ s) = \underline{6.00\times10^{-9}\ C}$

$E = \frac{\sigma}{\epsilon_0} = \frac{q}{\epsilon_0 A} = \frac{6.00\times10^{-9}\ C}{(8.854\times10^{-12}\ C^2/N\cdot m^2)(4.00\times10^{-4}\ m^2)} = \underline{1.69\times10^{6}\ V/m}$

$v = Ed = (1.69\times10^{6}\ V/m)(3.00\times10^{-3}\ m) = \underline{5.08\times10^{3}\ V}$

b) $E = \frac{q}{\epsilon_0 A}$

$\frac{dE}{dt} = \frac{dq/dt}{\epsilon_0 A} = \frac{i_c}{\epsilon_0 A} = \frac{1.20\times10^{-3}\ A}{(8.854\times10^{-12}\ C^2/N\cdot m^2)(4.00\times10^{-4}\ m^2)} = \underline{3.39\times10^{11}\ V/m\cdot s}$

Since i_c is constant $\frac{dE}{dt}$ does not vary in time.

c) $j_D = \epsilon_0 \frac{dE}{dt}$ (Eq. 29-37, with ϵ replaced by ϵ_0 since there is vacuum between the plates.)

$j_D = (8.854\times10^{-12}\ C^2/N\cdot m^2)(3.39\times10^{11}\ V/m\cdot s) = \underline{3.00\ A/m^2}$

$i_D = j_D A = (3.00\ A/m^2)(4.00\times10^{-4}\ m^2) = \underline{1.20\times10^{-3}\ A}$; $i_D = i_c$

29-39

a) $i_c = i_D$, so $j_D = \frac{i_D}{A} = \frac{i_c}{A} = \frac{0.450\ A}{\pi r^2} = \frac{0.450\ A}{\pi(0.0500\ m)^2} = \underline{57.3\ A/m^2}$

b) $j_D = \epsilon_0 \frac{dE}{dt} \Rightarrow \frac{dE}{dt} = \frac{j_D}{\epsilon_0} = \frac{57.3\ A/m^2}{8.854\times10^{-12}\ C^2/N\cdot m^2} = \underline{6.47\times10^{12}\ V/m\cdot s}$

c) Apply Ampere's law $\oint \vec{B}\cdot d\vec{l} = \mu_0(i_c + i_D)$ (Eq. 29-36) to a circular path with radius $r = 0.0250\ m$.

end view

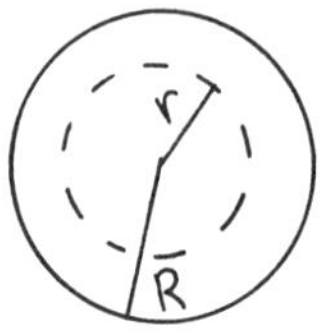

By symmetry the magnetic field is tangent to the path and constant around it $\Rightarrow \oint \vec{B}\cdot d\vec{l} = \oint B\,dl = B\int dl = B(2\pi r)$

$i_c = 0$ (no conduction current flows through the air space between the plates)

The displacement current enclosed by the path is $j_D \pi r^2$.

Thus $B(2\pi r) = \mu_0(j_D \pi r^2)$

$B = \frac{1}{2}\mu_0 j_D r = \frac{1}{2}(4\pi\times10^{-7}\ T\cdot m/A)(57.3\ A/m^2)(0.0250\ m) = \underline{9.00\times10^{-7}\ T}$

d) Apply Ampere's law $\oint \vec{B}\cdot d\vec{l} = \mu_0(i_c + i_D)$ to a circular path with radius $r = 0.100\ m$.

$\oint \vec{B}\cdot d\vec{l} = B(2\pi r)$ as in part (c)

$i_c = 0$ through path

The displacement current enclosed by the path is $j_D \pi R^2 = i_c$.

Thus $B(2\pi r) = \mu_0 i_c$.

$B = \frac{\mu_0 i_c}{2\pi r} = (2\times10^{-7}\ T\cdot m/A)\frac{0.450\ A}{0.100\ m} = \underline{9.00\times10^{-7}\ T}$

Problems

29-41

$\vec{B} = B_0\left(\frac{x}{a}\right)\hat{\imath}$

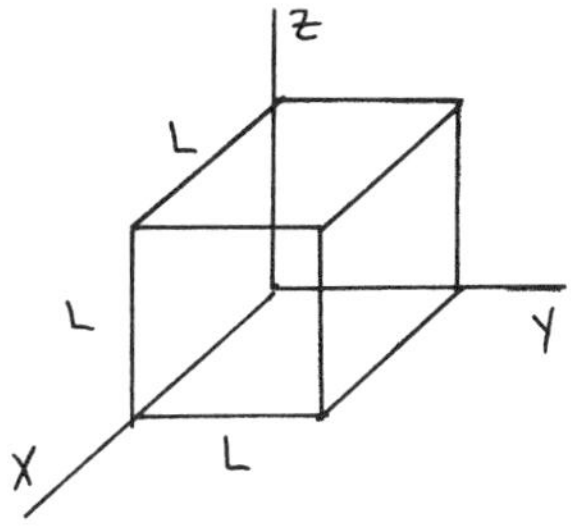

Apply Gauss's law for magnetic fields to a cube with side length L, one corner at the origin, and sides parallel to the x-, y-, and z-axes.

Since $\vec{B}$ is parallel to the x-axis the only sides that have nonzero flux are the front side (parallel to the yz-plane at $x = L$) and the back side (parallel to the yz-plane at $x = 0$).

<u>front</u>

$\Phi_B = \int \vec{B}\cdot d\vec{A}$

$d\vec{A} = dA\,\hat{\imath}$ so $\vec{B}\cdot d\vec{A} = B_0\left(\frac{x}{a}\right)dA\,(\hat{\imath}\cdot\hat{\imath}) = B_0\left(\frac{x}{a}\right)dA$

$x = L$ on this face so $\vec{B}\cdot d\vec{A} = B_0\left(\frac{L}{a}\right)dA$

$\Phi_B = B_0\left(\frac{L}{a}\right)\int dA = B_0\left(\frac{L}{a}\right)L^2 = B_0\frac{L^3}{a}$

<u>back</u>

On the back face $x = 0$ so $B = 0$ and $\Phi_B = 0$.

The total flux through the cubical Gaussian surface is $\Phi_B = B_0\frac{L^3}{a}$. But this violates Eq. (28-8), which says that $\Phi_B = 0$ for any closed surface. The claimed $\vec{B}$ is impossible because it has been shown to violate Gauss's law for magnetism.

29-43

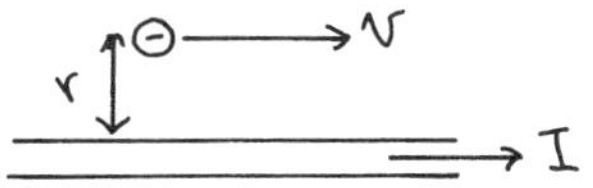

At the electron's position the magnetic field $\vec{B}$ due to the current in the wire has magnitude

$$B = \frac{\mu_0 I}{2\pi r} = \frac{(2\times10^{-7}\,\mathrm{T\cdot m/A})(1.50\,\mathrm{A})}{0.0800\,\mathrm{m}} = 3.75\times10^{-6}\,\mathrm{T}$$

By the right-hand rule, $\vec{B}$ is ⊙ (out of page) at this point.

$\vec{B}$ ⊙ ⊖ $\vec{v}$ → $\vec{v}\times\vec{B}$ ↓ But $\vec{F} = q\,\vec{v}\times\vec{B} = -e\,\vec{v}\times\vec{B}$ has the direction ↑ (away from the wire)

$F = |q|vB\sin\phi = evB = (1.602\times10^{-19}\,\mathrm{C})(4.00\times10^{4}\,\mathrm{m/s})(3.75\times10^{-6}\,\mathrm{T}) = 2.40\times10^{-20}\,\mathrm{N}$

The force has magnitude <u>2.40×10^{-20} N</u> and is directed <u>away from the wire</u>.

29-47

a)

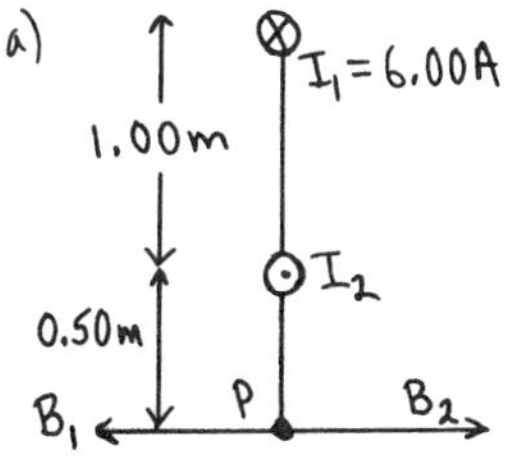

$\vec{B}_1$ and $\vec{B}_2$ must be equal and opposite for the resultant field at P to be zero ⇒ $\vec{B}_2$ to the right ⇒ I_2 <u>out of page</u>.

$$B_1 = \frac{\mu_0 I_1}{2\pi r_1} = \frac{\mu_0}{2\pi}\frac{6.00\,\mathrm{A}}{1.50\,\mathrm{m}}$$

$$B_2 = \frac{\mu_0 I_2}{2\pi r_2} = \frac{\mu_0}{2\pi}\frac{I_2}{0.50\,\mathrm{m}}$$

29-47 (cont)

$$B_1 = B_2 \Rightarrow \frac{\mu_0}{2\pi}\left(\frac{6.00\,A}{1.50\,m}\right) = \frac{\mu_0}{2\pi}\left(\frac{I_2}{0.50\,m}\right)$$

$$I_2 = \left(\frac{0.50\,m}{1.50\,m}\right)(6.00\,A) = \underline{2.00\,A}$$

b)

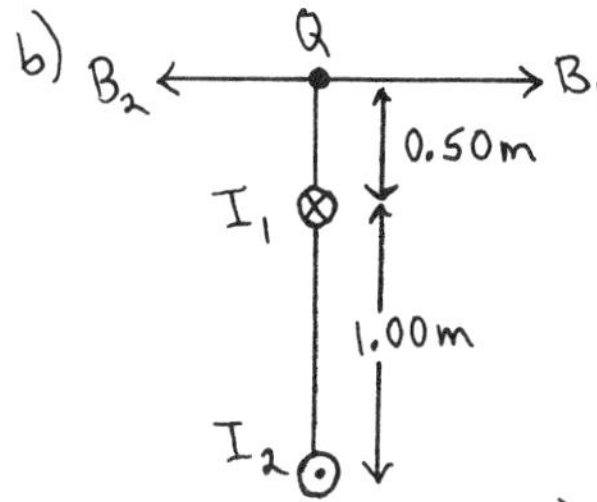

$$B_1 = \frac{\mu_0 I_1}{2\pi r_1} = (2\times10^{-7}\,T\cdot m/A)\left(\frac{6.00\,A}{0.50\,m}\right) = 2.40\times10^{-6}\,T$$

$$B_2 = \frac{\mu_0 I_2}{2\pi r_2} = (2\times10^{-7}\,T\cdot m/A)\left(\frac{2.00\,A}{1.50\,m}\right) = 2.67\times10^{-7}\,T$$

$\vec{B}_1$ and $\vec{B}_2$ are in opposite directions and $B_1 > B_2$

$\Rightarrow B = B_1 - B_2 = 2.40\times10^{-6}\,T - 2.67\times10^{-7}\,T = \underline{2.13\times10^{-6}\,T}$, and $\vec{B}$ is to the right.

c)

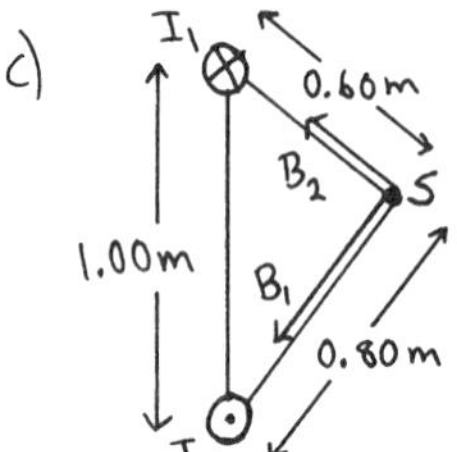

$$B_1 = \frac{\mu_0 I_1}{2\pi r_1} = (2\times10^{-7}\,T\cdot m/A)\frac{6.00\,A}{0.60\,m} = 2.00\times10^{-6}\,T$$

$$B_2 = \frac{\mu_0 I_2}{2\pi r_2} = (2\times10^{-7}\,T\cdot m/A)\frac{2.00\,A}{0.80\,m} = 5.00\times10^{-7}\,T$$

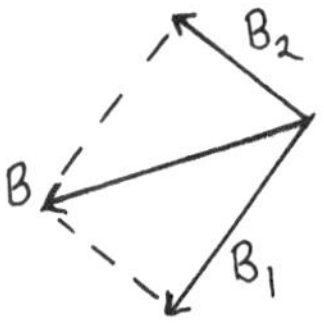

$\vec{B}_1$ and $\vec{B}_2$ are at right angles to each other, so the magnitude of their resultant is given by

$$B = \sqrt{B_1^2 + B_2^2} = \sqrt{(2.00\times10^{-6}T)^2 + (5.00\times10^{-7}T)^2}$$

$$B = \underline{2.06\times10^{-6}\,T}$$

29-49

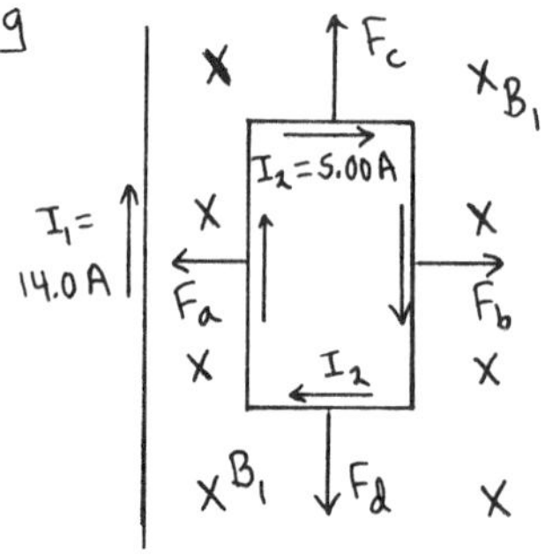

At the location of the loop the magnetic field $\vec{B}_1$ due to the current in the long, straight wire is directed into the page. The direction of the force on each side of the rectangle, given by the right-hand rule is shown in the sketch.

$\vec{F}_c$ and $\vec{F}_d$ are equal and opposite, so $\vec{F}_c + \vec{F}_d = 0$.

$F_a = I_2 L B_1$, where $L = 0.200m$ and B_1 is the magnetic field due to I_1 at a distance of 0.026 m from the wire.

$$B_1 = \frac{\mu_0 I_1}{2\pi r} = (2\times10^{-7}\,T\cdot m/A)\left(\frac{14.0\,A}{0.026\,m}\right) = 1.077\times10^{-4}\,T$$

$$F_a = I_2 L B_1 = (5.00A)(0.200m)(1.077\times10^{-4}T) = 1.077\times10^{-4}\,N$$

$F_b = I_2 L B_1$, where now B_1 is the magnetic field due to I_1 at a distance of 0.100m from the wire.

$$B_1 = \frac{\mu_0 I_1}{2\pi r} = (2\times10^{-7}\,T\cdot m/A)\frac{14.0\,A}{0.100\,m} = 2.80\times10^{-5}\,T$$

$$F_b = I_2 L B_1 = (5.00A)(0.200m)(2.80\times10^{-5}T) = 2.80\times10^{-5}\,N$$

$F_a > F_b$ and $\vec{F}_a$ and $\vec{F}_b$ are in opposite directions, so the resultant force $\vec{F}$ has magnitude $F = F_a - F_b = 1.077\times10^{-4}\,N - 2.80\times10^{-5}\,N = \underline{7.97\times10^{-5}\,N}$ and is directed toward the wire.

29-55

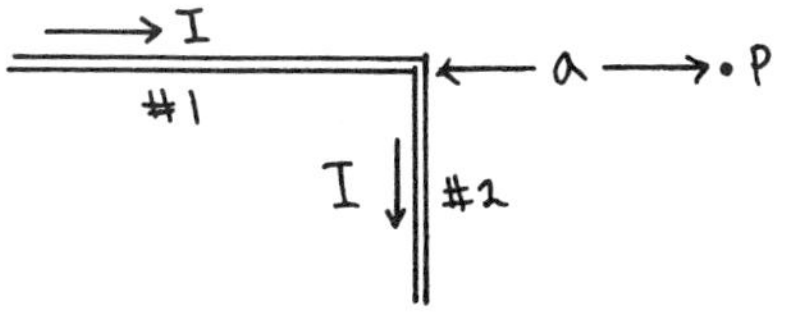

$$d\vec{B} = \frac{\mu_0}{4\pi}\frac{I\, d\vec{l}\times\vec{r}}{r^3}$$

Section #1

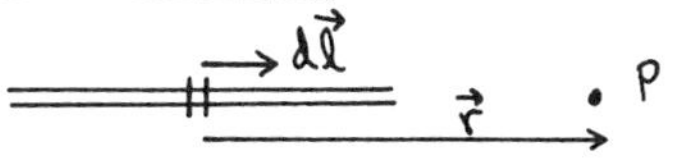

$d\vec{l}$ and $\vec{r}$ are parallel, so $d\vec{l}\times\vec{r} = 0$, $dB = 0$, and this section of the wire produces no magnetic field at P.

section #2

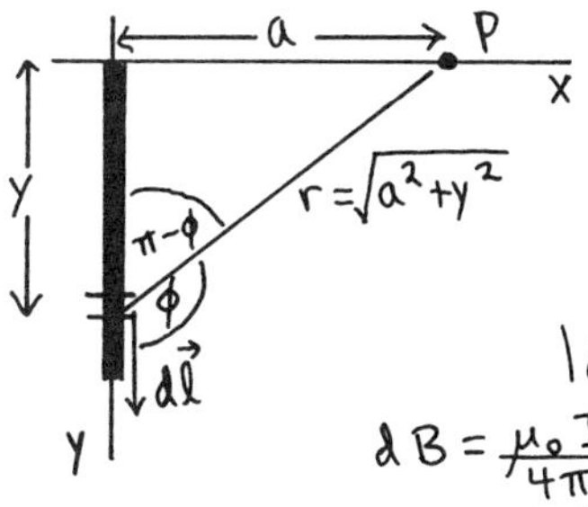

Divide this section of the wire up into short segments of length $dl = dy$. For each segment $d\vec{l}\times\vec{r}$ is directed out of the page ($\odot$) so the fields due to each segment all add and give a resultant field $\vec{B}$ directed out of the page.

$$|d\vec{l}\times\vec{r}| = r\sin\phi\, dl = r\sin(\pi-\phi)\, dl = \sqrt{a^2+y^2}\left(\frac{a}{\sqrt{a^2+y^2}}\right)dy = a\, dy$$

$$dB = \frac{\mu_0 I}{4\pi}\frac{|d\vec{l}\times\vec{r}|}{r^3} = \frac{\mu_0 I}{4\pi}\frac{a\, dy}{(a^2+y^2)^{3/2}}$$

$$B = \int dB = \frac{\mu_0 I a}{4\pi}\int_0^\infty \frac{dy}{(a^2+y^2)^{3/2}} = \frac{\mu_0 I a}{4\pi}\left[\frac{1}{a^2}\frac{y}{(a^2+y^2)^{1/2}}\Big|_0^\infty\right] = \underline{\frac{\mu_0 I}{4\pi a}}$$

29-59

a)

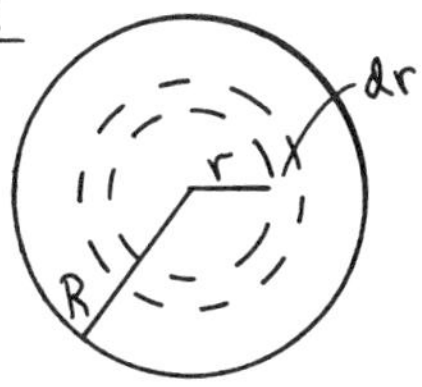

To integrate J over the cross section of the wire divide the wire cross section up into thin concentric rings of radius r and width dr. (We can't just take $I = JA = J\pi R^2$, since J varies across the cross section.)

The area of such a ring is dA, and the current through it is

$dI = J\, dA$; $dA = 2\pi r\, dr \Rightarrow dI = J\, dA = \alpha r(2\pi r\, dr) = 2\pi\alpha r^2\, dr$

$$I = \int dI = 2\pi\alpha\int_0^R r^2 dr = 2\pi\alpha\left(\tfrac{1}{3}R^3\right) = \frac{2\pi\alpha R^3}{3} \Rightarrow \alpha = \frac{3I}{2\pi R^3}$$

b) (i) $r \le R$

Apply Ampere's law to a circle of radius $r < R$:

$\oint \vec{B}\cdot d\vec{l} = \oint B\, dl = B\oint dl = B(2\pi r)$, by the symmetry and direction of $\vec{B}$.

The current passing through the path is $I_{encl} = \int dI$, where in the integral r goes from 0 to r.

$$I_{encl} = 2\pi\alpha\int_0^r r^2 dr = \frac{2\pi\alpha r^3}{3} = \frac{2\pi}{3}\left(\frac{3I}{2\pi R^3}\right)r^3 = \frac{Ir^3}{R^3}$$

Thus $\oint \vec{B}\cdot d\vec{l} = \mu_0 I_{encl} \Rightarrow B(2\pi r) = \mu_0\frac{Ir^3}{R^3} \Rightarrow \underline{B = \frac{\mu_0 I r^2}{2\pi R^3}}$

(ii) $r \ge R$

Apply Ampere's law to a circle of radius $r > R$:

$\oint \vec{B}\cdot d\vec{l} = \oint B\, dl = B\oint dl = B(2\pi r)$

$I_{encl} = I$; all the current in the wire passes through the path

Thus $\oint \vec{B}\cdot d\vec{l} = \mu_0 I_{encl} \Rightarrow B(2\pi r) = \mu_0 I$

$$\underline{B = \frac{\mu_0 I}{2\pi r}}$$

29-59 (cont)

Note: At $r=R$ the expression in (i) $(r \le R)$ gives $B = \frac{\mu_0 I}{2\pi R}$.
At $r=R$ the expression in (ii) $(r \ge R)$ gives $B = \frac{\mu_0 I}{2\pi R}$, which is the same.

29-61

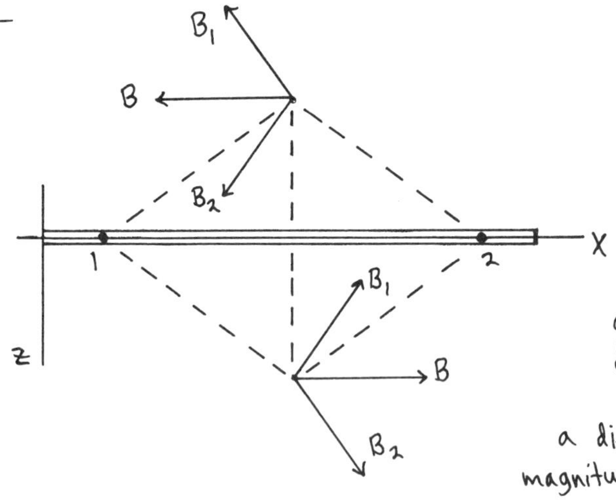

Consider the individual currents in pairs, where the currents in each pair are equidistant on either side of the point where $\vec{B}$ is being calculated.

The sketch shows that for each pair the z-components cancel, and that above the sheet the field is in the -x-direction and that below the sheet it is in the +x direction.

Also, by symmetry the magnitude of $\vec{B}$ a distance a above the sheet must equal the magnitude of $\vec{B}$ a distance a below the sheet.

Now that we have deduced the symmetry of $\vec{B}$, apply Ampere's law. Use a path that is a rectangle:

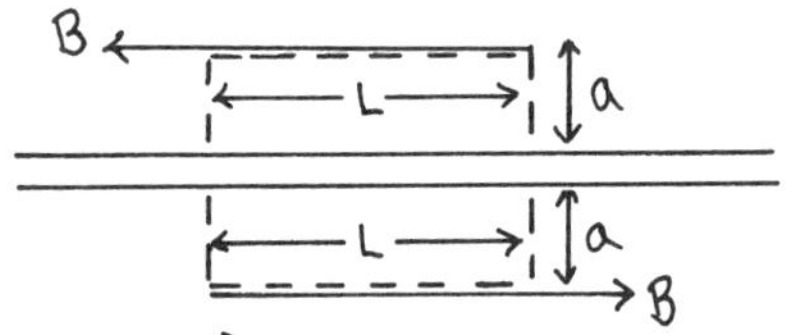

$\oint \vec{B} \cdot d\vec{l} = \mu_0 I_{encl}$

I is directed out of the page, so for I to be positive the integral around the path is taken in the counterclockwise direction.

Since $\vec{B}$ is parallel to the sheet, on the sides of the rectangle that have length 2a $\oint \vec{B} \cdot d\vec{l} = 0$. Thus $\oint \vec{B} \cdot d\vec{l} = 2BL$.

n conductors per unit length and current I out of the page in each gives $I_{encl} = InL$.

Ampere's law then gives $2BL = \mu_0 InL$

$$B = \tfrac{1}{2}\mu_0 In$$

Note that B is independent of the distance a from the sheet. Compare this result to the electric field due to an infinite sheet of charge (Example 23-7).

29-63

a) Divide the cross section of the cylinder into thin concentric rings of radius r and width dr. The current through each ring is $dI = J\,dA = J\,2\pi r\,dr$.

$$dI = \frac{2I_0}{\pi a^2}\left[1-(r/a)^2\right]2\pi r\,dr = \frac{4I_0}{a^2}\left[1-\left(\frac{r}{a}\right)^2\right]r\,dr$$

The total current I is obtained by integrating dI over the cross section

$\Rightarrow I = \int_0^a dI = \frac{4I_0}{a^2}\int_0^a \left(1 - r^2/a^2\right) r\,dr = \frac{4I_0}{a^2}\left(\frac{1}{2}r^2 - \frac{1}{4}\frac{r^4}{a^2}\right)\Big|_0^a = I_0$, as was to be shown.

29-63 (cont)

b)

Apply Ampere's law to a path that is a circle of radius $r > a$.

$\oint \vec{B} \cdot d\vec{l} = B(2\pi r)$

$I_{encl} = I_0$ (the path encloses the entire cylinder)

$\oint \vec{B} \cdot d\vec{l} = \mu_0 I_{encl} \Rightarrow B(2\pi r) = \mu_0 I_0$

$$B = \frac{\mu_0 I_0}{2\pi r}$$

c)

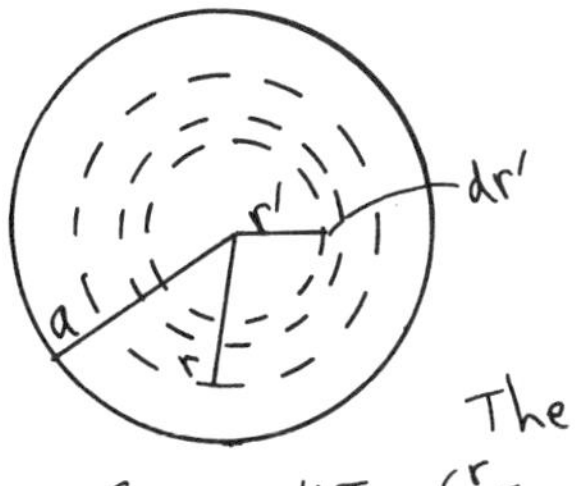

Divide the cross section of the cylinder into concentric rings of radius r' and width dr', as was done in part (a). The current dI through each ring is

$$dI = \frac{4I_0}{a^2}\left[1 - \left(\frac{r'}{a}\right)^2\right] r'\,dr'$$

The current I is obtained by integrating dI from $r' = 0$ to $r' = r$:

$$I = \int dI = \frac{4I_0}{a^2}\int_0^r \left[1 - \left(\frac{r'}{a}\right)^2\right] r'\,dr' = \frac{4I_0}{a^2}\left[\left(\tfrac{1}{2}r'^2 - \tfrac{1}{4}\frac{r'^4}{a^2}\right)\Big|_0^r\right] = \frac{4I_0}{a^2}\left(\frac{r^2}{2} - \frac{r^4}{4a^2}\right)$$

$$I = \frac{I_0 r^2}{a^2}\left(2 - \frac{r^2}{a^2}\right)$$

d)

Apply Ampere's law to a path that is a circle of radius $r < a$.

$\oint \vec{B} \cdot d\vec{l} = B(2\pi r)$

$I_{encl} = \frac{I_0 r^2}{a^2}\left(2 - \frac{r^2}{a^2}\right)$ (from part (c))

$$\oint \vec{B} \cdot d\vec{l} = \mu_0 I_{encl} \Rightarrow B(2\pi r) = \mu_0 \frac{I_0 r^2}{a^2}\left(2 - \frac{r^2}{a^2}\right)$$

$$B = \frac{\mu_0 I_0}{2\pi}\frac{r}{a^2}\left(2 - r^2/a^2\right)$$

result in part (b) evaluated at $r = a$: $B = \frac{\mu_0 I_0}{2\pi a}$

result in part (d) evaluated at $r = a$: $B = \frac{\mu_0 I_0}{2\pi}\frac{a}{a^2}\left(2 - \frac{a^2}{a^2}\right) = \frac{\mu_0 I_0}{2\pi a}$

The two results, one for $r > a$ and the other for $r < a$, agree at $r = a$.

29-67

Eq. (29-28): $M = \frac{\mu_{total}}{V}$, so $\mu_{total} = MV$

The average magnetic moment per atom is $\mu_{atom} = \frac{\mu_{total}}{N} = \frac{MV}{N}$, where N is the number of atoms in volume V.

The mass of volume V is $m = \rho V$, where ρ is the density ($\rho_{iron} = 7.8 \times 10^3\ \text{kg/m}^3$).

The number of moles of iron in volume V is $n = \frac{m}{55.847 \times 10^{-3}\ \text{kg/mol}} = \frac{\rho V}{55.847 \times 10^{-3}\ \text{kg/mol}}$, where 55.847×10^{-3} kg/mol is the atomic mass of iron from Appendix D.

$N = nN_A$, where $N_A = 6.022 \times 10^{23}$ atoms/mol is Avogadro's number.

$$N = nN_A = \frac{\rho V N_A}{55.847 \times 10^{-3}\ \text{kg/mol}}$$

$$\text{Thus } \mu_{atom} = \frac{MV}{N} = M\cancel{V}\frac{55.847 \times 10^{-3}\ \text{kg/mol}}{\rho \cancel{V} N_A} = \frac{(8.0 \times 10^4\ \text{A/m})(55.847 \times 10^{-3}\ \text{kg/mol})}{(7.8 \times 10^3\ \text{kg/m}^3)(6.022 \times 10^{23}\ \text{atoms/mol})}$$

$\mu_{atom} = \underline{9.5 \times 10^{-25}\ \text{A} \cdot \text{m}^2}$

$\mu_B = 9.274 \times 10^{-24}\ \text{A} \cdot \text{m}^2$, so $\mu_{atom} = \underline{0.10\,\mu_B}$

29-69

a)

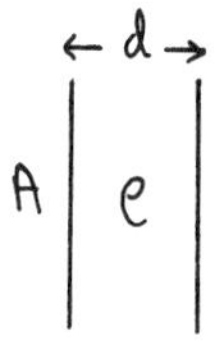

Apply Ohm's law to the dielectric:

$$i(t) = \frac{v(t)}{R}$$

$$v(t) = \frac{q(t)}{C} \quad \text{and} \quad C = K\frac{\epsilon_0 A}{d} \Rightarrow v = \frac{d}{K\epsilon_0 A}\, q(t)$$

The resistance R of the dielectric slab is $R = \frac{\rho d}{A}$.

Thus $i(t) = \frac{v(t)}{R} = \frac{q(t)\, d}{K\epsilon_0 A}\left(\frac{A}{\rho d}\right) = \frac{q(t)}{K\epsilon_0 \rho}$.

But the current $i(t)$ in the dielectric is related to the rate of change $\frac{dq}{dt}$ of the charge $q(t)$ on the plates by $i(t) = -\frac{dq}{dt}$ (a positive i in the direction from the + to the − plate of the capacitor corresponds to a decrease in the charge).

Use this in the above $\Rightarrow -\frac{dq}{dt} = \frac{1}{K\rho\epsilon_0}\, q(t)$

$$\frac{dq}{q} = -\frac{dt}{K\rho\epsilon_0}$$

Integrate both sides of this equation from $t=0$, where $q = Q_0$, to a later time t when the charge is $q(t)$

$$\int_{Q_0}^{q} \frac{dq}{q} = -\frac{1}{K\rho\epsilon_0}\int_0^t dt$$

$$\ln\left(\frac{q}{Q_0}\right) = -\frac{t}{K\rho\epsilon_0} \Rightarrow q(t) = Q_0\, e^{-t/K\rho\epsilon_0}$$

Then $i(t) = -\frac{dq}{dt} = \frac{Q_0}{K\rho\epsilon_0}\, e^{-t/K\rho\epsilon_0}$.

$$j_c(t) = \frac{i(t)}{A} = \frac{Q_0}{KA\rho\epsilon_0}\, e^{-t/K\rho\epsilon_0}$$

The conduction current flows from the positive to the negative plate of the capacitor.

b) $E(t) = \frac{q(t)}{\epsilon A} = \frac{q(t)}{K\epsilon_0 A}$

$$j_D(t) = \epsilon\frac{dE}{dt} = K\epsilon_0\frac{dE}{dt} = K\epsilon_0\frac{dq(t)/dt}{K\epsilon_0 A} = -\frac{i_c(t)}{A} = -j_c(t)$$

The minus sign means that $j_D(t)$ is directed from the negative towards the positive plate. $\vec{E}$ is from + to − but $\frac{dE}{dt}$ is negative (E decreases) so $j_D(t)$ is from − to +.

CHAPTER 30

Exercises 1, 3, 7, 9, 13, 19, 21, 23, 25, 29

Problems 31, 33, 37, 43, 47, 49

Exercises

30-1

The magnitude of the induced emf is $|\mathcal{E}| = \left|\frac{d\Phi_B}{dt}\right|$

$\Phi_B = BA$, so $\frac{d\Phi_B}{dt} = A\frac{dB}{dt}$

$|\mathcal{E}| = A\left|\frac{dB}{dt}\right| = (0.0900\,m^2)(0.190\,T/s) = 0.0171\,V$

$I = \frac{|\mathcal{E}|}{R} = \frac{0.0171\,V}{0.300\,\Omega} = \underline{0.0570\,A}$

Note that the induced emf does not depend on B, just on $\frac{dB}{dt}$.

30-3

The magnitude of the average induced emf is $|\mathcal{E}_{av}| = N\left|\frac{\Delta\Phi_B}{\Delta t}\right|$.

$\Delta\Phi_B = \Phi_{Bf} - \Phi_{Bi}$

initial

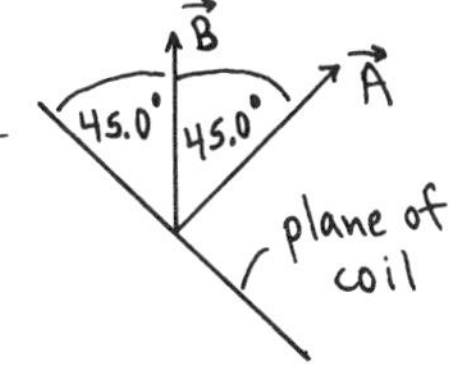

$\Phi_{Bi} = BA\cos\phi = (0.975T)(0.120\,m)(0.250\,m)\cos 45.0°$

$\Phi_{Bi} = 0.02068\,Wb$

final

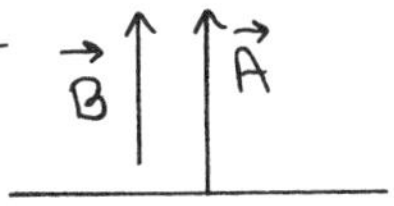

$\Phi_{Bf} = BA\cos\phi = (0.975T)(0.120\,m)(0.250\,m)\cos 0° = 0.02925\,Wb$

Then $|\mathcal{E}_{av}| = N\left|\frac{\Delta\Phi_B}{\Delta t}\right| = 50\left(\frac{0.02925\,Wb - 0.02068\,Wb}{0.0800\,s}\right) = \underline{5.36\,V}$.

30-7

The magnitude of the average emf induced in the coil is $|\mathcal{E}_{av}| = N\left|\frac{\Delta\Phi_B}{\Delta t}\right|$.

Initially, $\Phi_{Bi} = BA\cos\phi = BA$

The final flux is zero, so $|\mathcal{E}_{av}| = N\frac{|\Phi_{Bf} - \Phi_{Bi}|}{\Delta t} = \frac{NBA}{\Delta t}$.

The average induced current is $I = \frac{|\mathcal{E}_{av}|}{R} = \frac{NBA}{R\Delta t}$.

The total charge that flows through the coil is $Q = I\Delta t = \left(\frac{NBA}{R\Delta t}\right)\Delta t = \frac{NBA}{R}$.

30-9

Exercise 30-7 derived the result $Q = \frac{NBA}{R}$.

In this exercise the flux changes from its maximum value of $\Phi_B = BA$ to zero, so this equation applies. R is the total resistance so here $R = 50.0\,\Omega + 30.0\,\Omega = 80.0\,\Omega$.

$$Q = \frac{NBA}{R} \Rightarrow B = \frac{QR}{NA} = \frac{(7.28\times10^{-5}\,\text{C})(80.0\,\Omega)}{160(4.00\times10^{-4}\,\text{m}^2)} = \underline{0.0910\,\text{T}}.$$

30-13

a)

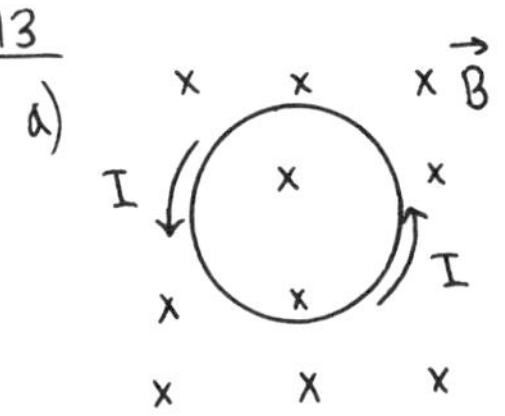

B increasing. The flux through the coil is directed into the paper and is increasing. By Lenz's law the field of the induced current, $\vec{B}_{induced}$, is directed out of the page (⊙) inside the loop, to oppose the increase in flux. This $\vec{B}_{induced}$ is produced by a <u>counterclockwise</u> current.

The sign convention for Faraday's law is an alternative means of determining the direction of I. Let $\vec{A}$ be directed into the page. Then Φ_B is positive. Since B is increasing the flux is increasing in magnitude and Φ_B positive gives $\frac{d\Phi_B}{dt} > 0$. $\mathcal{E} = -\frac{d\Phi_B}{dt}$ then gives $\mathcal{E} < 0$. By our choice of $\vec{A}$, $\mathcal{E} < 0$ means $\mathcal{E}$ is counterclockwise, so I is counterclockwise. This agrees with the direction we deduced from Lenz's law.

b)

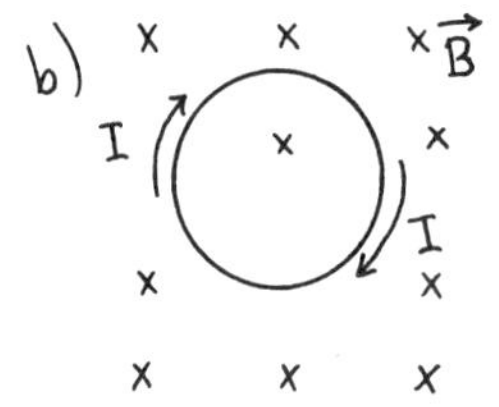

B decreasing. The flux through the coil is directed into the paper and is decreasing. By Lenz's law the field of the induced current, $\vec{B}_{induced}$, is directed into the paper (⊗) inside the loop, to oppose the decrease in flux. This $\vec{B}_{induced}$ is produced by a clockwise current.

Using Faraday's law: Let $\vec{A}$ be directed into the page. Then Φ_B is positive. Since B is decreasing in magnitude and with Φ_B positive, $\frac{d\Phi_B}{dt} < 0$. $\mathcal{E} = -\frac{d\Phi_B}{dt}$ then gives $\mathcal{E} > 0$. By our choice of $\vec{A}$, $\mathcal{E} > 0$ means $\mathcal{E}$ is clockwise, so I is clockwise. This agrees with the direction we deduced from Lenz's law.

30-19

a)

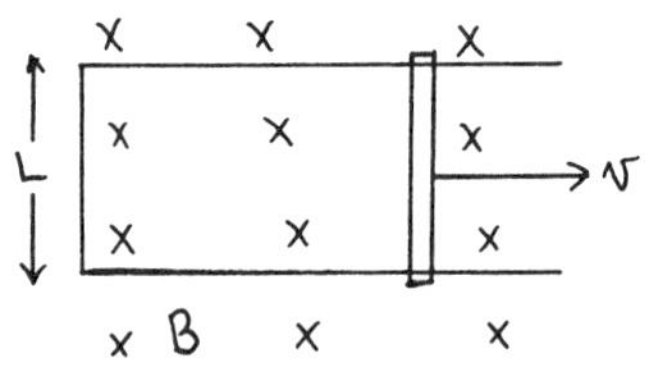

The magnitude of the induced emf is given by

$$|\mathcal{E}| = vLB \Rightarrow v = \frac{|\mathcal{E}|}{LB} = \frac{0.320\,\text{V}}{(0.0650\,\text{m})(1.20\,\text{T})} = \underline{4.10\,\text{m/s}}$$

b) $$I = \frac{|\mathcal{E}|}{R} = \frac{0.320\,\text{V}}{0.800\,\Omega} = \underline{0.400\,\text{A}}$$

c) The direction of the induced emf and induced current can be determined as discussed in Sect. 30-3:

Let positive $\vec{A}$ be into the page (in the same direction as $\vec{B}$).

30-19 (cont) $\vec{B}$ is in the positive direction (the direction of $\vec{A}$) so Φ_B is positive. The magnitude of the flux through the circuit is increasing as the area of the circuit is getting larger, so $d\Phi_B/dt$ is positive. Then by $\mathcal{E} = -d\Phi_B/dt$, $\mathcal{E}$ is negative. With our chosen direction of $\vec{A}$, clockwise is positive so a negative $\mathcal{E}$ is counterclockwise and I is counterclockwise.

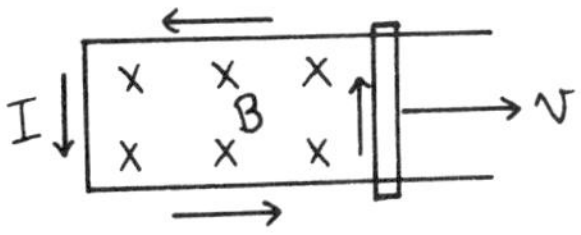

Force on the current in the rod: $\vec{F} = I\vec{l} \times \vec{B}$.

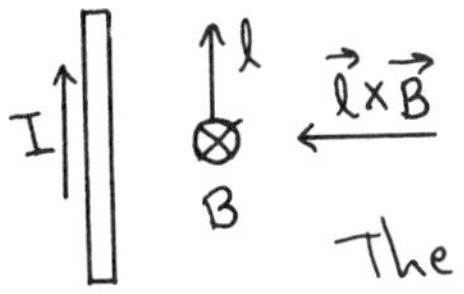

The force on the rod due to the induced currents is to the left.
$F = IlB\sin\phi = ILB = (0.400\ \text{A})(0.0650\ \text{m})(1.20\ \text{T}) = 0.0312\ \text{N}$.
The force on the rod has magnitude 0.0312 N and is directed to the left.

30-21

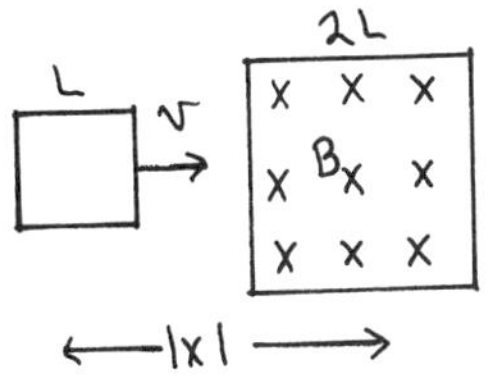

For $x < -\frac{3L}{2}$ or $x > \frac{3L}{2}$ the loop is completely outside the field region, $\Phi_B = 0$, and $\frac{d\Phi_B}{dt} = 0$. Thus $\mathcal{E} = 0$ and $I = 0$, so there is no force from the magnetic field and the external force F necessary to maintain constant velocity is zero.

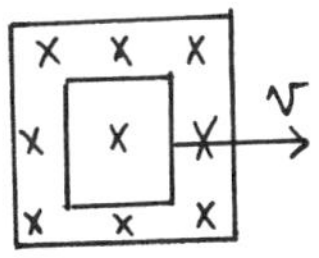

For $-\frac{L}{2} < x < \frac{L}{2}$ the loop is completely inside the field region and $\Phi_B = BL^2$. But $\frac{d\Phi_B}{dt} = 0$ so $\mathcal{E} = 0$ and $I = 0$. There is no force $\vec{F} = I\vec{l} \times \vec{B}$ from the magnetic field and the external force F necessary to maintain constant velocity is zero.

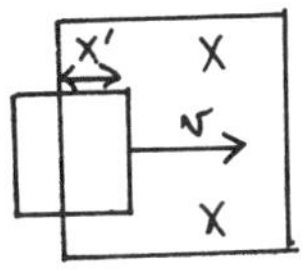

For $-\frac{3L}{2} < x < -\frac{L}{2}$ the loop is entering the field region. Let x' be the length of the loop that is within the field. Then $|\Phi_B| = BLx'$ and $\left|\frac{d\Phi_B}{dt}\right| = BLv$. The magnitude of the induced emf is $|\mathcal{E}| = \left|\frac{d\Phi_B}{dt}\right| = BLv$ and the induced current is $I = \frac{|\mathcal{E}|}{R} = \frac{BLv}{R}$.

Direction of I: Let $\vec{A}$ be directed into the plane of the figure. Then Φ_B is positive. The flux is positive and increasing in magnitude, so $\frac{d\Phi_B}{dt}$ is positive. Then by Faraday's law $\mathcal{E}$ is negative, and with our choice for direction of $\vec{A}$ a negative $\mathcal{E}$ is counterclockwise. The current induced in the loop is counterclockwise. Then

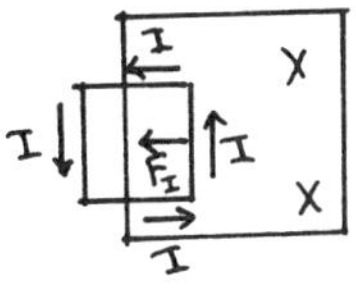

$\vec{F}_I = I\vec{l} \times \vec{B}$ gives that the force $\vec{F}_I$ exerted on the loop by the magnetic field is to the left and has magnitude $F_I = ILB = \left(\frac{BLv}{R}\right)LB = \frac{B^2L^2v}{R}$. The external force $\vec{F}$ needed to move the loop at constant speed is equal in magnitude and opposite in direction to F_I so is to the right.

30-21 (cont)

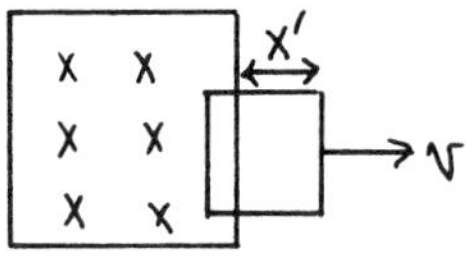

For $\frac{L}{2} < x < \frac{3L}{2}$ the loop is leaving the field region. Let x' be the length of the loop that is outside the field. Then $|\Phi_B| = BL(L-x')$ and $\left|\frac{d\Phi_B}{dt}\right| = BLv$. The magnitude of the induced emf is $|\mathcal{E}| = \left|\frac{d\Phi_B}{dt}\right| = BLv$ and the induced current is $I = \frac{|\mathcal{E}|}{R} = \frac{BLv}{R}$.

Direction of I:

Again let $\vec{A}$ be directed into the plane of the figure. Then Φ_B is positive and decreasing in magnitude, so $\frac{d\Phi_B}{dt}$ is negative. Then by Faraday's law $\mathcal{E}$ is positive, and with our choice for direction of $\vec{A}$ a positive $\mathcal{E}$ is clockwise. The current induced in the loop is clockwise. Then $\vec{F}_I = I\vec{l}\times\vec{B}$ gives that $\vec{F}$ is to the left and has magnitude $F_I = ILB = \left(\frac{BLv}{R}\right)LB = \frac{B^2L^2v}{R}$. The external force $\vec{F}$ is to the right and has this same magnitude.

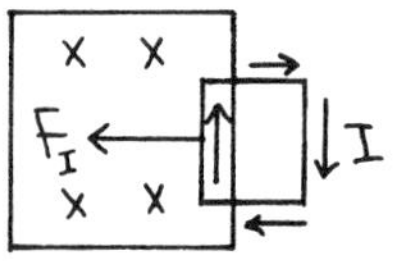

a)

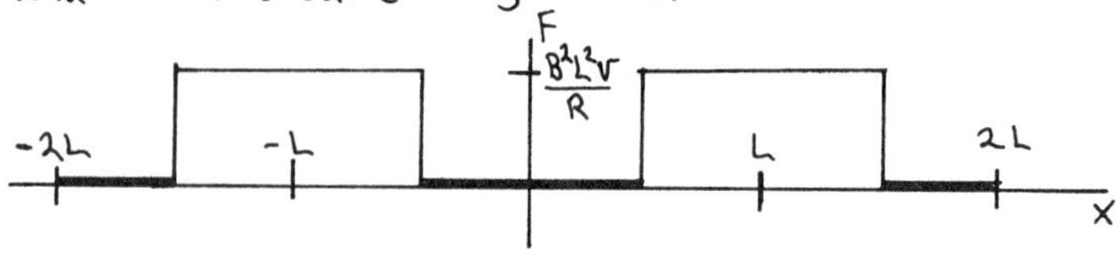

b)

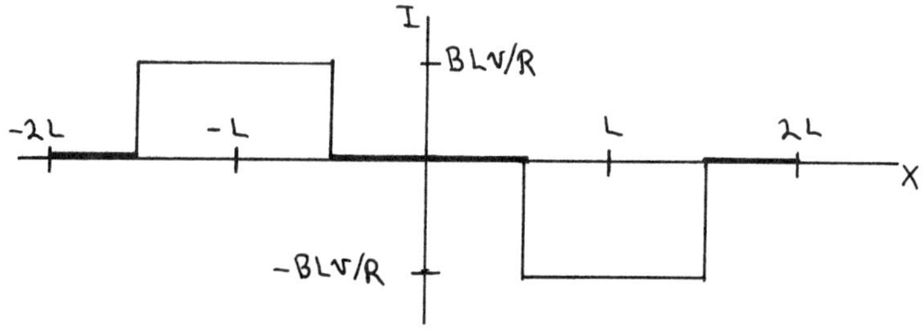

30-23

Away from the ends of the solenoid, $B = \mu_0 nI$ inside and $B = 0$ outside.

a) end view

Let R be the radius of the solenoid.

Apply $\oint \vec{E}\cdot d\vec{l} = -\frac{d\Phi_B}{dt}$ to an integration path that is a circle of radius r, where $r < R$. We need to calculate just the magnitude of E so can take absolute values.

$$\left|\oint \vec{E}\cdot d\vec{l}\right| = E(2\pi r)$$

$$\Phi_B = B\pi r^2, \quad \left|-\frac{d\Phi_B}{dt}\right| = \pi r^2\left|\frac{dB}{dt}\right|$$

$$\left|\oint \vec{E}\cdot d\vec{l}\right| = \left|-\frac{d\Phi_B}{dt}\right| \Rightarrow E(2\pi r) = \pi r^2\left|\frac{dB}{dt}\right|$$

$$E = \frac{1}{2}r\left|\frac{dB}{dt}\right|$$

$$B = \mu_0 nI, \text{ so } \frac{dB}{dt} = \mu_0 n\frac{dI}{dt}$$

$$E = \frac{1}{2}r\mu_0 n\frac{dI}{dt} = \frac{1}{2}(0.00400\text{ m})(4\pi\times10^{-7}\text{ T}\cdot\text{m/A})(800\text{ m}^{-1})(50.0\text{ A/s}) = \underline{1.01\times10^{-4}\text{ V/m}}$$

b) $r = 0.800$ cm is still inside the solenoid so the expression in part (a) applies

30-23 (cont)

$E = \frac{1}{2} r \mu_0 n \frac{dI}{dt} = \frac{1}{2}(0.00800\,m)(4\pi \times 10^{-7}\,T\cdot m/A)(800\,m^{-1})(50.0\,A/s) = \underline{2.02 \times 10^{-4}\,V/m}$

(Inside the solenoid E is proportional to r, so E doubles when r doubles.)

30-25

a) Because of the axial symmetry and the absence of any electric charge, the field lines are concentric circles.

b)

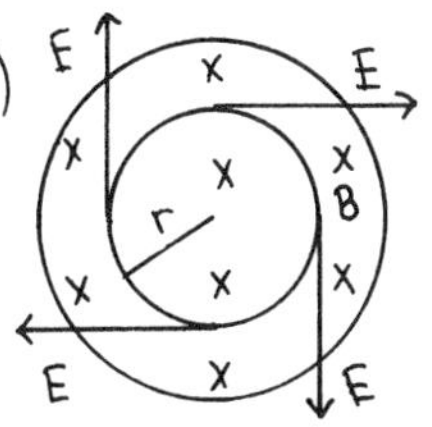

$\vec{E}$ is tangent to the ring. The direction of $\vec{E}$ (clockwise or counterclockwise) is the direction in which current will be induced in the ring. Use the sign convention for Faraday's law to deduce this direction. Let $\vec{A}$ be into the paper. Then Φ_B is positive. B decreasing then means $\frac{d\Phi_B}{dt}$ is negative, so by $\mathcal{E} = -\frac{d\Phi_B}{dt}$, $\mathcal{E}$ is positive and therefore clockwise. Thus $\vec{E}$ is <u>clockwise</u> around the ring.

To calculate E apply $\oint \vec{E}\cdot d\vec{l} = -\frac{d\Phi_B}{dt}$ to a circular path that coincides with the ring.

$\oint \vec{E}\cdot d\vec{l} = E(2\pi r)$

$\Phi_B = B\pi r^2$; $\left|\frac{d\Phi_B}{dt}\right| = \pi r^2 \left|\frac{dB}{dt}\right|$

$E(2\pi r) = \pi r^2 \left|\frac{dB}{dt}\right| \Rightarrow E = \frac{1}{2} r \left|\frac{dB}{dt}\right| = \frac{1}{2}(0.100\,m)(0.0450\,T/s) = \underline{2.25 \times 10^{-3}\,V/m}$

c) The induced emf has magnitude $\mathcal{E} = \oint \vec{E}\cdot d\vec{l} = E(2\pi r) = (2.25 \times 10^{-3}\,V/m)(2\pi)(0.100\,m)$

$\mathcal{E} = 1.414 \times 10^{-3}\,V$

Then $I = \frac{\mathcal{E}}{R} = \frac{1.414 \times 10^{-3}\,V}{2.00\,\Omega} = \underline{7.07 \times 10^{-4}\,A.}$

d) $\mathcal{E} = IR$ for any segment of the ring. Therefore, the potential difference between points a and b, or between any other points of the ring, is zero.

e) The induced emf $\mathcal{E}$ remains the same, but now there is no current and no IR potential drop. The potential difference will be $\mathcal{E} = 1.41 \times 10^{-3}\,V$ (from part (c)).

25-29

Apply Ampere's law to a circular path of radius $r < R$, where R is the radius of the wire.

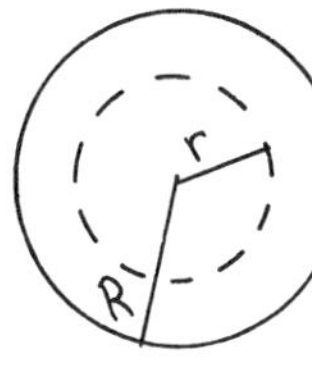

$\oint \vec{B}\cdot d\vec{l} = \mu_0 (I_c + \epsilon_0 \frac{d\Phi_E}{dt})$

There is no displacement current, so $\oint \vec{B}\cdot d\vec{l} = \mu_0 I_c$.

The magnetic field inside the superconducting material is zero, so $\oint \vec{B}\cdot d\vec{l} = 0$. But then Ampere's law says that $I_c = 0$; there can be no conduction current through the path. This same argument applies to any circular path with $r < R$, so all the current must be at the surface of the wire.

Problems

30-31

a)

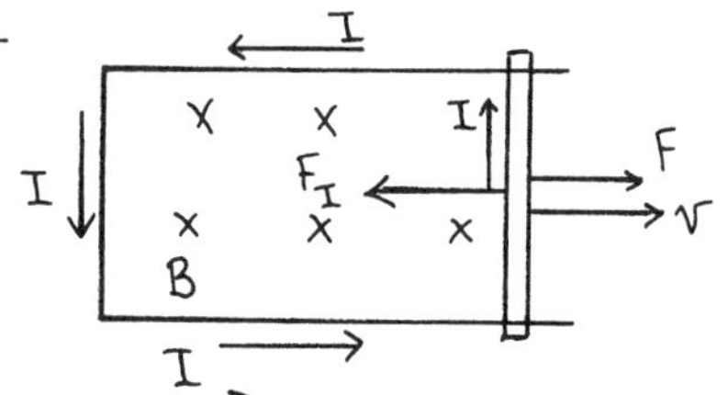

$$|\mathcal{E}| = \left|\frac{d\Phi_B}{dt}\right| = BLv$$

$$I = \frac{BLv}{R}$$

Use $\mathcal{E} = -\frac{d\Phi_B}{dt}$ to find the direction of I:

Let $\vec{A}$ be into the page. Then $\Phi_B > 0$. The area of the circuit is increasing, so $\frac{d\Phi_B}{dt} > 0$. Then $\mathcal{E} < 0$ and with our direction for $\vec{A}$ this means that $\mathcal{E}$ and I are counterclockwise, as shown on the sketch. The force F_I on the rod due to the induced current is given by $\vec{F}_I = I\vec{l} \times \vec{B}$. This gives $\vec{F}_I$ to the left with magnitude $F_I = ILB = \left(\frac{BLv}{R}\right)LB = \frac{B^2L^2v}{R}$. Note that $\vec{F}_I$ is directed to oppose the motion of the rod, as required by Lenz's law.

The net force on the rod is $F - F_I$, so its acceleration is $a = \frac{F - F_I}{m} = \frac{F - B^2L^2v/R}{m}$. The rod starts with $v = 0$ and $a = \frac{F}{m}$. As the speed v increases the acceleration a decreases. When $a = 0$ the rod has reached its terminal speed v_t.

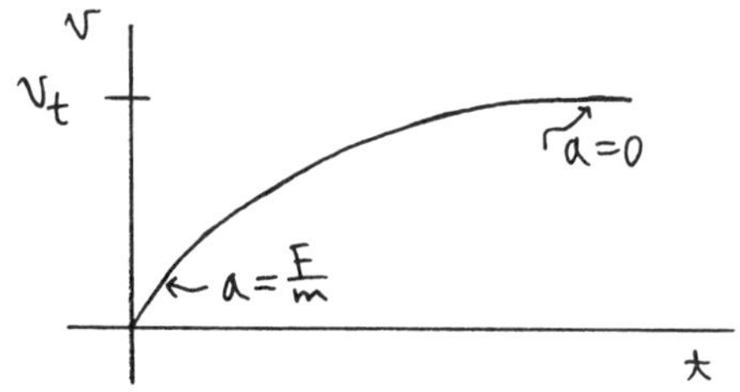

(Recall that a is the slope of the tangent to the v versus t curve.)

b) $v = v_t$ when $a = 0 \Rightarrow \frac{F - B^2L^2v_t/R}{m} = 0$

$$v_t = \frac{RF}{B^2L^2}$$

30-33

a)

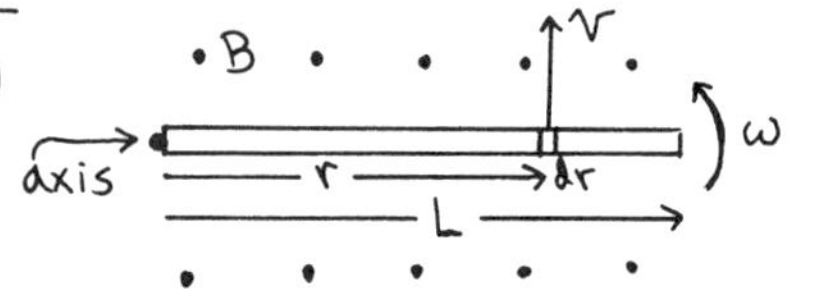

The emf induced in a thin slice is $d\mathcal{E} = \vec{v} \times \vec{B} \cdot d\vec{l}$

Assume that $\vec{B}$ is directed out of the page. Then $\vec{v} \times \vec{B}$ is directed radially outward and $dl = dr$, so $\vec{v} \times \vec{B} \cdot d\vec{l} = vB\,dr$

$$v = r\omega \Rightarrow d\mathcal{E} = \omega B r\,dr$$

The $d\mathcal{E}$ for all the thin slices that make up the rod are in series so they add:

$$\mathcal{E} = \int d\mathcal{E} = \int_0^L \omega B r\,dr = \tfrac{1}{2}\omega B L^2 = \tfrac{1}{2}(8.00\ \text{rad/s})(0.500\ \text{T})(0.190\ \text{m})^2 = \underline{0.0722\ \text{V}}$$

b) No current flows so there is no IR drop in potential. Thus the potential difference between the ends equals the emf of 0.0722 V calculated in part (a).

30-37

a)

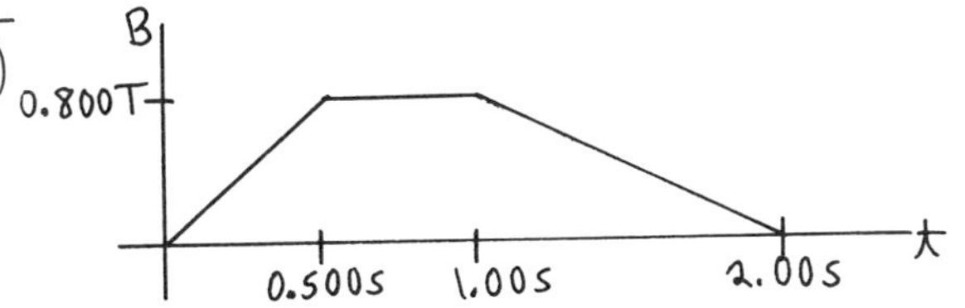

b)

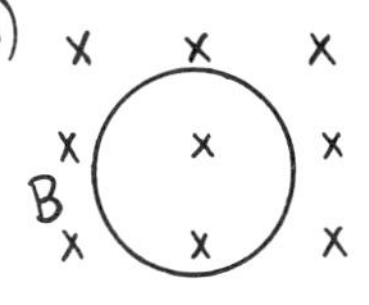

When $\vec{B}$ is increasing in magnitude ($t = 0$ to $t = 0.500\,\text{s}$), by Lenz's law $\vec{B}_{induced}$ is directed out of the page ($\odot$) inside the coil and I is counterclockwise.

When $\vec{B}$ is decreasing in magnitude ($t = 1.00\,\text{s}$ to $t = 2.00\,\text{s}$), $\vec{B}_{induced}$ is directed into the page ($\otimes$) inside the coil and I is clockwise.

When $\vec{B}$ is constant ($t = 0.500\,\text{s}$ to $t = 1.00\,\text{s}$), the induced current is zero.

Magnitude of I

$$|\mathcal{E}| = N\left|\frac{d\Phi_B}{dt}\right| = N\pi r^2\left|\frac{dB}{dt}\right|, \quad I = \frac{|\mathcal{E}|}{R}$$

$t = 0$ to $t = 0.500\,\text{s}$: $\left|\frac{dB}{dt}\right| = \frac{0.800\,\text{T}}{0.500\,\text{s}} = 1.60\,\text{T/s}$

$$|\mathcal{E}| = 20\pi(0.500\,\text{m})^2(1.60\,\text{T/s}) = 25.13\,\text{V}$$

$$I = \frac{|\mathcal{E}|}{R} = \frac{25.13\,\text{V}}{1.57\,\Omega} = 16.0\,\text{A}$$

$t = 1.00\,\text{s}$ to $t = 2.00\,\text{s}$: $\left|\frac{dB}{dt}\right| = \frac{0.800\,\text{T}}{1.00\,\text{s}} = 0.800\,\text{T/s}$

$$|\mathcal{E}| = 20\pi(0.500\,\text{m})^2(0.800\,\text{T/s}) = 12.57\,\text{V}$$

$$I = \frac{|\mathcal{E}|}{R} = \frac{12.57\,\text{V}}{1.57\,\Omega} = \underline{8.0\,\text{A}}$$

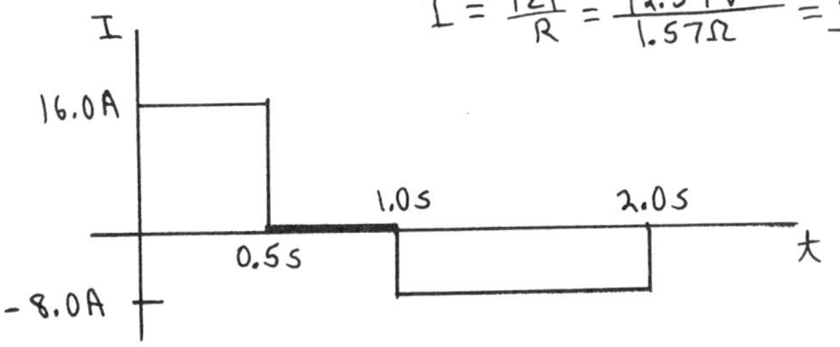

c) $N\oint \vec{E}\cdot d\vec{l} = \mathcal{E}$ $\quad \left(\oint \vec{E}\cdot d\vec{l} = \frac{\mathcal{E}}{N}\text{ is the emf in one turn of the coil}\right)$

$$N E(2\pi r) = \mathcal{E}$$

$$E = \frac{\mathcal{E}}{N 2\pi r}$$

E is maximum when $\mathcal{E}$ is maximum. From part (b) we have that $\mathcal{E}_{max} = 25.13\,\text{V}$.

$$E_{max} = \frac{\mathcal{E}_{max}}{N 2\pi r} = \frac{25.13\,\text{V}}{20(2\pi)(0.500\,\text{m})} = \underline{0.400\,\text{V/m}}$$

30-43

Calculate the induced electric field at each point and then use $\vec{F} = q\vec{E}$.

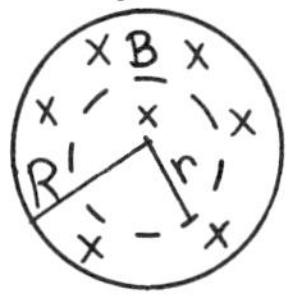

Apply $\oint \vec{E}\cdot d\vec{l} = -\frac{d\Phi_B}{dt}$ to a concentric circle of radius r. Take $\vec{A}$ to be into the page, in the direction of $\vec{B}$. B increasing then gives $\frac{d\Phi_B}{dt} > 0$, so $\oint \vec{E}\cdot d\vec{l}$ is negative. This means that $\vec{E}$ is tangent to the

30-43 (cont)

circle in the counterclockwise direction.

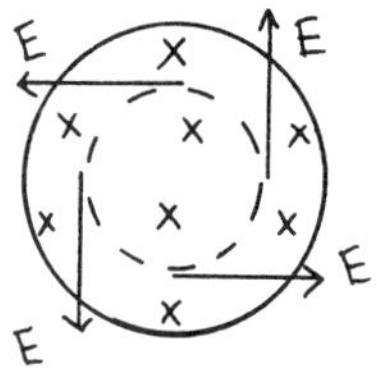

$$\oint \vec{E}\cdot d\vec{l} = -E(2\pi r)$$

$$-\frac{d\Phi_B}{dt} = -\pi r^2 \frac{dB}{dt}$$

$$-E(2\pi r) = -\pi r^2 \frac{dB}{dt} \Rightarrow E = \frac{1}{2} r \frac{dB}{dt}$$

point a

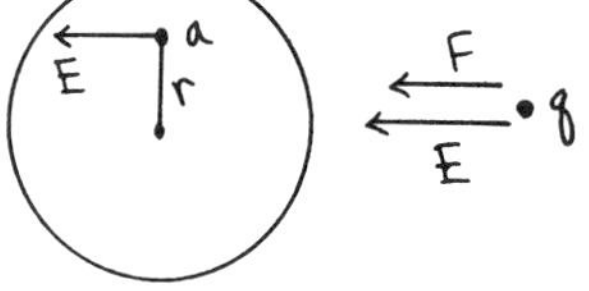

$F = qE = \frac{1}{2} q r \frac{dB}{dt}$

$\vec{F}$ is to the left

($\vec{F}$ is in the same direction as $\vec{E}$ since q is positive.)

point b

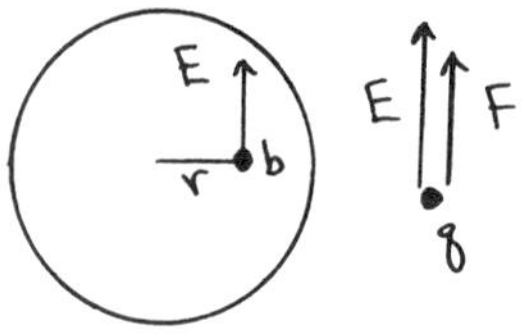

$F = qE = \frac{1}{2} q r \frac{dB}{dt}$

$\vec{F}$ is toward the top of the page.

point c

$r = 0$ here, so $E = 0$ and $F = 0$.

30-47

a) and b)

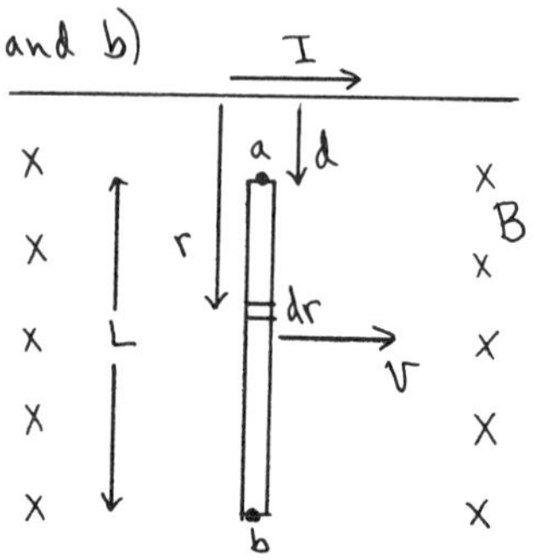

The magnetic field of the wire is given by $B = \frac{\mu_0 I}{2\pi r}$ and varies along the length of the bar. At every point along the bar $\vec{B}$ has direction into the page. Divide the bar up into thin slices.

The emf $d\mathcal{E}$ induced in each slice is given by $d\mathcal{E} = \vec{v}\times\vec{B}\cdot d\vec{l}$. $\vec{v}\times\vec{B}$ is directed toward the wire, so $d\mathcal{E} = -vB\,dr = -v\left(\frac{\mu_0 I}{2\pi r}\right)dr$. The total emf induced in the bar is

$$V_{ab} = \int_a^b d\mathcal{E} = -\int_d^{d+L} \frac{\mu_0 I v}{2\pi r}\,dr = -\frac{\mu_0 I v}{2\pi}\int_d^{d+L}\frac{dr}{r} = -\frac{\mu_0 I v}{2\pi}\left[\ln(r)\Big|_d^{d+L}\right]$$

$$V_{ab} = -\frac{\mu_0 I v}{2\pi}\left(\ln(d+L) - \ln(d)\right) = -\frac{\mu_0 I v}{2\pi}\ln\left(1+\frac{L}{d}\right)$$

The minus sign means that V_{ab} is negative; point a is at higher potential than point b. (The force $\vec{F} = q\vec{v}\times\vec{B}$ on positive charge carriers in the bar is towards a, so a is at higher potential.)

30-47 (cont)

c)

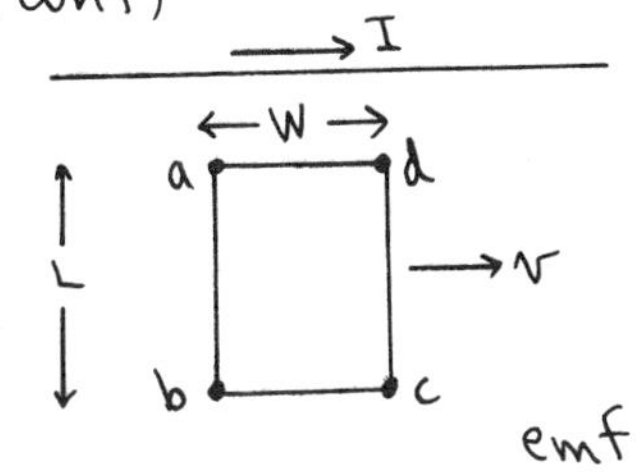

As the loop moves to the right the magnetic flux through it doesn't change. Thus $\mathcal{E} = -\frac{d\Phi_B}{dt} = 0$ and $I = 0$.

This result can also be understood as follows. The induced emf in section ab puts point a at higher potential; the induced emf in section dc puts point d at higher potential. If you travel around the loop then these two induced emf's sum to zero. There is no net emf in the loop and hence <u>no current</u>.

30-49

a)

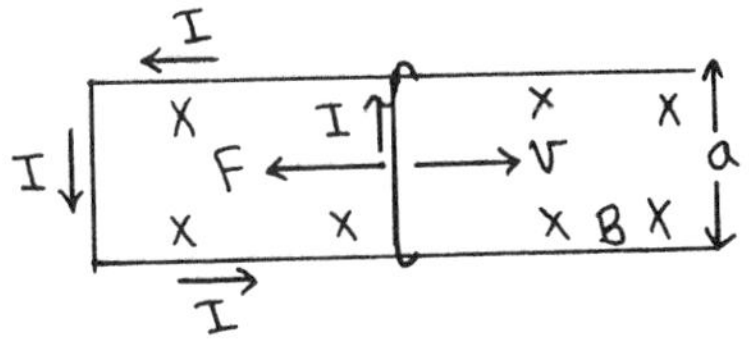

When the wire has speed v the induced emf is $\mathcal{E} = Bva$ and the induced current is $I = \frac{\mathcal{E}}{R} = \frac{Bva}{R}$. The induced current flows upward in the wire as shown, so the force $\vec{F} = I\vec{l} \times \vec{B}$ exerted by the magnetic field on the induced current is to the left. $\vec{F}$ opposes the motion of the wire, as it must by Lenz's law. The magnitude of the force is $F = IaB = \frac{B^2a^2v}{R}$.

b) Apply $\sum \vec{F} = m\vec{a}$ to the wire. Take $+x$ to be toward the right and let the origin be at the location of the wire at $t = 0$, so $x_0 = 0$.

$$\sum F_x = ma_x$$

$$-F = ma_x \Rightarrow a_x = -\frac{F}{m} = -\frac{B^2a^2v}{mR}$$

$$a_x = \frac{dv}{dt} = -\frac{B^2a^2v}{mR}$$

$$\frac{dv}{v} = -\frac{B^2a^2}{mR}\,dt$$

$$\int_{v_0}^{v} \frac{dv}{v} = -\frac{B^2a^2}{mR}\int_0^t dt$$

$$\ln v - \ln v_0 = -\frac{B^2a^2t}{mR}$$

$$\ln\left(\frac{v}{v_0}\right) = -\frac{B^2a^2t}{mR} \Rightarrow v = v_0 e^{-B^2a^2t/mR}$$

Note: $t = 0 \Rightarrow v = v_0$ and $v \to 0$ when $t \to \infty$

$$v = \frac{dx}{dt} = v_0 e^{-B^2a^2t/mR} \Rightarrow dx = v_0 e^{-B^2a^2t/mR}\,dt$$

$$\int_0^x dx = \int_0^t v_0 e^{-B^2a^2t/mR}\,dt$$

$$x = v_0\left(-\frac{mR}{B^2a^2}\right)\left[e^{-B^2a^2t/mR}\right]\Big|_0^t = \frac{mRv_0}{B^2a^2}\left(1 - e^{-B^2a^2t/mR}\right)$$

Comes to rest $\Rightarrow v \to 0$. This happens when $t \to \infty$.

$t \to \infty$ gives $x = \frac{mRv_0}{B^2a^2}$. Thus the distance the wire travels before coming to rest is given by $\frac{mRv_0}{B^2a^2}$, as was to be shown.

CHAPTER 31

Exercises 3, 5, 7, 11, 13, 15, 21, 23, 29, 31, 33

Problems 39, 45, 47, 49, 51, 53, 55, 57, 59

Exercises

31-3

a) $|\mathcal{E}_2| = M\left|\frac{di_1}{dt}\right| = (0.275\text{ H})(0.0500\text{ A/s}) = \underline{0.0138\text{ V}}$; yes, it is constant.

b) $|\mathcal{E}_1| = M\left|\frac{di_2}{dt}\right|$; M is a property of the pair of coils so it is the same as in part (a)
$\Rightarrow |\mathcal{E}_1| = \underline{0.0138\text{ V}}$

31-5

a) $M = \frac{N_2 \Phi_{B2}}{i_1} = \frac{300\,(0.0280\text{ Wb})}{3.55\text{ A}} = \underline{2.37\text{ H}}$

b) $M = \frac{N_1 \Phi_{B1}}{i_2} \Rightarrow \Phi_{B1} = \frac{M i_2}{N_1} = \frac{(2.37\text{ H})(1.60\text{ A})}{800} = \underline{0.00474\text{ Wb}}$

31-7

a) $|\mathcal{E}| = L\left|\frac{di}{dt}\right| \Rightarrow L = \frac{|\mathcal{E}|}{|di/dt|} = \frac{0.0180\text{ V}}{0.0600\text{ A/s}} = \underline{0.300\text{ H}}$

b) $N\Phi_B = Li \Rightarrow \Phi_B = \frac{Li}{N} = \frac{(0.300\text{ H})(0.800\text{ A})}{300} = \underline{8.00\times10^{-4}\text{ Wb}}$

31-11

$$L = \frac{N\Phi_B}{i}$$

If the magnetic field is uniform inside the solenoid $\Phi_B = BA$.

Eq. (29-23): $B = \mu_0 n i = \mu_0\left(\frac{N}{l}\right)i \Rightarrow \Phi_B = \frac{\mu_0 N i A}{l}$

$$L = \frac{N}{i}\left(\frac{\mu_0 N i A}{l}\right) = \frac{\mu_0 N^2 A}{l}$$

31-13

a) The energy density (energy per unit volume) in a magnetic field (in vacuum) is given by $u = \frac{U}{V} = \frac{B^2}{2\mu_0}$ (Eq. 31-10).

$\Rightarrow V = \frac{2\mu_0 U}{B^2} = \frac{2(4\pi\times10^{-7}\text{ T}\cdot\text{m/A})(3.60\times10^6\text{ J})}{(0.500\text{ T})^2} = \underline{36.2\text{ m}^3}$

b) $u = \frac{U}{V} = \frac{B^2}{2\mu_0}$

$B = \sqrt{\frac{2\mu_0 U}{V}} = \sqrt{\frac{2(4\pi\times10^{-7}\text{ T}\cdot\text{m/A})(3.60\times10^6\text{ J})}{0.125\text{ m}^3}} = \underline{8.51\text{ T}}$

31-15

$$U = \tfrac{1}{2}LI^2 \Rightarrow L = \frac{2U}{I^2} = \frac{2(0.350\ \text{J})}{(12.5\ \text{A})^2} = 0.00448\ \text{H}$$

Example 31-3 gives the inductance of a toroidal solenoid to be $L = \dfrac{\mu_0 N^2 A}{2\pi r}$

$$N = \sqrt{\frac{2\pi r L}{\mu_0 A}} = \sqrt{\frac{2\pi(0.120\text{m})(0.00448\ \text{H})}{(4\pi\times10^{-7}\ \text{T}\cdot\text{m/A})(4.00\times10^{-4}\text{m}^2)}} = \underline{2592}$$

31-21

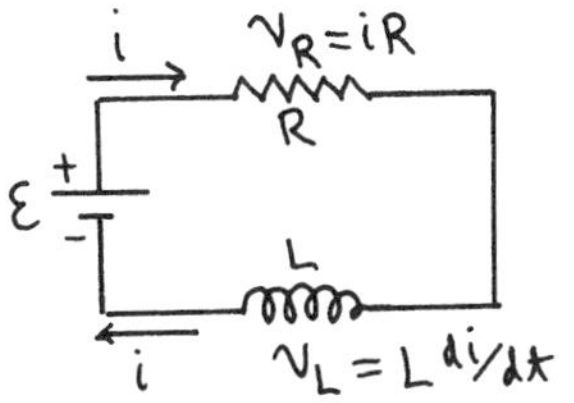

$\frac{di}{dt}$ is positive as the current increases from its initial value of zero.

$$\mathcal{E} - v_R - v_L = 0$$
$$\mathcal{E} - iR - L\frac{di}{dt} = 0 \Rightarrow i = \frac{\mathcal{E}}{R}\left(1 - e^{-(R/L)t}\right)$$

a) Initially $(t=0)$ $i = 0 \Rightarrow \mathcal{E} - L\frac{di}{dt} = 0$

$$\frac{di}{dt} = \frac{\mathcal{E}}{L} = \frac{12.0\ \text{V}}{3.00\ \text{H}} = \underline{4.00\ \text{A/s}}$$

b) $\mathcal{E} - iR - L\frac{di}{dt} = 0$

(Use this equation rather than Eq. (31-14) since i rather than t is given.)

$$\frac{di}{dt} = \frac{\mathcal{E} - iR}{L} = \frac{12.0\text{V} - (1.00\text{A})(7.00\Omega)}{3.00\ \text{H}} = \underline{1.67\ \text{A/s}}$$

c) $i = \frac{\mathcal{E}}{R}\left(1 - e^{-(R/L)t}\right) = \frac{12.0\text{V}}{7.00\,\Omega}\left(1 - e^{-\left(\frac{7.00\Omega}{3.00\text{H}}\right)(0.200\text{s})}\right) = 1.714\ \text{A}\left(1 - e^{-0.467}\right) = \underline{0.639\ \text{A}}$

d) Final steady state $\Rightarrow t \to \infty$ and $\frac{di}{dt} \to 0$, so $\mathcal{E} - iR = 0$.

$$i = \frac{\mathcal{E}}{R} = \frac{12.0\text{V}}{7.00\Omega} = \underline{1.71\ \text{A}}$$

31-23

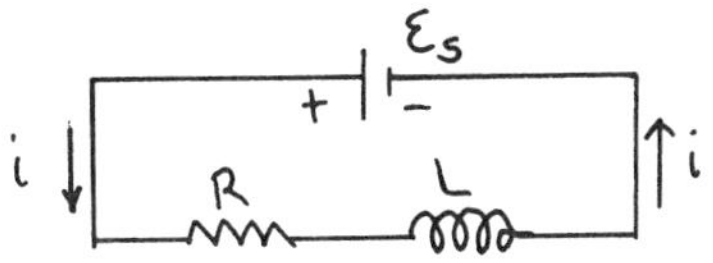

$$\mathcal{E}_S - iR - L\frac{di}{dt} = 0$$

constant current is established $\Rightarrow \frac{di}{dt} = 0$

$$i = \frac{\mathcal{E}_S}{R} = \frac{120\text{V}}{400\Omega} = 0.300\ \text{A}$$

a)

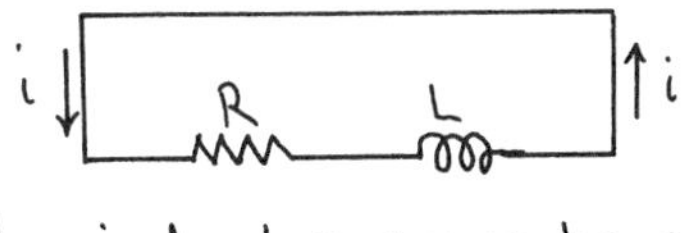

$$i = I_0 e^{-(R/L)t}$$
$$t = 0 \Rightarrow i = I_0 = 0.300\text{A}$$

The inductor prevents an instantaneous change in the current; the current in the inductor just after S_2 is closed and S_1 is opened equals the current in the inductor just before this is done.

b) $i = I_0 e^{-(R/L)t} = (0.300\text{A})e^{-\left(\frac{400\Omega}{0.200\text{H}}\right)(2.00\times10^{-4}\text{s})} = (0.300\text{A})e^{-0.400} = \underline{0.201\ \text{A}}$

31-23 (cont)

c)

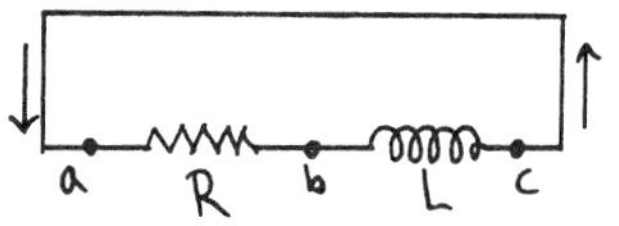

If we trace around the loop in the direction of the current the potential falls as we travel through the resistor so it must rise as we pass through the inductor: $V_{ab} > 0$ and $V_{bc} < 0$. So point c is at higher potential than point b.

$V_{ab} + V_{bc} = 0 \Rightarrow V_{bc} = -V_{ab}$,

or $V_{cb} = V_{ab} = iR = (0.201\,A)(400\,\Omega) = \underline{80.4\,V}$

d) $i = I_0 e^{-(R/L)t}$

$i = \frac{1}{2} I_0 \Rightarrow \frac{1}{2} I_0 = I_0 e^{-(R/L)t} \Rightarrow \frac{1}{2} = e^{-(R/L)t}$

Take natural logs of both sides of this equation $\Rightarrow \ln\left(\frac{1}{2}\right) = -\frac{R}{L}t$

$t = \frac{L}{R}\ln 2$ (since $\ln\left(\frac{1}{2}\right) = -\ln 2$)

$t = \left(\frac{0.200\,H}{400\,\Omega}\right)\ln 2 = \underline{3.47 \times 10^{-4}\,s}$

31-29

a) Eq. (31-22) $\omega = \frac{1}{\sqrt{LC}} = \frac{1}{\sqrt{(3.00\,H)(6.00 \times 10^{-4}\,F)}} = \underline{23.6\,rad/s}$

The period is given by $T = \frac{2\pi}{\omega} = \frac{2\pi}{23.6\,rad/s} = \underline{0.266\,s}$

b)

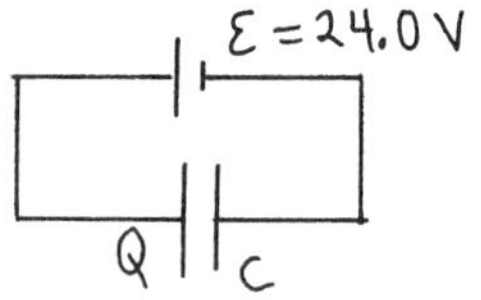

$\mathcal{E} - \frac{Q}{C} = 0$

$Q = \mathcal{E}C = (24.0\,V)(6.00 \times 10^{-4}\,F) = \underline{0.0144\,C}$

c) $U = \frac{1}{2}CV^2 = \frac{1}{2}(6.00 \times 10^{-4}\,F)(24.0\,V)^2 = \underline{0.173\,J}$

d) $q = Q\cos(\omega t + \phi)$ (Eq. 31-21)

$q = Q$ at $t = 0 \Rightarrow \phi = 0$

$q = Q\cos\omega t = (0.0144\,C)\cos([23.6\,rad/s][0.0444\,s]) = \underline{0.00719\,C}$

e) $i = -\omega Q\sin(\omega t + \phi)$ (Eq. 31-23)

$i = -(23.6\,rad/s)(0.0144\,C)\sin([23.6\,rad/s][0.0444\,s]) = \underline{-0.294\,A}$

or

$\frac{1}{2}Li^2 + \frac{q^2}{2C} = \frac{Q^2}{2C} \Rightarrow i = \pm\sqrt{\frac{1}{LC}}\sqrt{Q^2 - q^2}$ (Eq. 31-25)

$i = \pm(23.6\,rad/s)\sqrt{(0.0144\,C)^2 - (0.00719\,C)^2} = \pm 0.294\,A$, which checks

f) $U_C = \frac{q^2}{2C} = \frac{1}{2}\frac{(0.00719\,C)^2}{6.00 \times 10^{-4}\,F} = \underline{0.0431\,J}$

$U_L = \frac{1}{2}Li^2 = \frac{1}{2}(3.00\,H)(0.294\,A)^2 = \underline{0.130\,J}$

Note: $U_C + U_L = 0.0431\,J + 0.130\,J = 0.173\,J$. This agrees with the total energy initially stored in the capacitor, $U = \frac{Q^2}{2C} = \frac{(0.0144\,C)^2}{2(6.00 \times 10^{-4}\,F)} = 0.173\,J$.

31-31

a) 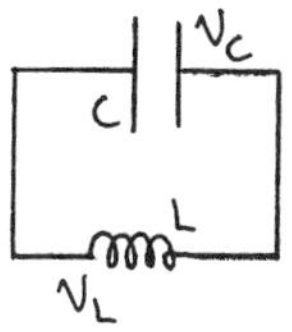

$$v_L = v_c$$

$$L\frac{di}{dt} = q/C$$

$$q = LC\frac{di}{dt} = (0.750\ \text{H})(18.0\times10^{-6}\ \text{F})(3.40\ \text{A/s}) = \underline{4.59\times10^{-5}\ \text{C}}$$

b) $v_c = \frac{q}{C} = \frac{4.20\times10^{-4}\ \text{C}}{18.0\times10^{-6}\ \text{F}} = 23.3\ \text{V}$

$v_L = v_c = \underline{23.3\ \text{V}}$

31-33

a) $\omega = \frac{1}{\sqrt{LC}} = \frac{1}{\sqrt{(0.500\text{H})(6.00\times10^{-4}\text{F})}} = \underline{57.7\ \text{rad/s}}$

b) $\omega' = \sqrt{\frac{1}{LC} - \frac{R^2}{4L^2}}$

10% decrease $\Rightarrow \omega' = 0.90\,\omega = 0.90\frac{1}{\sqrt{LC}}$

$$\sqrt{\frac{1}{LC} - \frac{R^2}{4L^2}} = (0.90)\frac{1}{\sqrt{LC}}$$

Square both sides of the equation: $\frac{1}{LC} - \frac{R^2}{4L^2} = 0.81\frac{1}{LC}$

$\frac{R^2}{4L^2} = \frac{0.19}{LC} \Rightarrow R = \sqrt{0.76\frac{L}{C}} = \sqrt{0.76\frac{0.500\text{H}}{6.00\times10^{-4}\text{F}}} = \underline{25.2\ \Omega}$

Problems

31-39

a) $|\mathcal{E}| = L\left|\frac{di}{dt}\right| \Rightarrow L = \frac{|\mathcal{E}|}{|di/dt|}$

$\frac{di}{dt} = \frac{\Delta i}{\Delta t}$ (since rate of increase is constant) $= \frac{50.0\ \text{A}}{10.0\ \text{s}} = 5.00\ \text{A/s}$

$L = \frac{60.0\ \text{V}}{5.00\ \text{A/s}} = \underline{12.0\ \text{H}}$

b) $N\Phi_B = Li = (12.0\ \text{H})(50.0\ \text{A}) = \underline{600\ \text{Wb}}$

c) Rate at which electrical energy is being dissipated by the resistance is
$P_R = i^2R = (50.0\text{A})^2(25.0\Omega) = 6.25\times10^4\ \text{W}.$
Rate at which electrical energy is being stored in the magnetic field of the inductor is
$P_L = |\mathcal{E}_L|i = Li\left|\frac{di}{dt}\right| = (12.0\text{H})(50.0\text{A})(5.00\ \text{A/s}) = 3.00\times10^3\ \text{W}.$

The ratio is $\frac{P_L}{P_R} = \frac{3.00\times10^3\ \text{W}}{6.25\times10^4\ \text{W}} = \underline{0.0480}.$

31-45

a) end view

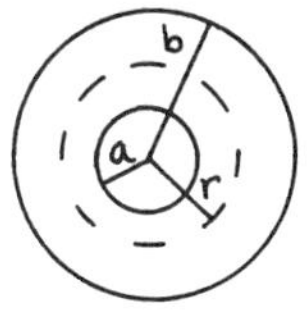

Apply Ampere's law to a circular path of radius r.

$$\oint \vec{B}\cdot d\vec{l} = \mu_0 I_{encl}$$

$$\oint \vec{B}\cdot d\vec{l} = B(2\pi r)$$

$I_{encl} = i$, the current in the inner conductor

Thus $B(2\pi r) = \mu_0 i \Rightarrow B = \frac{\mu_0 i}{2\pi r}$

31-45 (cont)

b) $u = \frac{B^2}{2\mu_0}$

$dU = u\,dV$, where $dV = 2\pi r l\,dr$

$dU = \frac{1}{2\mu_0}\left(\frac{\mu_0 i}{2\pi r}\right)^2 (2\pi r l)\,dr = \frac{\mu_0 i^2 l}{4\pi r}\,dr$

c) $U = \int dU = \frac{\mu_0 i^2 l}{4\pi}\int_a^b \frac{dr}{r} = \frac{\mu_0 i^2 l}{4\pi} \ln r\Big|_a^b = \frac{\mu_0 i^2 l}{4\pi}(\ln b - \ln a) = \frac{\mu_0 i^2 l}{4\pi}\ln\left(\frac{b}{a}\right)$

d) Eq. (31-9): $U = \frac{1}{2}Li^2$

Part (c): $U = \frac{\mu_0 i^2 l}{4\pi}\ln\left(\frac{b}{a}\right)$

$\Rightarrow \frac{1}{2}Li^2 = \frac{\mu_0 i^2 l}{4\pi}\ln\left(\frac{b}{a}\right)$

$L = \frac{\mu_0 l}{2\pi}\ln\left(\frac{b}{a}\right)$, the same result as calculated in part (d) of Problem 31-44.

31-47

Energy density in the electric field: $u_E = \frac{1}{2}\epsilon_0 E^2$ (Eq. 25-11)

Energy density in the magnetic field: $u_B = \frac{1}{2\mu_0}B^2$ (Eq. 31-10)

$u_E = u_B \Rightarrow \frac{1}{2}\epsilon_0 E^2 = \frac{1}{2\mu_0}B^2$

$B = \sqrt{\mu_0\epsilon_0}\,E = \sqrt{(4\pi\times10^{-7}\,\text{T}\cdot\text{m/A})(8.854\times10^{-12}\,\text{C}^2/\text{N}\cdot\text{m}^2)}\,(900\,\text{V/m}) = \underline{3.00\times10^{-6}\,\text{T}}$

31-49

$L = 3.00\,\text{H}$, $R = 7.00\,\Omega$, $\mathcal{E} = 12.0\,\text{V}$

$i = \frac{\mathcal{E}}{R}(1 - e^{-t/\tau})$, $\tau = \frac{L}{R}$

a) Eq. (31-9): $U_L = \frac{1}{2}Li^2$

$t = \tau \Rightarrow i = \frac{\mathcal{E}}{R}(1 - e^{-1}) = \frac{12.0\,\text{V}}{7.00\,\Omega}(1 - e^{-1}) = 1.084\,\text{A}$

$U_L = \frac{1}{2}Li^2 = \frac{1}{2}(3.00\,\text{H})(1.084\,\text{A})^2 = \underline{1.76\,\text{J}}$

Exercise 31-25 (c): $P_L = \frac{dU_L}{dt} = Li\frac{di}{dt}$

$i = \frac{\mathcal{E}}{R}(1 - e^{-t/\tau})$; $\frac{di}{dt} = \frac{\mathcal{E}}{L}e^{-(R/L)t} = \frac{\mathcal{E}}{L}e^{-t/\tau}$

$\Rightarrow P_L = L\left(\frac{\mathcal{E}}{R}(1 - e^{-t/\tau})\right)\left(\frac{\mathcal{E}}{L}e^{-t/\tau}\right) = \frac{\mathcal{E}^2}{R}(e^{-t/\tau} - e^{-2t/\tau})$

$U_L = \int_0^\tau P_L\,dt = \frac{\mathcal{E}^2}{R}\int_0^\tau (e^{-t/\tau} - e^{-2t/\tau})\,dt = \frac{\mathcal{E}^2}{R}\left[-\tau e^{-t/\tau} + \frac{\tau}{2}e^{-2t/\tau}\right]\Big|_0^\tau$

$U_L = -\frac{\mathcal{E}^2}{R}\tau\left[e^{-t/\tau} - \frac{1}{2}e^{-2t/\tau}\right]\Big|_0^\tau = \frac{\mathcal{E}^2}{R}\tau\left[1 - \frac{1}{2} - e^{-1} + \frac{1}{2}e^{-2}\right] = \frac{\mathcal{E}^2}{2R}\left(\frac{L}{R}\right)(1 - 2e^{-1} + e^{-2})$

$U_L = \frac{1}{2}\left(\frac{\mathcal{E}}{R}\right)^2 L(1 - 2e^{-1} + e^{-2}) = \frac{1}{2}\left(\frac{12.0\,\text{V}}{7.00\,\Omega}\right)^2(3.00\,\text{H})(0.3996) = 1.76\,\text{J}$, which checks

b) Exercise 31-25 (a): The rate at which the battery supplies energy is

$P_\mathcal{E} = \mathcal{E}i = \mathcal{E}\left(\frac{\mathcal{E}}{R}(1 - e^{-t/\tau})\right) = \frac{\mathcal{E}^2}{R}(1 - e^{-t/\tau})$

$U_\mathcal{E} = \int_0^\tau P_\mathcal{E}\,dt = \frac{\mathcal{E}^2}{R}\int_0^\tau (1 - e^{-t/\tau})\,dt = \frac{\mathcal{E}^2}{R}\left[t + \tau e^{-t/\tau}\right]\Big|_0^\tau = \frac{\mathcal{E}^2}{R}(\tau + \tau e^{-1} - \tau)$

$U_\mathcal{E} = \frac{\mathcal{E}^2}{R}\tau e^{-1} = \frac{\mathcal{E}^2}{R}\left(\frac{L}{R}\right)e^{-1} = \left(\frac{\mathcal{E}}{R}\right)^2 Le^{-1} = \left(\frac{12.0\,\text{V}}{7.00\,\Omega}\right)^2(3.00\,\text{H})(0.3679) = \underline{3.24\,\text{J}}$

c) $P_R = i^2R = \frac{\mathcal{E}^2}{R}(1 - e^{-t/\tau})^2 = \frac{\mathcal{E}^2}{R}(1 - 2e^{-t/\tau} + e^{-2t/\tau})$

31-49 (cont)

$$U_R = \int_0^\tau P_R\,dt = \frac{\mathcal{E}^2}{R}\int_0^\tau (1-2e^{-t/\tau}+e^{-2t/\tau})\,dt = \frac{\mathcal{E}^2}{R}\left[t+2\tau e^{-t/\tau}-\frac{\tau}{2}e^{-2t/\tau}\right]\Big|_0^\tau$$

$$U_R = \frac{\mathcal{E}^2}{R}\left[\tau+2\tau e^{-1}-\frac{\tau}{2}e^{-2}-2\tau+\frac{\tau}{2}\right] = \frac{\mathcal{E}^2}{R}\left[-\frac{\tau}{2}+2\tau e^{-1}-\frac{\tau}{2}e^{-2}\right] = \frac{\mathcal{E}^2}{2R}\tau\left[-1+4e^{-1}-e^{-2}\right]$$

$$U_R = \frac{\mathcal{E}^2}{2R}\left(\frac{L}{R}\right)\left[-1+4e^{-1}-e^{-2}\right] = \left(\frac{\mathcal{E}}{R}\right)^2\frac{1}{2}L\left[-1+4e^{-1}-e^{-2}\right] = \left(\frac{12.0\text{ V}}{7.00\,\Omega}\right)^2\frac{1}{2}(3.00\text{ H})(0.3362) = \underline{1.48\text{ J}}$$

d) $U_{\mathcal{E}} = U_R + U_L$ $\quad (3.24\text{ J} = 1.48\text{ J} + 1.76\text{ J})$

The energy supplied by the battery equals the sum of the energy stored in the magnetic field of the inductor and the energy dissipated in the resistance of the inductor.

31-51

The equation is $-iR - L\frac{di}{dt} - \frac{q}{C} = 0.$

Multiply by $-i \Rightarrow i^2R + Li\frac{di}{dt} + \frac{qi}{C} = 0.$

$\frac{d}{dt}U_L = \frac{d}{dt}\left(\frac{1}{2}Li^2\right) = \frac{1}{2}L\frac{d}{dt}(i^2) = \frac{1}{2}L\left(2i\frac{di}{dt}\right) = Li\frac{di}{dt}$, the second term

$\frac{d}{dt}U_C = \frac{d}{dt}\left(\frac{q^2}{2C}\right) = \frac{1}{2C}\frac{d}{dt}(q^2) = \frac{1}{2C}2q\frac{dq}{dt} = \frac{qi}{C}$, the third term

$i^2R = P_R$, the rate at which electrical energy is dissipated in the resistance.

$\frac{d}{dt}U_L = P_L$, the rate at which the amount of energy stored in the inductor is changing.

$\frac{d}{dt}U_C = P_C$, the rate at which the amount of energy stored in the capacitor is changing

The equation says that $P_R + P_L + P_C = 0$; the net rate of change of energy in the circuit is zero.

Note: At any given time one of P_C or P_L is negative. If the current and U_L are increasing the charge on the capacitor and U_C are decreasing, and vice versa.

31-53

Need $U_C = 2.00\times10^{-4}\text{ J}$ and $V_C = 50.0\text{ V}$

$$U_C = \frac{1}{2}CV^2 \Rightarrow C = \frac{2U_C}{V^2} = \frac{2(2.00\times10^{-4}\text{ J})}{(50.0\text{ V})^2} = 1.60\times10^{-7}\text{ F} = \underline{0.160\,\mu\text{F}}$$

$$L = \frac{1}{C\omega^2} = \frac{1}{(1.60\times10^{-7}\text{ F})(8.00\times10^4\text{ rad/s})^2} = 9.77\times10^{-4}\text{ H} = \underline{0.977\text{ mH}}$$

31-55

With S closed the circuit is

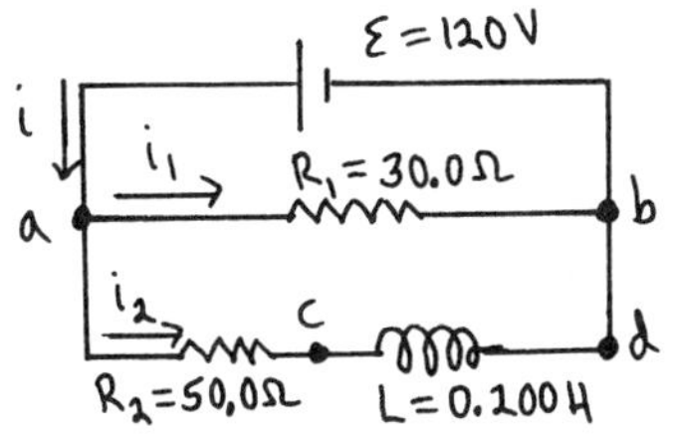

The rate of change of the current through the inductor is limited by the induced emf. Just after the switch is closed the current in the inductor has not had time to increase from zero, so $i_2 = 0$.

31-55 (cont)

a) $\mathcal{E} - V_{ab} = 0$, so $V_{ab} = 120\text{ V}$

b) The voltage drops across R_1 as travel through the resistor in the direction of the current, so point a is at higher potential.

c) $i_2 = 0 \Rightarrow V_{R_2} = i_2 R_2 = 0$

$\mathcal{E} - V_{R_2} - V_L = 0 \Rightarrow V_L = \mathcal{E} = 120\text{ V}$

d) The voltage rises when go from b to a through the emf, so it must drop when go from a to b through the inductor. Point c must be at higher potential than point d.

e) After the switch is closed a long time $\frac{di_2}{dt} \to 0$ so $V_L = 0$. Then $\mathcal{E} - V_{R_2} = 0$ and $i_2 R_2 = \mathcal{E} \Rightarrow i_2 = \frac{\mathcal{E}}{R_2} = \frac{120\text{ V}}{50.0\,\Omega} = 2.40\text{ A}$.

The rate of change of the current through the inductor is limited by the induced emf. Just after the switch is opened again the current through the inductor hasn't had time to change and is still $i_2 = 2.40\text{ A}$. The circuit is

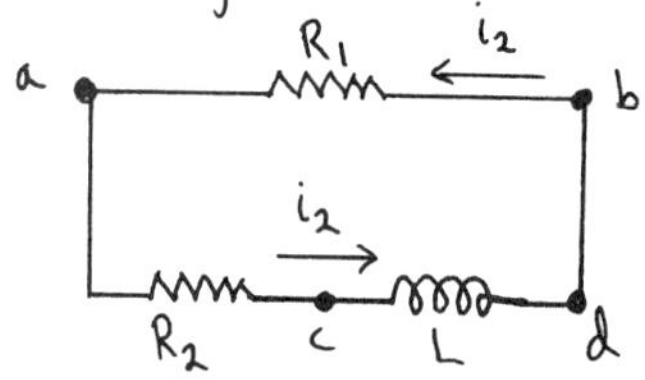

The current through R_1 is $i_2 = 2.40\text{A}$, in the direction b to a. Thus $V_{ab} = -i_2 R_1 = -(2.40\text{A})(30.0\,\Omega) = -72\text{ V}$

f) Point where current enters resistor is at higher potential; point b is at higher potential.

g) $V_L - V_{R_1} - V_{R_2} = 0$

$V_L = V_{R_1} + V_{R_2}$

$V_{R_1} = -V_{ab} = 72\text{ V}$

$V_{R_2} = i_2 R_2 = (2.40\text{A})(50.0\,\Omega) = 120\text{V}$

$V_L = V_{R_1} + V_{R_2} = 72\text{V} + 120\text{ V} = 192\text{V}$

As you travel counterclockwise around the circuit in the direction of the current, the voltage drops across each resistor, so it must rise across the inductor and point d is at higher potential than point c. The current is decreasing, so the induced emf in the inductor is directed in the direction of the current.

31-57

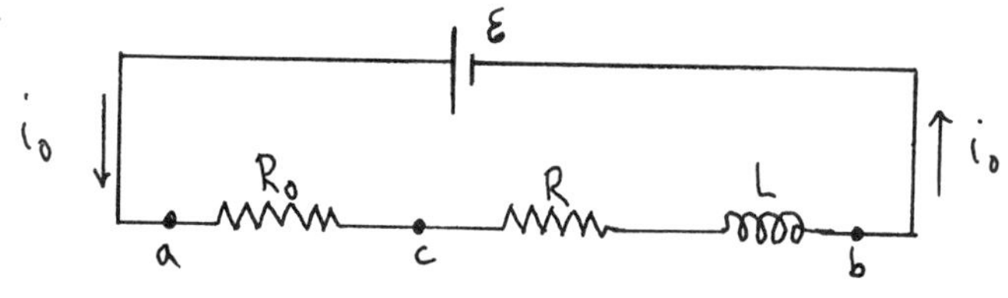

31-57 (cont)

a) At time $t=0$, $i=0$ so $V_{ac} = iR_0 = 0$. By the loop rule $\mathcal{E} - V_{ac} - V_{cb} = 0$, so $V_{cb} = \mathcal{E} - V_{ac} = \mathcal{E} = 36.0\ \text{V}$ ($iR = 0$ so this potential difference of 36.0 V is across the inductor and is an induced emf produced by the changing current.)

b) After a long time $\frac{di}{dt} \to 0$ so the potential $-L\frac{di}{dt}$ across the inductor becomes zero. The loop rule gives $\mathcal{E} - i_0(R_0 + R) = 0$.

$$i_0 = \frac{\mathcal{E}}{R_0 + R} = \frac{36.0\ \text{V}}{50.0\ \Omega + 150\ \Omega} = 0.180\ \text{A}$$

$$V_{ac} = i_0 R_0 = (0.180\ \text{A})(50.0\ \Omega) = 9.0\ \text{V}$$

$$V_{cb} = i_0 R + L\frac{di}{dt}^{\to 0} = (0.180\ \text{A})(150\ \Omega) = 27.0\ \text{V}$$ (Note that $V_{ac} + V_{cb} = \mathcal{E}$.)

c)
$$\mathcal{E} - V_{ac} - V_{cb} = 0$$

$$\mathcal{E} - iR_0 - iR - L\frac{di}{dt} = 0$$

$$L\frac{di}{dt} = \mathcal{E} - i(R_0 + R)$$

$$-\left(\frac{L}{R+R_0}\right)\frac{di}{dt} = i - \frac{\mathcal{E}}{R+R_0}$$

$$\frac{di}{i - \frac{\mathcal{E}}{R+R_0}} = -\left(\frac{R+R_0}{L}\right)dt$$

Integrate from $t=0$, when $i=0$, to t, when $i=i_0$

$$\int_0^{i_0} \frac{di}{i - \mathcal{E}/R+R_0} = -\frac{R+R_0}{L}\int_0^t dt$$

$$\ln\left(i - \frac{\mathcal{E}}{R+R_0}\right)\Big|_0^{i_0} = -\left(\frac{R+R_0}{L}\right)t \Rightarrow \ln\left(i_0 - \frac{\mathcal{E}}{R+R_0}\right) - \ln\left(-\frac{\mathcal{E}}{R+R_0}\right) = -\left(\frac{R+R_0}{L}\right)t$$

$$\ln\left(\frac{i_0 - \frac{\mathcal{E}}{R+R_0}}{-\mathcal{E}/R+R_0}\right) = -\left(\frac{R+R_0}{L}\right)t$$

Take exponentials of both sides $\Rightarrow \dfrac{i_0 - \frac{\mathcal{E}}{R+R_0}}{-\mathcal{E}/R+R_0} = e^{-\left(\frac{R+R_0}{L}\right)t}$

$$i_0 = \frac{\mathcal{E}}{R+R_0}\left(1 - e^{-\left(\frac{R+R_0}{L}\right)t}\right)$$

Substituting in the numerical values gives

$$i_0 = \frac{36.0\ \text{V}}{50\ \Omega + 150\ \Omega}\left(1 - e^{-\left(\frac{200\ \Omega}{4.00\ \text{H}}\right)t}\right) = (0.180\ \text{A})\left(1 - e^{-t/0.020\text{s}}\right)$$

At $t \to 0$, $i_0 = (0.180\ \text{A})(1-1) = 0$ (agrees with part (a))

At $t \to \infty$, $i_0 = (0.180\ \text{A})(1-0) = 0.180\ \text{A}$ (agrees with part (b))

$$V_{ac} = i_0 R_0 = \frac{\mathcal{E}R_0}{R+R_0}\left(1 - e^{-\left(\frac{R+R_0}{L}\right)t}\right) = 9.0\ \text{V}\left(1 - e^{-t/0.020\text{s}}\right)$$

$$V_{cb} = \mathcal{E} - V_{ac} = 36.0\ \text{V} - (9.0\ \text{V})\left(1 - e^{-t/0.020\text{s}}\right) = 9.0\ \text{V}\left(3.00 + e^{-t/0.020\text{s}}\right)$$

At $t \to 0$, $V_{ac} = 0$, $V_{cb} = 36.0\ \text{V}$ (agrees with part (a)).

At $t \to \infty$, $V_{ac} = 9.0\ \text{V}$, $V_{cb} = 27.0\ \text{V}$ (agrees with part (b)).

31-57 (cont)

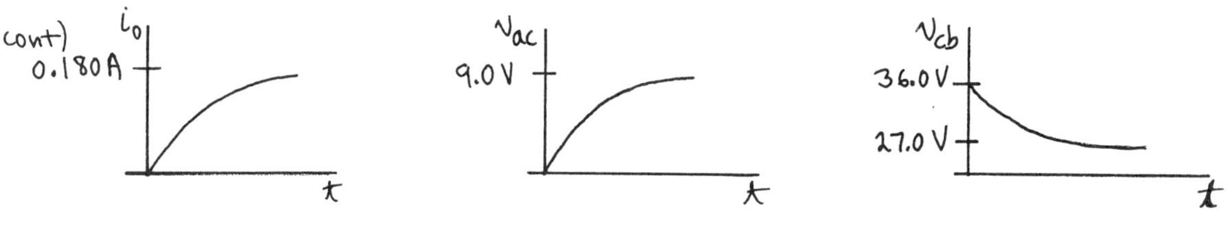

31-59

a) With switch S closed the circuit is

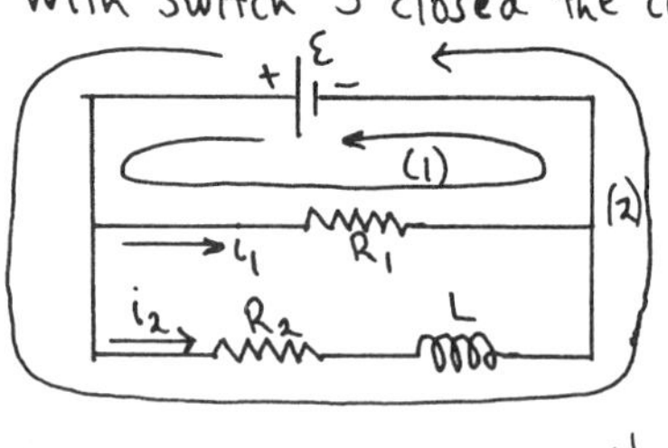

Loop (1)

$\mathcal{E} - i_1 R_1 = 0$

$i_1 = \frac{\mathcal{E}}{R_1}$ (independent of t)

Loop (2)

$\mathcal{E} - i_2 R_2 - L\frac{di_2}{dt} = 0$

This is in the form of equation (31-12), so the solution is analogous to Eq. (31-14): $i_2 = \frac{\mathcal{E}}{R_2}(1 - e^{-R_2 t/L})$

b) The expressions derived in part (a) give that as $t \to \infty$, $i_1 = \frac{\mathcal{E}}{R_1}$, $i_2 = \frac{\mathcal{E}}{R_2}$ (Since $\frac{di}{dt} \to 0$ at steady-state the inductance has no effect on the circuit.)

c) The circuit now becomes

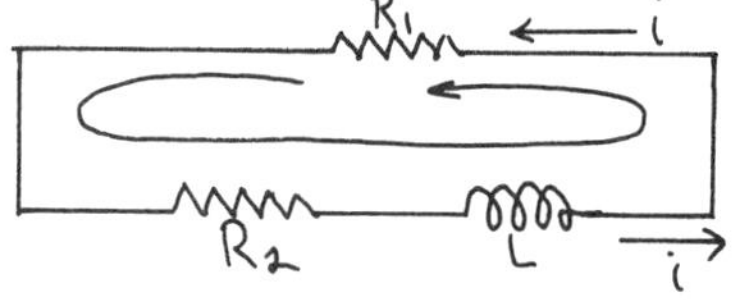

Let $t = 0$ now be when S is opened. At $t = 0$, $i = \frac{\mathcal{E}}{R_2}$. The loop rule applied to the single current loop gives $-i(R_1+R_2) - L\frac{di}{dt} = 0$ (Now $\frac{di}{dt}$ is negative.)

$L\frac{di}{dt} = -i(R_1+R_2) \Rightarrow \frac{di}{i} = -\left(\frac{R_1+R_2}{L}\right)dt$

Integrate from $t=0$, when $i = I_0 = \frac{\mathcal{E}}{R_2}$, to t.

$\int_{I_0}^{i} \frac{di}{i} = -\left(\frac{R_1+R_2}{L}\right)\int_0^t dt$

$\ln\left(\frac{i}{I_0}\right) = -\left(\frac{R_1+R_2}{L}\right)t$

Take exponentials of both sides of this equation $\Rightarrow i = I_0 e^{-(R_1+R_2)t/L} = \frac{\mathcal{E}}{R_2}e^{-(R_1+R_2)t/L}$

d) $L = 52.0\ \text{H}$

$P_{R_1} = \frac{V^2}{R_1} = 60.0\ \text{W} \Rightarrow R_1 = \frac{V^2}{P_{R_1}} = \frac{(120\ \text{V})^2}{60.0\ \text{W}} = 240\ \Omega$

We are asked to find R_2 and $\mathcal{E}$. Use the expression derived in part (c).

$I_0 = 0.900\ \text{A} \Rightarrow \frac{\mathcal{E}}{R_2} = 0.900\ \text{A}$

$i = 0.300\ \text{A}$ when $t = 0.200\ \text{s}$, so $i = \frac{\mathcal{E}}{R_2}e^{-(R_1+R_2)t/L}$ gives

$0.300\ \text{A} = 0.900\ \text{A}\ e^{-(R_1+R_2)t/L}$

$\frac{1}{3} = e^{-(R_1+R_2)t/L} \Rightarrow \ln 3 = \frac{(R_1+R_2)t}{L}$

$R_2 = \frac{L\ln 3}{t} - R_1 = \frac{(52.0\ \text{H})\ln 3}{0.200\ \text{s}} - 240\ \Omega = 285.6\ \Omega - 240\ \Omega = \underline{45.6\ \Omega}$

Then $\frac{\mathcal{E}}{R_2} = 0.900\ \text{A} \Rightarrow \mathcal{E} = (0.900\ \text{A})R_2 = (0.900\ \text{A})(45.6\ \Omega) = \underline{41.1\ \text{V}}$

31-59 (cont)

e) Use the expressions derived in part (a). The current through the light bulb before the switch is opened is $i_1 = \frac{\mathcal{E}}{R_1} = \frac{41.1\text{ V}}{240\,\Omega} = 0.171\text{ A}$

When the switch is opened the current through the light bulb jumps from 0.171 A to 0.900 A.

CHAPTER 32

Exercises 3, 7, 9, 11, 15, 21, 23, 25, 27, 29

Problems 33, 35, 37, 39, 41, 43, 47, 49, 51

Exercises

32-3

a) $X_L = \omega L = 2\pi f L = 2\pi(50.0\text{ Hz})(2.00\text{ H}) = \underline{628\,\Omega}$

b) $X_L = 2\pi f L \Rightarrow L = \dfrac{X_L}{2\pi f} = \dfrac{2.00\,\Omega}{2\pi(50.0\text{ Hz})} = \underline{6.37\times10^{-3}\text{ H}}$

c) $X_C = \dfrac{1}{\omega C} = \dfrac{1}{2\pi f C} = \dfrac{1}{2\pi(50.0\text{Hz})(2.00\times10^{-6}\text{F})} = \underline{1.59\times10^{3}\,\Omega}$

d) $X_C = \dfrac{1}{2\pi f C} \Rightarrow C = \dfrac{1}{2\pi f X_C} = \dfrac{1}{2\pi(50.0\text{Hz})(2.00\,\Omega)} = \underline{1.59\times10^{-3}\text{ F}}$

32-7

a) $v_R = (2.50\text{V})\cos[(950\text{ rad/s})t]$

$v_R = iR$, so $i = \dfrac{v_R}{R} = \left(\dfrac{2.50\text{V}}{300\,\Omega}\right)\cos[(950\text{ rad/s})t] = (8.33\text{ mA})\cos[(950\text{ rad/s})t]$

b) $X_L = \omega L$

$\omega = 950$ rad/s, $L = 0.800$H, so $X_L = \omega L = (950\text{ rad/s})(0.800\text{ H}) = \underline{760\,\Omega}$

c) If $i = I\cos\omega t$ then $v_L = V_L\cos(\omega t + 90°)$ (from Eq. 32-10).

$V_L = I\omega L = IX_L = (0.00833\text{ A})(760\,\Omega) = 6.33\text{ V}$

$v_L = (6.33\text{ V})\cos[(950\text{ rad/s})t + 90°]$

But $\cos(a+90°) = -\sin a$ (Appendix B), so $v_L = -(6.33\text{V})\sin(950\text{ rad/s})t$.

32-9

$V = IX_C \Rightarrow X_C = \dfrac{V}{I} = \dfrac{170\text{ V}}{1.40\text{A}} = 121.4\,\Omega$

$X_C = \dfrac{1}{\omega C} = \dfrac{1}{2\pi f C} \Rightarrow C = \dfrac{1}{2\pi f X_C} = \dfrac{1}{2\pi(60.0\text{Hz})(121.4\,\Omega)} = 2.18\times10^{-5}\text{F} = \underline{21.8\,\mu\text{F}}$

32-11

$R = 300\,\Omega$, $L = 0.250$H, $C = 8.00\times10^{-6}$ F

a) $\omega = 500$ rad/s

$X_L = \omega L = (500\text{ rad/s})(0.250\text{ H}) = 125\,\Omega$

$X_C = \dfrac{1}{\omega C} = \dfrac{1}{(500\text{ rad/s})(8.00\times10^{-6}\text{F})} = 250\,\Omega$

$Z = \sqrt{R^2 + (X_L - X_C)^2} = \sqrt{(300\,\Omega)^2 + (125\,\Omega - 250\,\Omega)^2} = \underline{325\,\Omega}$

$\tan\phi = \dfrac{X_L - X_C}{R} = \dfrac{125\,\Omega - 250\,\Omega}{300\,\Omega} = -0.4167 \Rightarrow \phi = \underline{-22.6°}$

32-11 (cont)

The phase angle is negative, so the source voltage lags behind the current.

Since $X_C > X_L$ the phasor diagram is qualitatively

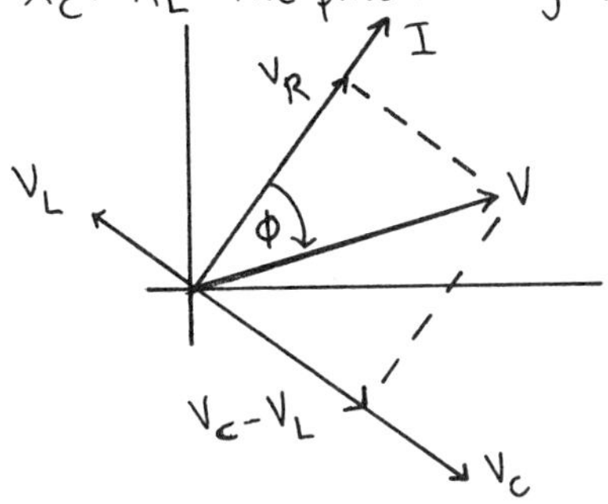

b) $\omega = 1000$ rad/s

$X_L = \omega L = (1000\text{ rad/s})(0.250\text{ H}) = 250\,\Omega$

$X_C = \frac{1}{\omega C} = \frac{1}{(1000\text{ rad/s})(8.00\times10^{-6}\text{ F})} = 125\,\Omega$

$Z = \sqrt{R^2 + (X_L - X_C)^2} = \sqrt{(300\,\Omega)^2 + (250\,\Omega - 125\,\Omega)^2} = \underline{325\,\Omega}$

$\tan\phi = \frac{X_L - X_C}{R} = \frac{250\,\Omega - 125\,\Omega}{300\,\Omega} = +0.4167 \Rightarrow \phi = \underline{22.6^\circ}$

The phase angle is positive, so the source voltage leads the current.
Since $X_L > X_C$ the phasor diagram is qualitatively

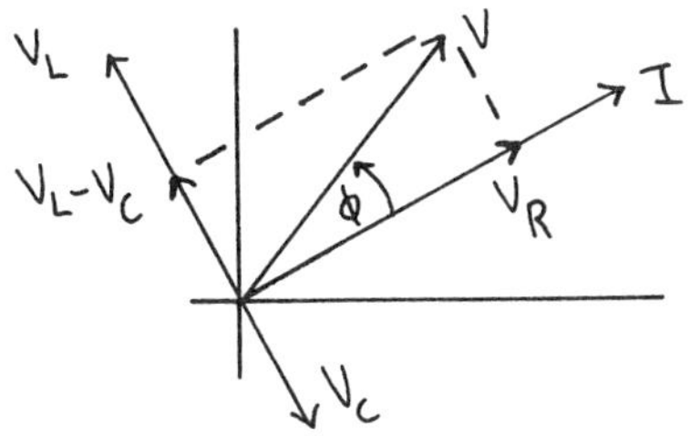

29-15

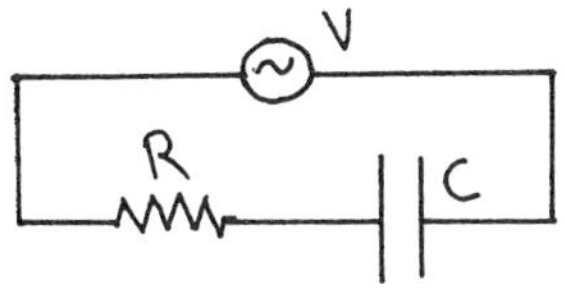

No inductor $\Rightarrow X_L = 0$

$R = 300\,\Omega$, $C = 4.00\times10^{-6}$ F, $V = 70.0$ V, $\omega = 200$ rad/s

a) $X_C = \frac{1}{\omega C} = \frac{1}{(200\text{ rad/s})(4.00\times10^{-6}\text{ F})} = 1250\,\Omega$

$Z = \sqrt{R^2 + (X_L - X_C)^2} = \sqrt{R^2 + X_C^2} = \sqrt{(300\,\Omega)^2 + (1250\,\Omega)^2} = \underline{1290\,\Omega}$

b) $I = \frac{V}{Z} = \frac{70.0\text{ V}}{1290\,\Omega} = 0.0543\text{ A} = \underline{54.3\text{ mA}}$

c) The voltage amplitude across the resistor is
$V_R = IR = (0.0543\text{ A})(300\,\Omega) = \underline{16.3\text{ V}}$

The voltage amplitude across the capacitor is
$V_C = IX_C = (0.0543\text{ A})(1250\,\Omega) = \underline{67.9\text{ V}}$

32-15 (cont)

d) $\tan\phi = \dfrac{X_L - X_C}{R} = \dfrac{0 - 1250\Omega}{300\Omega} = -4.167 \Rightarrow \phi = \underline{-76.5^\circ}$

The phase angle is negative, so the source voltage <u>lags</u> behind the current.

e) The phasor diagram is qualitatively

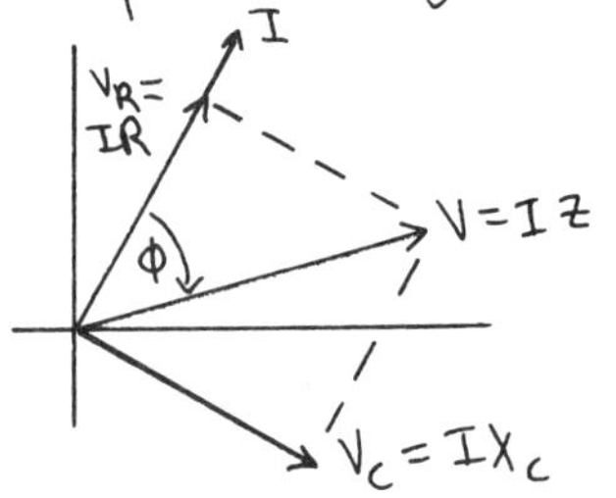

32-21

$R = 300\Omega$, $L = 0.250\text{ H}$, $C = 8.00 \times 10^{-6}\text{ F}$, $\omega = 1000\text{ rad/s}$, $I_{rms} = 0.325\text{ A}$

a) $P_{av} = V_{rms} I_{rms} \cos\phi$ (Eq. 32-30)

$X_L = \omega L = (1000\text{ Hz})(0.250\text{ H}) = 250\Omega$

$X_C = \dfrac{1}{\omega C} = \dfrac{1}{(1000\text{ Hz})(8.00\times10^{-6}\text{ F})} = 125\Omega$

$Z = \sqrt{R^2 + (X_L - X_C)^2} = \sqrt{(300\Omega)^2 + (250\Omega - 125\Omega)^2} = 325\Omega$

$V_{rms} = I_{rms} Z = (0.325\text{ A})(325\Omega) = 105.6\text{ V}$

$\tan\phi = \dfrac{X_L - X_C}{R} = \dfrac{250\Omega - 125\Omega}{300\Omega} = 0.4167 \Rightarrow \phi = 22.62^\circ$

Then $P_{av} = V_{rms} I_{rms} \cos\phi = (105.6\text{ V})(0.325\text{ A})\cos(22.62^\circ) = \underline{31.7\text{ W}}$

b) In the resistor the current and voltage are in phase, so
$P_{av} = V_{R,rms} I_{rms} = (I_{rms} R) I_{rms} = I_{rms}^2 R = (0.325\text{ A})^2 (300\Omega) = \underline{31.7\text{ W}}$

c) For the capacitor the instantaneous voltage and current are 90° out of phase, and the average power consumed is zero. Electrical energy is periodically stored and released in the capacitor, but not converted to other forms and thereby lost to the circuit.

d) For the inductor the instantaneous voltage and current are 90° out of phase, and the average power consumed is zero. Electrical energy is periodically stored and released in the magnetic field of the inductor, but not converted to other forms and thereby lost to the circuit.

e) part (a): $P_{av,source} = 31.7\text{ W}$

part (b): $P_{av,R} = 31.7\text{ W}$

part (c): $P_{av,C} = 0$

part (d): $P_{av,L} = 0$

$P_{av,source} = P_{av,R} + P_{av,C} + P_{av,L}$; the average power delivered by the source equals the sum of the average power consumed in the resistor, capacitor, and inductor.

32-23

a)

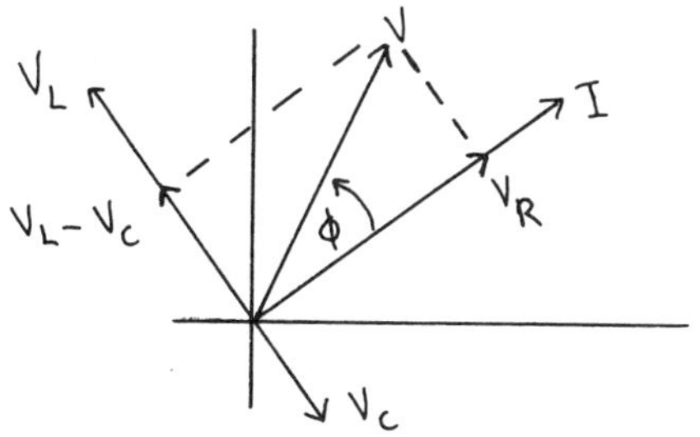

From the diagram $\cos\phi = \frac{V_R}{V} = \frac{IR}{IZ} = \frac{R}{Z}$, as was to be shown

b) $P_{av} = V_{rms} I_{rms} \cos\phi = V_{rms} I_{rms}\left(\frac{R}{Z}\right) = \frac{V_{rms}}{Z} I_{rms} R$

But $\frac{V_{rms}}{Z} = I_{rms}$, so $P_{av} = I_{rms}^2 R$

32-25

$R = 400\,\Omega$, $L = 0.900\,H$, $C = 2.50\,\mu F$

a) The resonant angular frequency is given by Eq. (32-31)

$\omega_0 = \frac{1}{\sqrt{LC}} = \frac{1}{\sqrt{(0.900H)(2.50\times10^{-6}F)}} = \underline{667\text{ rad/s}}$

b) At resonance $X_L = X_C$, so $\tan\phi = \frac{X_L - X_C}{R} = 0$ and $\phi = 0$. The source voltage and the current in the circuit are in phase.

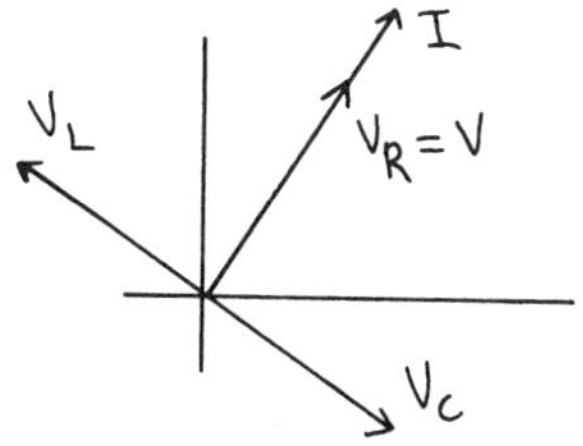

c) $X_L = \omega L = (667\text{ rad/s})(0.900H) = 600\,\Omega$

$X_C = \frac{1}{\omega C} = \frac{1}{(667\text{ rad/s})(2.50\times10^{-6}F)} = 600\,\Omega$

$X_L = X_C \Rightarrow Z = \sqrt{R^2 + (X_L - X_C)^2} = R = 400\,\Omega$ (with $X_L - X_C = 0$)

$I = \frac{V}{Z} = \frac{90.0\,V}{400\,\Omega} = 0.225\,A$; $I_{rms} = \frac{I}{\sqrt{2}} = 0.159\,A$

V_1 reads $V_{R,rms} = I_{rms} R = (0.159\,A)(400\,\Omega) = 63.6\,V$

V_2 reads $V_{L,rms} = I_{rms} X_L = (0.159\,A)(600\,\Omega) = 95.4\,V$

V_3 reads $V_{C,rms} = I_{rms} X_C = (0.159\,A)(600\,\Omega) = 95.4\,V$

V_4 reads $(V_L - V_C)_{rms} = I_{rms}(X_L - X_C) = 0$. At any instant the voltages across the inductor and capacitor are equal and opposite and sum to zero.

V_5 reads $V_{rms} = \frac{90.0\,V}{\sqrt{2}} = 63.6\,V$, the same as V_1.

d) $\omega_0 = \frac{1}{\sqrt{LC}}$ is independent of R, so $\omega_0 = 667$ rad/s, the same as when $R = 400\,\Omega$.

e) At resonance $X_L = X_C$ and $Z = R$.

$I_{rms} = \frac{V_{rms}}{Z} = \frac{V_{rms}}{R} = \frac{63.6\,V}{100\,\Omega} = \underline{0.636\,A}$.

32-25 (cont)

The current at resonance is inversely proportional to the resistance so increases by a factor of 4 when the resistance is decreased by a factor of 4.

32-27

$R = 300\,\Omega$, $L = 0.800\,H$, $C = 0.0400\,\mu F$, $V = 300\,V$

a) At the resonance frequency $X_L = X_C$ and from $\tan\phi = \frac{X_L - X_C}{R}$ we have that $\phi = 0°$ and the power factor is $\cos\phi = 1$.

b) $P_{av} = \frac{1}{2} V I \cos\phi$ (Eq. 32-30)

At the resonance frequency $Z = R$, so $I = \frac{V}{Z} = \frac{V}{R}$

$$P_{av} = \frac{1}{2} V \left(\frac{V}{R}\right) \cos\phi^{\nearrow 1} = \frac{1}{2} \frac{V^2}{R} = \frac{1}{2} \frac{(300\,V)^2}{300\,\Omega} = \underline{150\,W}$$

c) Nothing changes in $P_{av} = \frac{1}{2}\frac{V^2}{R}$, so the average power is unchanged: $P_{av} = \underline{150\,W}$.
(The resonance frequency changes but since $Z = R$ at resonance, the current doesn't change.)

32-29

a) Eq. (32-34): $\frac{V_2}{V_1} = \frac{N_2}{N_1} \Rightarrow \frac{N_1}{N_2} = \frac{V_1}{V_2} = \frac{120\,V}{6.00\,V} = \underline{20}$

b) $I_2 = \frac{V_2}{R} = \frac{6.00\,V}{4.00\,\Omega} = \underline{1.50\,A}$

c) $P_{av} = I_2^2 R = (1.50\,A)^2 (4.00\,\Omega) = \underline{9.00\,W}$

d) The power drawn from the line by the transformer is the 9.00 W that is delivered to the load.

$$P_{av} = \frac{V^2}{R} \Rightarrow R = \frac{V^2}{P_{av}} = \frac{(120\,V)^2}{9.00\,W} = \underline{1600\,\Omega}$$

And $\left(\frac{N_1}{N_2}\right)^2 (4.00\,\Omega) = (20)^2 (4.00\,\Omega) = 1600\,\Omega$, as was to be shown.

Problems

32-33

a) From Fig. 32-2b, the rectified current is zero at the same values of t for which the sinusoidal current is zero $\Rightarrow \cos\omega t = 0$ and $\omega t = \pm\frac{\pi}{2}, \pm\frac{3\pi}{2}, \ldots$
The two smallest positive times are

$$t_1 = \frac{\pi}{2\omega}, \quad t_2 = \frac{3\pi}{2\omega}$$

b) $A = \int_{t_1}^{t_2} i\,dt = \int_{t_1}^{t_2} I \cos\omega t\,dt = I\left(\frac{1}{\omega}\sin\omega t \Big|_{t_1}^{t_2}\right) = \frac{I}{\omega}(\sin\omega t_2 - \sin\omega t_1)$

$$\sin\omega t_1 = \sin\left[\omega\left(\frac{\pi}{2\omega}\right)\right] = \sin\frac{\pi}{2} = 1$$

$$\sin\omega t_2 = \sin\left[\omega\left(\frac{3\pi}{2\omega}\right)\right] = \sin\left(\frac{3\pi}{2}\right) = -1$$

32-33 (cont)

$$A = \frac{I}{\omega}\left(1-(-1)\right) = \frac{2I}{\omega}$$

c) $I_{rav}(t_2 - t_1) = \frac{2I}{\omega}$

$$I_{rav} = \frac{2I}{\omega(t_2 - t_1)} = \frac{2I}{\omega\left(\frac{3\pi}{2\omega} - \frac{\pi}{2\omega}\right)} = \frac{2I}{\pi}$$, which is Eq. (32-3)

32-35

The voltage across the coil leads the current in it by 38.4° $\Rightarrow \phi = +38.4°$.

$\tan\phi = \frac{X_L - X_C}{R}$. But there is no capacitance in the circuit so $X_C = 0$.

$\tan\phi = \frac{X_L}{R} \Rightarrow X_L = R\tan\phi = (40.0\,\Omega)\tan 38.4° = 31.7\,\Omega$

$X_L = \omega L = 2\pi f L \Rightarrow L = \frac{X_L}{2\pi f} = \frac{31.7\,\Omega}{2\pi\,(100\,Hz)} = \underline{0.0505\ H}$

32-37

a) $V_C = IX_C \Rightarrow I = \frac{V_C}{X_C} = \frac{720\,V}{600\,\Omega} = \underline{1.20\,A}$

b) $V = IZ \Rightarrow Z = \frac{V}{I} = \frac{240\,V}{1.20\,A} = \underline{200\,\Omega}$

c) $Z^2 = R^2 + (X_L - X_C)^2$

$X_L - X_C = \pm\sqrt{Z^2 - R^2} \Rightarrow X_L = X_C \pm \sqrt{Z^2 - R^2} = 600\,\Omega \pm \sqrt{(200\,\Omega)^2 - (120\,\Omega)^2}$

$X_L = 600\,\Omega \pm 160\,\Omega$

$X_L = \underline{760\,\Omega \text{ or } 440\,\Omega}$

32-39

$P_{av} = V_{rms} I_{rms} \cos\phi$

But from the phasor diagram (Fig. 32-8) $V_{rms}\cos\phi = V_R = I_{rms} R$

$\Rightarrow P_{av} = I_{rms}^2 R$

$R = \frac{P_{av}}{I_{rms}^2} = \frac{16.0\,J/s}{(0.500\,A)^2} = 64.0\,\Omega$

Then $Z = \sqrt{R^2 + (X_L - X_C)^2} = \sqrt{(64.0\,\Omega)^2 + (25.0\,\Omega - 0)^2} = \underline{68.7\,\Omega}$

32-41

a) Source voltage lags current $\Rightarrow X_C > X_L \Rightarrow$ add an <u>inductor</u> in series with the circuit.

b) power factor $\cos\phi$ equals $1 \Rightarrow \phi = 0 \Rightarrow X_C = X_L$

Calculate the present value of $X_C - X_L$ to see how much more X_L is needed:

From the phasor diagram $\cos\phi = \frac{R}{Z}$ (Exercise 32-23).

$R = Z\cos\phi = (50.0\,\Omega)(0.630) = 31.5\,\Omega$

32-41 (cont)

$$\tan\phi = \frac{X_L - X_C}{R} \Rightarrow X_L - X_C = R\tan\phi$$

$\cos\phi = 0.630 \Rightarrow \phi = -50.95°$ (ϕ is negative since the voltage lags the current)

$X_L - X_C = R\tan\phi = (31.5\ \Omega)\tan(-50.95°) = -38.8\ \Omega$

Therefore need to add 38.8 Ω of X_L.

$X_L = \omega L = 2\pi f L \Rightarrow L = \frac{X_L}{2\pi f} = \frac{38.8\ \Omega}{2\pi\,(60.0\ \text{Hz})} = 0.103\ \text{H}$, amount of inductance to add

32-43

source voltage lags current $\Rightarrow \phi = -32.0°$

$X_C = 400\ \Omega,\ R = 200\ \Omega,\ P_{av} = 210\ \text{W}$

a) $\tan\phi = \frac{X_L - X_C}{R}$

$X_L = R\tan\phi + X_C = (200\ \Omega)\tan(-32.0°) + 400\ \Omega = -125\ \Omega + 400\ \Omega = \underline{275\ \Omega}$

b) $P_{av} = V_{rms} I_{rms}\cos\phi = I_{rms}^2 R$ (Exercise 32-23).

$I_{rms} = \sqrt{\frac{P_{av}}{R}} = \sqrt{\frac{210\ \text{W}}{200\ \Omega}} = \underline{1.02\ \text{A}}$ (1.025 A before rounding to three significant figures)

c) $P_{av} = V_{rms} I_{rms}\cos\phi$

$V_{rms} = \frac{P_{av}}{I_{rms}\cos\phi} = \frac{210\ \text{W}}{(1.025\ \text{A})\cos(-32.0°)} = \underline{242\ \text{V}}$

or

$Z = \sqrt{R^2 + (X_L - X_C)^2} = \sqrt{(200\ \Omega)^2 + (275\ \Omega - 400\ \Omega)^2} = 236\ \Omega$

$V_{rms} = I_{rms} Z = (1.025\ \text{A})(236\ \Omega) = 242\ \text{V}$, which agrees

32-47

$R = 400\ \Omega,\ L = 0.500\ \text{H},\ C = 0.600\times10^{-6}\ \text{F},\ V = 120\ \text{V}$

a) The resonant angular frequency of the parallel L-R-C circuit is the same as for the series L-R-C circuit (Problem 32-45):

$\omega_0 = \frac{1}{\sqrt{LC}} = \frac{1}{\sqrt{(0.500\text{H})(0.600\times10^{-6}\text{F})}} = \underline{1830\ \text{rad/s}}$

b) The phasor diagram is like that drawn in Problem 32-44, but with $I_C = I_L$, so $\phi = 0$.

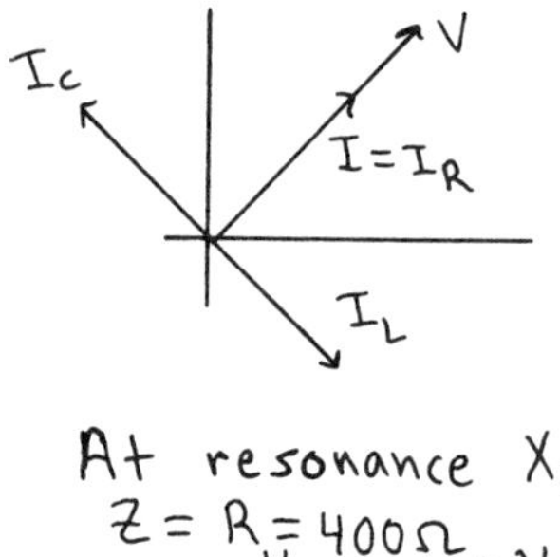

c) At resonance $X_L = X_C$ so $\frac{1}{Z} = \sqrt{\frac{1}{R^2} + \left(\frac{1}{X_C} - \frac{1}{X_L}\right)^2} = \frac{1}{R}$

$Z = R = 400\ \Omega$

$I = \frac{V}{Z} = \frac{120\ \text{V}}{400\ \Omega} = \underline{0.300\ \text{A}}$

32-47 (cont)

d) For the parallel circuit the voltage across each circuit element equals the source voltage.

$$X_L = \omega L = (1830\ \text{rad/s})(0.500\text{H}) = 915\,\Omega$$
$$X_C = \frac{1}{\omega C} = \frac{1}{(1830\ \text{rad/s})(0.600\times10^{-6}\text{F})} = 915\,\Omega$$

The amplitude of the current through the resistor is $I_R = \frac{V}{R} = \frac{120\text{ V}}{400\,\Omega} = \underline{0.300\text{A}}$.
The amplitude of the current through the inductor is $I_L = \frac{V}{X_L} = \frac{120\text{ V}}{915\,\Omega} = \underline{0.131\text{A}}$.
The amplitude of the current through the capacitor branch is $I_C = \frac{V}{X_C} = \frac{120\text{V}}{915\,\Omega} = \underline{0.131\text{ A}}$.

(At any instant in time the inductor and capacitor currents are equal in magnitude and opposite in direction, so their sum is zero. At resonance then the current through the source equals the current through the resistor.)

32-49

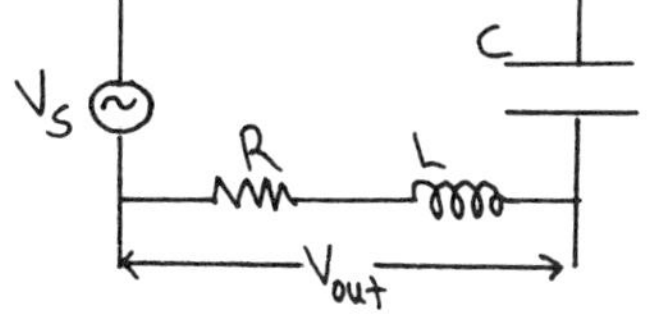

$$X_L = \omega L, \quad X_C = \frac{1}{\omega C}$$
$$Z = \sqrt{R^2 + \left(\omega L - \frac{1}{\omega C}\right)^2}$$
$$I = \frac{V_S}{Z} = \frac{V_S}{\sqrt{R^2 + \left(\omega L - \frac{1}{\omega C}\right)^2}}$$

The voltages across the resistor and the inductor are 90° out of phase, so

$$V_{out} = \sqrt{V_R^2 + V_L^2} = I\sqrt{R^2 + X_L^2} = I\sqrt{R^2 + \omega^2 L^2} = V_S\sqrt{\frac{R^2 + \omega^2 L^2}{R^2 + \left(\omega L - \frac{1}{\omega C}\right)^2}}$$
$$\frac{V_{out}}{V_S} = \sqrt{\frac{R^2 + \omega^2 L^2}{R^2 + \left(\omega L - \frac{1}{\omega C}\right)^2}}$$

$\underline{\omega \text{ small}}$

As ω gets small, $R^2 + \left(\omega L - \frac{1}{\omega C}\right)^2 \to \frac{1}{\omega^2 C^2}$, $R^2 + \omega^2 L^2 \to R^2$

Therefore $\frac{V_{out}}{V_S} \xrightarrow[\omega\text{ small}]{} \sqrt{\frac{R^2}{\frac{1}{\omega^2 C^2}}} = \omega RC.$

$\underline{\omega \text{ large}}$

As ω gets large, $R^2 + \omega^2 L^2 \to \omega^2 L^2$, $R^2 + \left(\omega L - \frac{1}{\omega C}\right)^2 \to R^2 + \omega^2 L^2 \to \omega^2 L^2$

Therefore $\frac{V_{out}}{V_S} \xrightarrow[\omega\text{ large}]{} \sqrt{\frac{\omega^2 L^2}{\omega^2 L^2}} = 1.$

32-51

Consider the cycle of the repeating current from $t_1 = \frac{\tau}{2}$ to $t_2 = \frac{3\tau}{2}$. In this interval $i = \frac{2I_0}{\tau}(t - \tau)$.

$$I_{av} = \frac{1}{t_2 - t_1}\int_{t_1}^{t_2} i\,dt = \frac{1}{\tau}\int_{\tau/2}^{3\tau/2} \frac{2I_0}{\tau}(t-\tau)\,dt = \frac{2I_0}{\tau^2}\left(\frac{1}{2}t^2 - \tau t\right)\Big|_{\tau/2}^{3\tau/2}$$
$$I_{av} = \frac{2I_0}{\tau^2}\left(\frac{9\tau^2}{8} - \frac{3\tau^2}{2} - \frac{\tau^2}{8} + \frac{\tau^2}{2}\right) = (2I_0)\frac{1}{8}(9 - 12 - 1 + 4) = \frac{I_0}{4}(13-13) = 0.$$

32-51 (cont)

$$I_{rms}^2 = (I^2)_{av} = \frac{1}{t_2 - t_1}\int_{t_1}^{t_2} i^2\,dt = \frac{1}{T}\int_{T/2}^{3T/2} \frac{4I_0^2}{T^2}(t-T)^2\,dt$$

$$I_{rms}^2 = \frac{4I_0^2}{T^3}\int_{T/2}^{3T/2}(t-T)^2\,dt = \frac{4I_0^2}{T^3}\,\frac{1}{3}(t-T)^3\Big|_{T/2}^{3T/2} = \frac{4I_0^2}{3T^3}\left[\left(\frac{T}{2}\right)^3 - \left(-\frac{T}{2}\right)^3\right]$$

$$I_{rms}^2 = \frac{I_0^2}{6}[1+1] = \frac{1}{3}I_0^2$$

$$I_{rms} = \sqrt{I_{rms}^2} = \frac{I_0}{\sqrt{3}}$$

CHAPTER 33

Exercises 3, 5, 7, 11, 15, 17, 19, 21, 23, 25

Problems 31, 35, 37, 43, 45, 47, 49

Exercises

33-3

a) $c = f\lambda \Rightarrow \lambda = \frac{c}{f} = \frac{3.00\times10^{8}\ \text{m/s}}{90.9\times10^{6}\ \text{Hz}} = \underline{3.30\ \text{m}}$

b) Eq. (33-4): $E = cB = (3.00\times10^{8}\ \text{m/s})(5.56\times10^{-11}\ \text{T}) = \underline{0.0167\ \text{V/m}}$

33-5

$$\omega = 2\pi f = 2\pi\,(4.00\times10^{13}\ \text{Hz}) = 2.51\times10^{14}\ \text{rad/s}$$

$$k = \frac{2\pi}{\lambda} = \frac{2\pi f}{c} = \frac{\omega}{c} = \frac{2.51\times10^{14}\ \text{rad/s}}{3.00\times10^{8}\ \text{m/s}} = 8.37\times10^{5}\ \text{rad/m}$$

$$B_{max} = 7.30\times10^{-4}\ \text{T}$$

$$E_{max} = cB_{max} = (3.00\times10^{8}\ \text{m/s})(7.30\times10^{-4}\ \text{T}) = 2.19\times10^{5}\ \text{V/m}$$

$\vec{B}$ is along the x-axis. $\vec{E}\times\vec{B}$ is in the direction of propagation (the +y-direction). From this we can deduce the direction of $\vec{E}$.

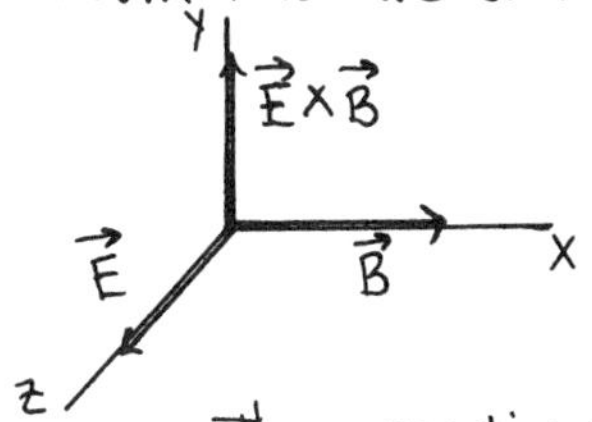

$\vec{E}$ is along the z-axis.

The equations are in the form of Eqs. (33-17):

$$\vec{E} = E_{max}\,\hat{k}\,\sin(\omega t - ky) = (2.19\times10^{5}\ \text{V/m})\,\hat{k}\,\sin[(2.51\times10^{14}\ \text{rad/s})t - (8.37\times10^{5}\ \text{rad/m})\,y]$$

$$\vec{B} = B_{max}\,\hat{\imath}\,\sin(\omega t - ky) = (7.30\times10^{-4}\ \text{T})\,\hat{\imath}\,\sin[(2.51\times10^{14}\ \text{rad/s})t - (8.37\times10^{5}\ \text{rad/m})\,y]$$

33-7

a) $\vec{E}(x,t) = -(2.30\times10^{5}\ \text{V/m})\,\hat{k}\,\sin[(1.45\times10^{14}\ \text{rad/s})t - kx]$

Compare to Eq. (33-16) $\Rightarrow \omega = 1.45\times10^{14}\ \text{rad/s}$

$$\omega = 2\pi f = \frac{2\pi c}{\lambda} \Rightarrow \lambda = \frac{2\pi c}{\omega} = \frac{2\pi\,(2.998\times10^{8}\ \text{m/s})}{1.45\times10^{14}\ \text{rad/s}} = 1.30\times10^{-5}\ \text{m} = \underline{13.0\ \mu\text{m}}$$

b)

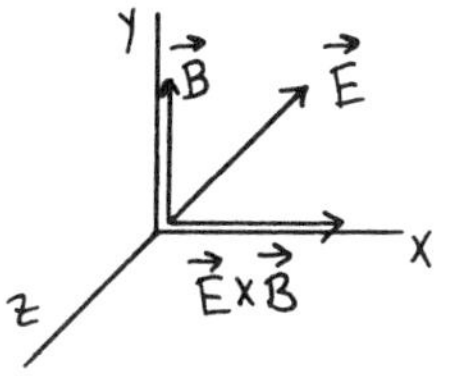

$\vec{E}\times\vec{B}$ must be in the +x-direction (the direction in which the wave is traveling). When $\vec{E}$ is in the −z-direction then $\vec{B}$ must be in the +y-direction.

$$k = \frac{2\pi}{\lambda} = \frac{\omega}{c} = \frac{1.45\times10^{14}\ \text{rad/s}}{2.998\times10^{8}\ \text{m/s}} = 4.84\times10^{5}\ \text{rad/m}$$

$$E_{max} = 2.30\times10^{5}\ \text{V/m}$$

33-7 (cont) $B_{max} = \frac{E_{max}}{c} = \frac{2.30\times10^{5}\text{ V/m}}{2.998\times10^{8}\text{ m/s}} = 7.67\times10^{-4}\text{ T}$

Using Eq. (33-16) and the fact that $\vec{B}$ is in the $+\hat{j}$ direction when $\vec{E}$ is in the $-\hat{k}$ direction,

$$\vec{B} = (7.67\times10^{-4}\text{ T})\,\hat{j}\,\sin[(1.45\times10^{14}\text{ rad/s})t - (4.84\times10^{5}\text{ rad/m})x]$$

33-11

a) $E_{max} = cB_{max} \Rightarrow B_{max} = \frac{E_{max}}{c} = \frac{0.0800\text{ V/m}}{3.00\times10^{8}\text{ m/s}} = \underline{2.67\times10^{-10}\text{ T}}$

b) $I = S_{av} = \frac{E_{max}B_{max}}{2\mu_0} = \frac{(0.0800\text{ V/m})(2.67\times10^{-10}\text{ T})}{2(4\pi\times10^{-7}\text{ T}\cdot\text{m/A})} = 8.50\times10^{-6}\text{ W/m}^2$

I gives the power per unit area. As in Example 33-3 surround the antenna with a sphere of radius $R = 50.0\times10^{3}$ m. The upper half of this sphere has area $A = 2\pi R^2$. All the power radiated passes through this surface, so the power per unit area at this surface is given by $I = \frac{P}{2\pi R^2}$.

Thus $P = 2\pi R^2 I = 2\pi(50.0\times10^{3}\text{ m})^2(8.50\times10^{-6}\text{ W/m}^2) = 1.34\times10^{5}\text{ W} = \underline{134\text{ kW}}$.

c) The discussion in part (b) shows that since $P = I\,2\pi R^2$, IR^2 is constant

$\Rightarrow I_1R_1^2 = I_2R_2^2$

But $I = \frac{E_{max}^2}{2\mu_0 c}$, so $E_{max,1}^2 R_1^2 = E_{max,2}^2 R_2^2 \Rightarrow E_{max,1}R_1 = E_{max,2}R_2$

$R_2 = R_1\frac{E_{max,1}}{E_{max,2}} = 50.0\text{ km}\left(\frac{0.0800\text{ V/m}}{0.0400\text{ V/m}}\right) = \underline{100.0\text{ km}}$

33-15

a) By Eq. (33-27) the average momentum density is $\frac{dp}{dV} = \frac{S_{av}}{c^2} = \frac{I}{c^2}$

$\frac{dp}{dV} = \frac{0.78\times10^{3}\text{ W/m}^2}{(3.00\times10^{8}\text{ m/s})^2} = 8.7\times10^{-15}\,\frac{\text{kg}\cdot\text{m/s}}{\text{m}^3} = \underline{8.7\times10^{-15}\text{ kg/m}^2\cdot\text{s}}$

b) By Eq. (33-28) the average momentum flow rate per unit area is

$\frac{S_{av}}{c} = \frac{I}{c} = \frac{0.78\times10^{3}\text{ W/m}^2}{3.00\times10^{8}\text{ m/s}} = \underline{2.6\times10^{-6}\text{ Pa}}$

33-17

a) Eq. (33-29): $p_{rad} = \frac{I}{c} = \frac{935\text{ W/m}^2}{2.998\times10^{8}\text{ m/s}} = \underline{3.12\times10^{-6}\text{ Pa}}\left(\frac{1\text{ atm}}{1.013\times10^{5}\text{ Pa}}\right) = \underline{3.08\times10^{-11}\text{ atm}}$

b) Eq. (33-30): $p_{rad} = \frac{2I}{c} = 2(3.12\times10^{-6}\text{ Pa}) = \underline{6.24\times10^{-6}\text{ Pa}} = \underline{6.16\times10^{-11}\text{ atm}}$

33-19

$K = 1.83$, $K_m = 1.36$, $B_{max} = 4.80\times10^{-9}\text{ T}$

$E_{max} = vB_{max}$

By Eq. (33-32) $v = \frac{c}{\sqrt{KK_m}}$, so $E_{max} = \frac{cB_{max}}{\sqrt{KK_m}} = \frac{(2.998\times10^{8}\text{ m/s})(4.80\times10^{-9}\text{ T})}{\sqrt{(1.83)(1.36)}} = \underline{0.912\text{ V/m}}$

33-21

a) $\lambda = \frac{v}{f} = \frac{2.48\times10^8\ \text{m/s}}{20.0\times10^6\ \text{Hz}} = \underline{12.4\ \text{m}}$

b) $\lambda = \frac{c}{f} = \frac{3.00\times10^8\ \text{m/s}}{20.0\times10^6\ \text{Hz}} = \underline{15.0\ \text{m}}$

c) $n = \frac{c}{v} = \frac{3.00\times10^8\ \text{m/s}}{2.48\times10^8\ \text{m/s}} = \underline{1.21}$

d) $n = \sqrt{KK_m} \cong \sqrt{K} \Rightarrow K = n^2 = (1.21)^2 = \underline{1.46}$

33-23

a) By Eq. (33-41) we see that the nodal planes of the $\vec{B}$ field are a distance $\frac{\lambda}{2}$ apart, so $\frac{\lambda}{2} = 4.00\ \text{mm} \Rightarrow \lambda = \underline{8.00\ \text{mm}}$.

b) By Eq. (33-40) we see that the nodal planes of the $\vec{E}$ field are also a distance $\frac{\lambda}{2}$ $= \underline{4.00\ \text{mm}}$ apart.

c) $v = f\lambda = (2.00\times10^{10}\ \text{Hz})(8.00\times10^{-3}\ \text{m}) = \underline{1.60\times10^8\ \text{m/s}}$

33-25

a) The distance between adjacent nodal planes of $\vec{B}$ is $\frac{\lambda}{2}$. There is an antinodal plane of $\vec{B}$ midway between any two adjacent nodal planes, so the distance between a nodal plane and an adjacent antinodal plane is $\frac{\lambda}{4}$.

$\lambda = \frac{v}{f} = \frac{2.40\times10^8\ \text{m/s}}{1.70\times10^{11}\ \text{Hz}} = 1.412\times10^{-3}\ \text{m}$

$\frac{\lambda}{4} = \frac{1.412\times10^{-3}\ \text{m}}{4} = 3.53\times10^{-4}\ \text{m} = \underline{0.353\ \text{mm}}$

b) The nodal planes of $\vec{E}$ are at $x = 0, \frac{\lambda}{2}, \lambda, \frac{3\lambda}{2}, \ldots$, so the antinodal planes of $\vec{E}$ are at $x = \frac{\lambda}{4}, \frac{3\lambda}{4}, \frac{5\lambda}{4}, \ldots$. The nodal planes of $\vec{B}$ are at $\frac{\lambda}{4}, \frac{3\lambda}{4}, \frac{5\lambda}{4}, \ldots$, so the antinodal planes of $\vec{B}$ are at $\frac{\lambda}{2}, \lambda, \ldots$.

The distance between adjacent antinodal planes of $\vec{E}$ and antinodal planes of $\vec{B}$ is therefore $\frac{\lambda}{4} = \underline{0.353\ \text{mm}}$.

c) From Eqs. (33-40) and (33-41) the distance between adjacent nodal planes of $\vec{E}$ and $\vec{B}$ is $\frac{\lambda}{4} = \underline{0.353\ \text{mm}}$.

Problems

33-31

Eq. (33-12): $\frac{\partial E_y}{\partial x} = -\frac{\partial B_z}{\partial t}$

Take $\frac{\partial}{\partial t} \Rightarrow \frac{\partial^2 E_y}{\partial x\,\partial t} = -\frac{\partial^2 B_z}{\partial t^2}$

Eq. (33-14): $-\frac{\partial B_z}{\partial x} = \epsilon_0 \mu_0 \frac{\partial E_y}{\partial t}$

33-31 (cont)

Take $\frac{\partial}{\partial x} \Rightarrow -\frac{\partial^2 B_z}{\partial x^2} = \epsilon_0 \mu_0 \frac{\partial^2 E_y}{\partial t \partial x} \Rightarrow \frac{\partial^2 E_y}{\partial t \partial x} = -\frac{1}{\epsilon_0 \mu_0}\frac{\partial^2 B_z}{\partial x^2}$

But $\frac{\partial^2 E_y}{\partial x \partial t} = \frac{\partial^2 E_y}{\partial t \partial x}$ (the order in which the partial derivatives are taken doesn't change the result)

So $-\frac{\partial^2 B_z}{\partial t^2} = -\frac{1}{\epsilon_0 \mu_0}\frac{\partial^2 B_z}{\partial x^2}$

$\frac{\partial^2 B_z}{\partial x^2} = \epsilon_0 \mu_0 \frac{\partial^2 B_z}{\partial t^2}$, as was to be shown

33-35

a) I gives the energy flow per unit time per unit area

$I = \frac{1}{A}\frac{dU}{dt} \Rightarrow \frac{dU}{dt} = AI$

$I = \frac{E_{max}^2}{2\mu_0 c} = \frac{(0.0350\ \text{V/m})^2}{2(4\pi \times 10^{-7}\ \text{T·m/A})(3.00 \times 10^8\ \text{m/s})} = 1.625 \times 10^{-6}\ \text{W/m}^2$

$\frac{dU}{dt} = AI = (6.00 \times 10^{-4}\ \text{m}^2)(1.625 \times 10^{-6}\ \text{W/m}^2) = 9.75 \times 10^{-10}\ \text{W}$

The energy incident on the mirror in 1.00 s is $(9.75 \times 10^{-10}\ \text{W})(1.00\ \text{s}) = \underline{9.75 \times 10^{-10}\ \text{J.}}$

b) The light is reflected by the mirror, so the average pressure is

$\frac{2I}{c} = \frac{2(1.625 \times 10^{-6}\ \text{W/m}^2)}{3.00 \times 10^8\ \text{m/s}} = \underline{1.08 \times 10^{-14}\ \text{Pa}}$

c) Surround the light bulb with a spherical surface of radius $R = 4.00\ \text{m}$ and surface area $A = 4\pi R^2$. All the power radiated by the bulb passes through this surface, so $I = \frac{P}{A} = \frac{P}{4\pi R^2}$.

$P = 4\pi R^2 I = 4\pi (4.00\ \text{m})^2 (1.625 \times 10^{-6}\ \text{W/m}^2) = \underline{3.27 \times 10^{-4}\ \text{W}}$

33-37

a) The intensity is the power per unit area:

$I = \frac{P}{A} = \frac{3.60 \times 10^{-3}\ \text{W}}{\pi (2.00 \times 10^{-3}\ \text{m})^2} = 286.5\ \text{W/m}^2$

$I = \frac{E_{max}^2}{2\mu_0 c} \Rightarrow E_{max} = \sqrt{2\mu_0 c I} = \sqrt{2(4\pi \times 10^{-7}\ \text{T·m/A})(3.00 \times 10^8\ \text{m/s})(286.5\ \text{W/m}^2)} = \underline{465\ \text{V/m}}$

$B_{max} = \frac{E_{max}}{c} = \frac{465\ \text{V/m}}{3.00 \times 10^8\ \text{m/s}} = \underline{1.55 \times 10^{-6}\ \text{T}}$

b) The energy density in the electric field is (Eq. 25-11) $u_E = \frac{1}{2}\epsilon_0 E^2$. $E = E_{max}\sin(\omega t - kx)$ and the average value of $\sin^2(\omega t - kx)$ is $\frac{1}{2}$. The average energy density in the electric field then is $u_{E,av} = \frac{1}{4}\epsilon_0 E_{max}^2 = \frac{1}{4}(8.85 \times 10^{-12}\ \text{C}^2/\text{N·m}^2)(465\ \text{V/m})^2 = 4.78 \times 10^{-7}\ \text{J/m}^3$.

The energy density in the magnetic field is (Eq. 31-10) $u_B = \frac{B^2}{2\mu_0}$. The average value is $u_{B,av} = \frac{B_{max}^2}{4\mu_0} = \frac{(1.55 \times 10^{-6}\ \text{T})^2}{4(4\pi \times 10^{-7}\ \text{T·m/A})} = 4.78 \times 10^{-7}\ \text{J/m}^3$.

33-37 (cont)

c) The total energy in this length of beam is the total energy density $u_{av} = u_{E,av} + u_{B,av} = 9.56 \times 10^{-7}\ \mathrm{J/m^3}$ times the volume of this part of the beam

$\Rightarrow u = u_{av} LA = (9.56 \times 10^{-7}\ \mathrm{J/m^3})(0.500\ \mathrm{m})\pi(2.00 \times 10^{-3}\ \mathrm{m})^2 = \underline{6.00 \times 10^{-12}\ \mathrm{J}}$

This quantity can also be calculated as the power output times the time it takes the light to travel $L = 0.500$ m:

$U = P\left(\frac{L}{c}\right) = (3.60 \times 10^{-3}\ \mathrm{W})\left(\frac{0.500\ \mathrm{m}}{3.00 \times 10^8\ \mathrm{m/s}}\right) = 6.00 \times 10^{-12}\ \mathrm{J}$, which checks.

33-43

a) The direction of $\vec{E}$ is parallel to the axis of the cylinder, in the direction of the current. From Eq. 26-7, $E = \rho J = \rho \frac{I}{\pi a^2}$. (E is uniform across the cross section of the conductor.)

b) Cross-sectional view of the conductor; take the current to be coming out of the page.

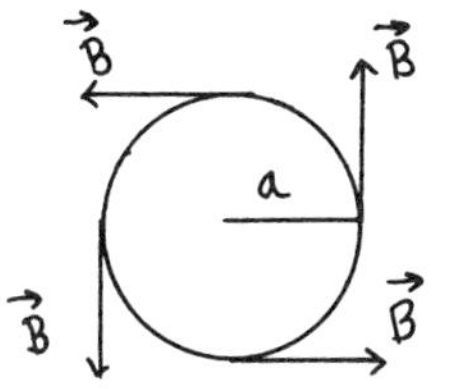

Apply Ampere's law to a circle of radius a.

$\oint \vec{B} \cdot d\vec{l} = B(2\pi a)$

$I_{encl} = I$

$\oint \vec{B} \cdot d\vec{l} = \mu_0 I_{encl} \Rightarrow B(2\pi a) = \mu_0 I$

$B = \frac{\mu_0 I}{2\pi a}$

The direction of $\vec{B}$ is counterclockwise around the circle.

c)

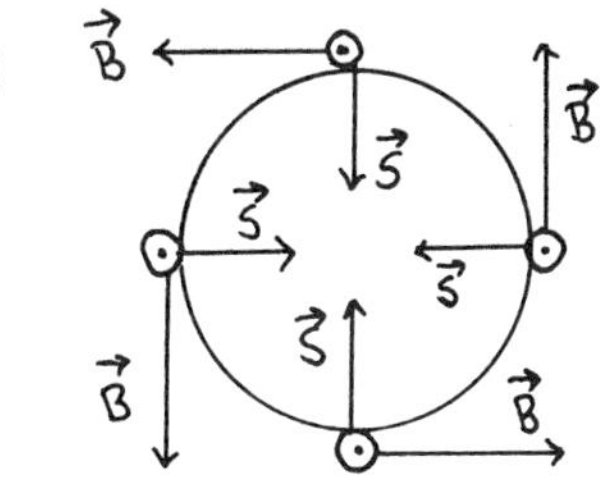

The direction of $\vec{S} = \frac{1}{\mu_0} \vec{E} \times \vec{B}$ is radially inward.

$S = \frac{1}{\mu_0} EB = \frac{1}{\mu_0}\left(\frac{\rho I}{\pi a^2}\right)\left(\frac{\mu_0 I}{2\pi a}\right) = \frac{\rho I^2}{2\pi^2 a^3}$

d) Since S is constant over the surface of the conductor, the rate of energy flow P is given by S times the surface area of a length l of the conductor:

$P = SA = S(2\pi a l) = \frac{\rho I^2}{2\pi^2 a^3}(2\pi a l) = \frac{\rho l I^2}{\pi a^2}$.

But $R = \frac{\rho l}{\pi a^2}$, so the result from the Poynting vector is $P = RI^2$. This agrees with $P_R = I^2 R$, the rate at which electrical energy is being dissipated by the resistance of the wire.

Since $\vec{S}$ is radially inward at the surface of the wire and has magnitude equal to the rate at which electrical energy is being dissipated in the wire, this energy can be thought of as entering through the cylindrical sides of the conductor.

33-45

The magnitude of the induced emf is given by Faraday's law: $|\mathcal{E}| = \left|\frac{d\Phi_B}{dt}\right|$.
$\Phi_B = B\pi R^2$, where $R = 0.200\text{ m}$ is the radius of the loop. (This assumes that the magnetic field is uniform across the loop, an excellent approximation.)

$|\mathcal{E}| = \pi R^2 \left|\frac{dB}{dt}\right|$

$B = B_{max}\sin(\omega t - kx) \Rightarrow \frac{dB}{dt} = B_{max}\,\omega\cos(\omega t - kx)$

The maximum value of $\frac{dB}{dt}$ is $B_{max}\omega$, so $|\mathcal{E}|_{max} = \pi R^2 B_{max}\,\omega$.

$R = 0.200\text{ m}$, $\omega = 2\pi f = 2\pi(10.0\times10^6\text{ Hz}) = 6.283\times10^7\text{ rad/s}$

Calculate the intensity I at this distance from the source, and from that the magnetic field amplitude B_{max}: $I = \frac{P}{4\pi r^2} = \frac{2.75\times10^6\text{ W}}{4\pi(500\text{ m})^2} = 0.8754\text{ W/m}^2$.

$$I = \frac{E_{max}^2}{2\mu_0 c} = \frac{(cB_{max})^2}{2\mu_0 c} = \frac{c}{2\mu_0}B_{max}^2$$

$$B_{max} = \sqrt{\frac{2\mu_0 I}{c}} = \sqrt{\frac{2(4\pi\times10^{-7}\text{ T}\cdot\text{m/A})(0.8754\text{ W/m}^2)}{3.00\times10^8\text{ m/s}}} = 8.564\times10^{-8}\text{ T}.$$

Then $|\mathcal{E}|_{max} = \pi R^2 B_{max}\,\omega = \pi(0.200\text{ m})^2(8.564\times10^{-8}\text{ T})(6.283\times10^7\text{ rad/s}) = \underline{0.676\text{ V}}$

33-47

The average time rate of change of the momentum of the light emitted by the flashlight (the momentum flow rate in the beam) is $\frac{dp}{dt} = \frac{AI}{c}$, where A is the area of the beam and I is the intensity. But $IA = P$ is the power output of the flashlight, so $\frac{dp}{dt} = \frac{P}{c}$.

By Newton's second law $\frac{dp}{dt}$ equals the force F exerted by the light on the astronaut, so $F_{av} = \frac{P}{c}$. Then $F = ma$ gives $ma = \frac{P}{c}$.

$$a = \frac{P}{mc} = \frac{100\text{ W}}{(200\text{ kg})(3.00\times10^8\text{ m/s})} = 1.67\times10^{-9}\text{ m/s}^2$$

Now that we have the acceleration we can use a constant-acceleration kinematic equation to find the time: $x - x_0 = \cancel{v_0 t}^{\,0} + \frac{1}{2}at^2$

$$t = \sqrt{\frac{2(x-x_0)}{a}} = \sqrt{\frac{2(12.0\text{ m})}{1.67\times10^{-9}\text{ m/s}^2}} = (1.199\times10^5\text{ s})\left(\frac{1\text{ h}}{3600\text{ s}}\right) = \underline{33.3\text{ h}}$$

33-49

a) The gravitational force is $F_g = G\frac{mM}{r^2}$. The mass of the dust particle is $m = \rho V = \rho\frac{4}{3}\pi R^3$. Thus $F_g = \frac{4\rho G\pi MR^3}{3r^2}$.

b) For a totally absorbing surface $p_{rad} = \frac{I}{c}$.

If L is the power output of the sun, the intensity of the solar radiation a distance r from the sun is $I = \frac{L}{4\pi r^2}$.

$p_{rad} = \frac{L}{4\pi c r^2}$

The force F_{rad} that corresponds to p_{rad} is in the direction of propagation of the radiation, so $F_{rad} = p_{rad}A_\perp$, where $A_\perp = \pi R^2$ is the component of area of

33-49 (cont)

the particle perpendicular to the radiation propagation direction. Thus

$$F_{rad} = \left(\frac{L}{4\pi c r^2}\right)(\pi R^2) = \frac{LR^2}{4cr^2}.$$

c) $F_g = F_{rad}$

$$\frac{4\rho G\pi M R^3}{3r^2} = \frac{LR^2}{4cr^2}$$

$$\left(\frac{4\rho G\pi M}{3}\right)R = \frac{L}{4c}$$

$$R = \frac{3L}{16 c\rho G\pi M} = \frac{3\,(3.9\times10^{26}\text{ W})}{16(2.998\times10^8\text{ m/s})(3000\text{ kg/m}^3)(6.673\times10^{-11}\text{ N}\cdot\text{m}^2/\text{kg}^2)\pi(1.99\times10^{30}\text{ kg})} = 2.0\times10^{-7}\text{ m} = \underline{0.20\ \mu\text{m}}$$

The gravitation force and the radiation force both have a r^{-2} dependence on the distance from the sun, so this distance divides out in the calculation of R.

d) $$\frac{F_{rad}}{F_g} = \left(\frac{LR^2}{4cr^2}\right)\left(\frac{3r^2}{4\rho G\pi M R^3}\right) = \frac{3L}{16 c\rho G\pi M R}$$

If $R < 0.20\ \mu\text{m}$ then $F_{rad} > F_g$ and the radiation force will drive the particles out of the solar system.

CHAPTER 34

Exercises 3, 7, 9, 11, 13, 17, 19

Problems 29, 31, 33, 35, 37, 39, 43, 45, 47

Exercises

34-3

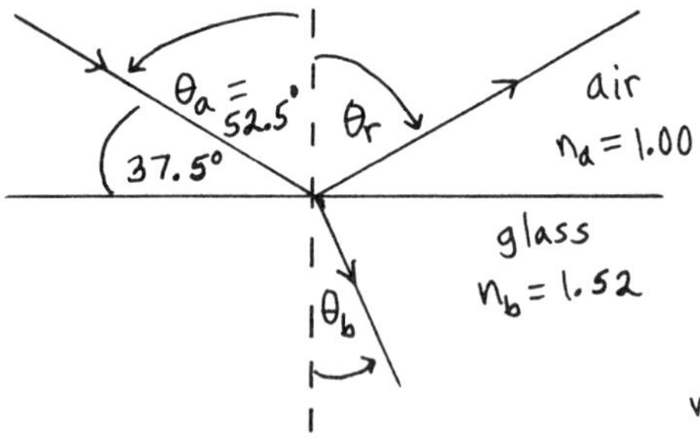

a) $\theta_r = \theta_a = 52.5°$

$\Rightarrow$ The reflected ray makes an angle of $90.0° - \theta_r = \underline{37.5°}$ with the surface of the glass.

b) $n_a \sin\theta_a = n_b \sin\theta_b$, where the angles are measured from the normal to the interface.

$$\sin\theta_b = \frac{n_a \sin\theta_a}{n_b} = \frac{(1.00)(\sin 52.5°)}{1.52} = 0.5219$$

$$\theta_b = 31.5°$$

The refracted rays makes an angle of $90.0° - \theta_b = \underline{58.5°}$ with the surface of the glass.

34-7

a) Eq. (34-1): $n = \frac{c}{v} \Rightarrow v = \frac{c}{n} = \frac{2.998 \times 10^8 \text{ m/s}}{1.62} = \underline{1.85 \times 10^8 \text{ m/s}}$

b) Eq. (34-5): $\lambda = \frac{\lambda_0}{n} = \frac{500 \text{ nm}}{1.62} = \underline{309 \text{ nm}}$

34-9

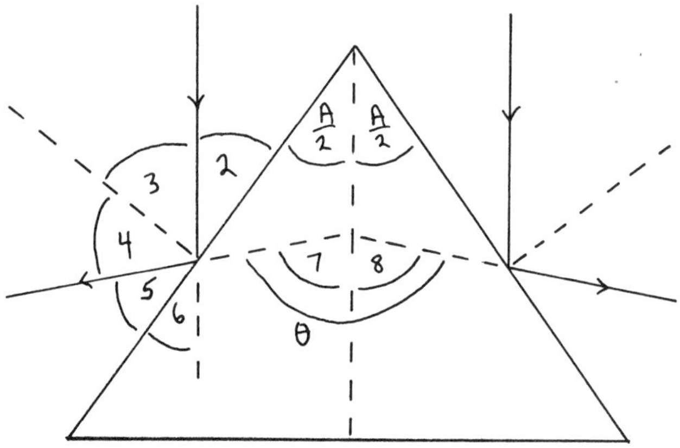

Angle θ is angle 7 + angle 8. We are asked to show that $\theta = 2A$. By symmetry angle 7 = angle 8, so we need to show that angle 7 = A.

From the sketch we see that angle 5 + angle 6 = angle 7 and that angle 2 = $\frac{A}{2}$.
But angle 2 = angle 6, so angle 6 = $\frac{A}{2}$.

The law of reflection $\theta_a = \theta_r$ says that angle 3 = angle 4. But then angle 2 = angle 5. Thus angle 5 = $\frac{A}{2}$.

Then angle 7 = angle 5 + angle 6 = $\frac{A}{2} + \frac{A}{2} = A$, which completes the proof.

34-11

a) Define the index of refraction of air for sound waves to be 1.00. Then $n_{water} = \frac{v_{air}}{v_{water}}$ (Eq. 34-1) $\Rightarrow n_{water} = \frac{344\ m/s}{1320\ m/s} = 0.2606$. $n_{water} < n_{air}$; air has the larger index of refraction for sound waves since sound travels slower in air than it does in water.

b) $n_a \sin\theta_a = n_b \sin\theta_b$

$\theta_a = \theta_{crit}$ when $\sin\theta_b = 1 \Rightarrow \sin\theta_{crit} = \frac{n_b \sin 90°}{n_a} = \frac{n_{water}}{n_{air}} = \frac{0.2606}{1.00} = 0.2606$

$\theta_{crit} = \underline{15.1°}$

34-13

Use the critical angle to find the index of refraction of the liquid. Total internal reflection requires that the light be incident in the material with the larger n, in this case the liquid. Apply $n_a \sin\theta_a = n_b \sin\theta_b$ with a = liquid and b = air, so $n_a = n_{liq}$ and $n_b = 1.0$.

$\theta_a = \theta_{crit}$ when $\theta_b = 90°$, so $n_{liq} \sin\theta_{crit} = (1.0)\sin 90°$

$n_{liq} = \frac{1}{\sin\theta_{crit}} = \frac{1}{\sin 37.0°} = 1.662$

a) $n_a \sin\theta_a = n_b \sin\theta_b$ (a = liquid, b = air)

$\sin\theta_b = \frac{n_a \sin\theta_a}{n_b} = \frac{(1.662)\sin 24.0°}{1.0} = 0.6760 \Rightarrow \theta_b = \underline{42.5°}$

b) Now $n_a \sin\theta_a = n_b \sin\theta_b$ with a = air, b = liquid

$\sin\theta_b = \frac{n_a \sin\theta_a}{n_b} = \frac{(1.0)\sin 24.0°}{1.662} = 0.2447 \Rightarrow \theta_b = \underline{14.2°}$

34-17

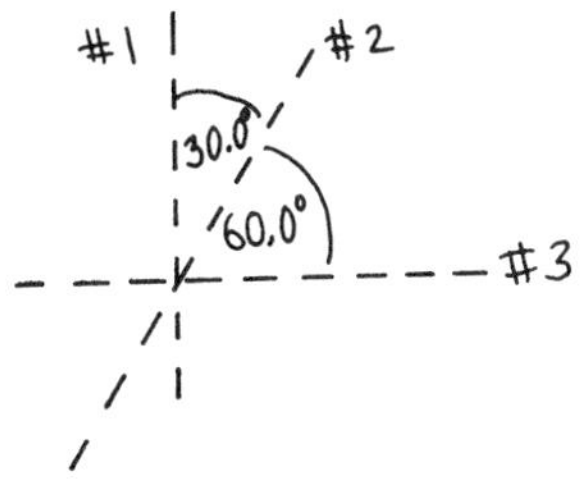

$I = I_{max}\cos^2\phi$ (Eq. 34-7)

After the first filter the intensity is $I_1 = \frac{1}{2}I_0$ and the light is linearly polarized along the axis of the first polarizer.

After the second filter the intensity is

$I_2 = I_1\cos^2\phi = (\frac{1}{2}I_0)(\cos 30.0°)^2 = 0.375\,I_0$

and the light is linearly polarized along the axis of the second polarizer.

After the third filter the intensity is

$I_3 = I_2\cos^2\phi = (0.375\,I_0)(\cos 60.0°)^2 = 0.0938\,I_0$

and the light is linearly polarized along the axis of the third polarizer.

b)

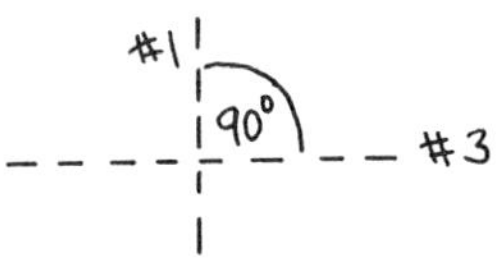

After the first filter the intensity is $I_1 = \frac{1}{2}I_0$ and the light is linearly polarized along the axis of the first polarizer.

34-17 (cont)

After the next filter the intensity is $I_3 = I_1 \cos^2\phi = \frac{1}{2} I_0 (\cos 90°)^2 = 0.$
No light is passed.

34-19

a) Reflected beam completely linearly polarized ⇒ the angle of incidence equals the polarizing angle ⇒ $\theta_p = 56.8°$

$\tan\theta_p = \frac{n_b}{n_a} \Rightarrow n_{glass} = n_{air} \tan\theta_p = (1.00)\tan 56.8° = \underline{1.53}$

b) $n_a \sin\theta_a = n_b \sin\theta_b$

$\sin\theta_b = \frac{n_a \sin\theta_a}{n_b} = \frac{1.00(\sin 56.8°)}{1.53} = 0.5469 \Rightarrow \theta_b = \underline{33.2°}$

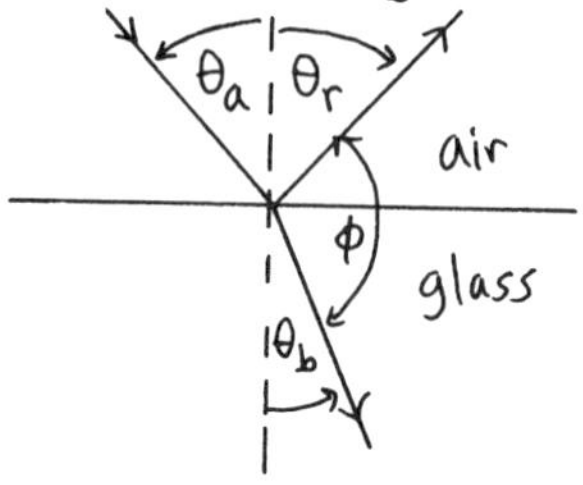

Note: $\phi = 180.0° - \theta_r - \theta_b$ and $\theta_r = \theta_a$
⇒ $\phi = 180.0° - 56.8° - 33.2° = 90.0°$; the reflected ray and the refracted ray are perpendicular to each other.

Problems

34-29

Before the liquid is poured in:

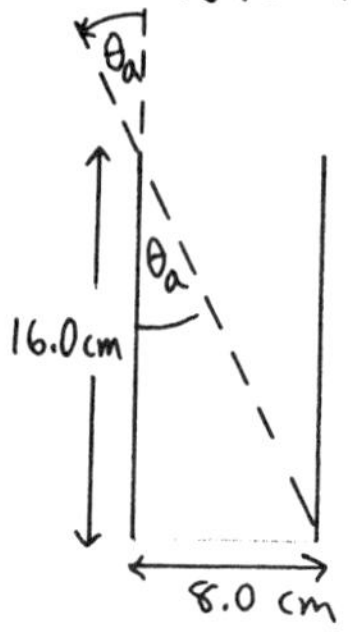

$\tan\theta_a = \frac{8.0\text{ cm}}{16.0\text{ cm}} = 0.500 \Rightarrow \theta_a = 26.57°$

After the liquid is poured in, θ_a is the same and the refracted ray passes through the center of the bottom of the glass:

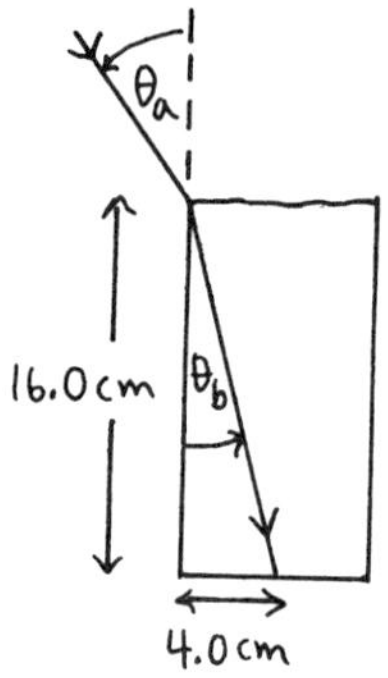

$\tan\theta_b = \frac{4.0\text{cm}}{16.0\text{cm}} = 0.250 \Rightarrow \theta_b = 14.04°$

Then can use Snell's law to find n_b, the refractive index of the liquid:

$$n_a \sin\theta_a = n_b \sin\theta_b$$

$$n_b = \frac{n_a \sin\theta_a}{\sin\theta_b} = \frac{(1.00)(\sin 26.57°)}{\sin 14.04°} = \underline{1.84}$$

34-31

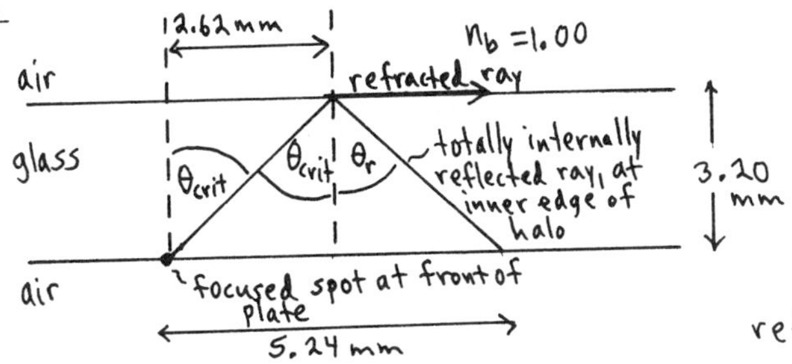

From the distances given in the sketch

$$\tan\theta_{crit} = \frac{2.62\text{ mm}}{3.20\text{ mm}} = 0.8188$$

$$\theta_{crit} = 39.3°$$

Apply Snell's law to the total internal reflection to find the refractive index of the glass:

$$n_a \sin\theta_a = n_b \sin\theta_b$$

$$n_{glass}\sin\theta_{crit} = 1.00 \sin 90°$$

$$n_{glass} = \frac{1}{\sin\theta_{crit}} = \frac{1}{\sin 39.3°} = \underline{1.58}$$

34-33

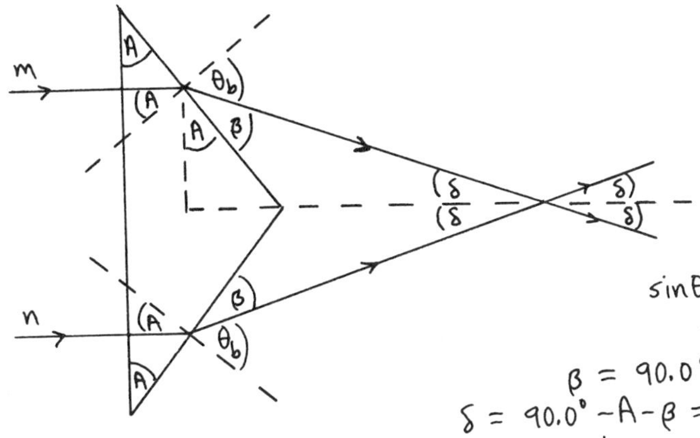

Apply Snell's law to the refraction of each ray as it emerges from the glass. The angle of incidence equals the angle $A = 30.0°$.

$$n_a \sin\theta_a = n_b \sin\theta_b$$

$$n_{glass}\sin 30.0° = 1.00 \sin\theta_b$$

$$\sin\theta_b = n_{glass}\sin 30.0° = 1.58 \sin 30.0° = 0.790$$

$$\Rightarrow \theta_b = 52.19°$$

$$\beta = 90.0° - \theta_b = 37.81°$$

$$\delta = 90.0° - A - \beta = 90.0° - 30.0° - 37.81° = 22.19°$$

The angle between the two rays is $2\delta = 44.4°$.

34-35

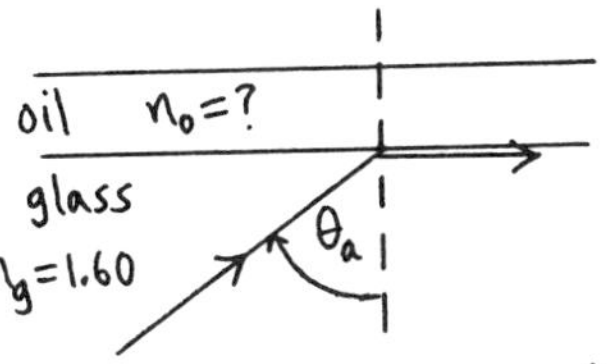

$$n_a \sin\theta_a = n_b \sin\theta_b$$

$$n_a = n_g = 1.60$$

$$n_b = n_o = ?$$

$$\theta_a = 51.0°$$

Maximum n_o for total reflection $\Rightarrow \theta_b = 90°$

$$(1.60)\sin 51.0° = n_o \sin 90°$$

$$n_o = (1.60)\sin 51.0° = \underline{1.24}$$

34-37

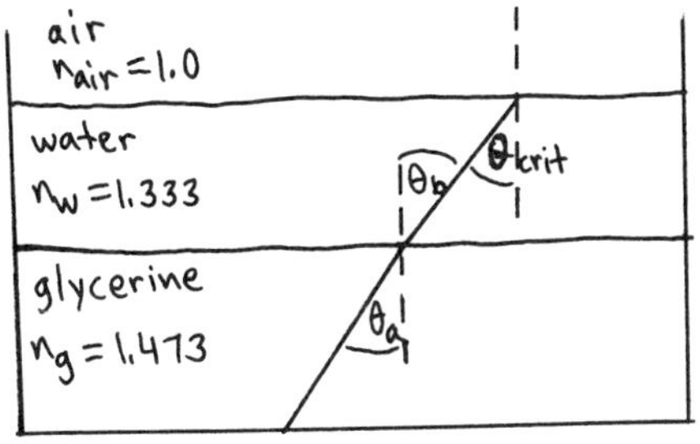

We want to find the incident angle θ_a at the glycerine – water interface that causes the incident angle at the water-air interface to be the critical angle.

water–air interface:

$$n_w \sin\theta_{crit} = 1.0 \sin 90°$$

34-37 (cont)

$n_w \sin\theta_{crit} = 1.0 \Rightarrow \sin\theta_{crit} = \frac{1}{n_w}$

But from the diagram we see that $\theta_b = \theta_{crit}$, so $\sin\theta_b = \frac{1}{n_w}$

glycerine – water interface:

$n_g \sin\theta_a = n_w \sin\theta_b$

But $\sin\theta_b = \frac{1}{n_w}$ so $n_g \sin\theta_a = 1.0$

$\sin\theta_a = \frac{1}{n_g} = \frac{1}{1.473} = 0.6789 \Rightarrow \underline{\theta_a = 42.8^\circ}$

Note: This is the critical angle for a glycerine – air interface; the answer would be the same if the water layer wasn't there!

34-39

a)

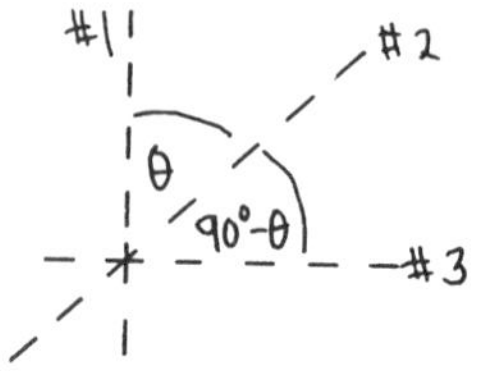

1st filter: $I_1 = \frac{1}{2} I_0$

2nd filter: $I_2 = I_1 (\cos\theta)^2 = \frac{1}{2} I_0 \cos^2\theta$

3rd filter: $I_3 = I_2 (\cos(90^\circ - \theta))^2 = \frac{1}{2} I_0 \cos^2\theta \cos^2(90^\circ - \theta)$

$\cos(90^\circ - \theta) = \cos(\theta - 90^\circ) = \sin\theta$ (Using the trig identities in Appendix B.)

Therefore, $I_3 = \frac{1}{2} I_0 \cos^2\theta \sin^2\theta$

But $\cos\theta \sin\theta = \frac{1}{2}\sin 2\theta$ (Appendix B again)

$\Rightarrow I_3 = \frac{1}{8} I_0 (\sin 2\theta)^2$

b) I_3 maximum $\Rightarrow \sin 2\theta = 1 \Rightarrow \underline{\theta = 45^\circ}$

34-43

a)

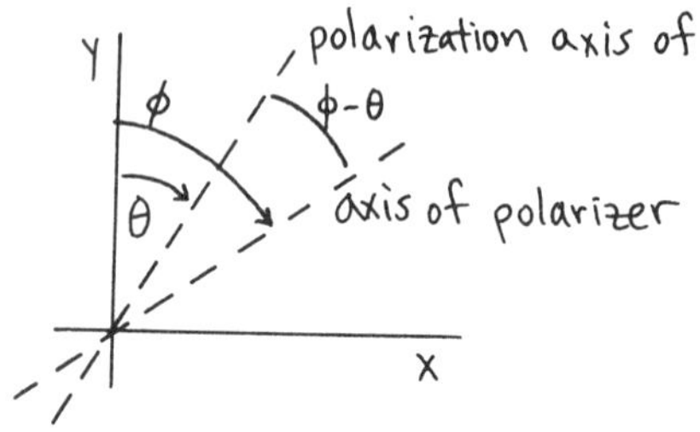

The polarizer passes $\frac{1}{2}$ of the intensity of the unpolarized component, independent of ϕ. Out of the intensity I_p of the polarized component the polarizer passes intensity $I_p \cos^2(\phi - \theta)$, where $\phi - \theta$ is the angle between the plane of polarization and the axis of the polarizer.

The total transmitted intensity is $I = \frac{1}{2} I_0 + I_p \cos^2(\phi - \theta)$. This is maximum when $\theta = \phi$ and from the table of data this occurs for ϕ between 30° and 40°, say at $35^\circ \Rightarrow \underline{\theta = 35^\circ}$.

Alternatively, the total transmitted intensity is minimum when $\phi - \theta = 90^\circ$ and from the data this occurs for $\phi = 125^\circ \Rightarrow \theta = \phi - 90^\circ = 125^\circ - 90^\circ = 35^\circ$, in agreement with the above.

b) $I = \frac{1}{2} I_0 + I_p \cos^2(\phi - \theta)$

Use data at two values of ϕ to determine the two constants I_0 and I_p. Use data where the I_p term is large ($\phi = 30^\circ$) and where it is small ($\phi = 130^\circ$) to have the greatest sensitivity to both I_0 and I_p:

34-43 (cont)

$\phi = 30° \Rightarrow 24.8\ W/m^2 = \frac{1}{2} I_o + I_p \cos^2(30° - 35°)$

$24.8\ W/m^2 = 0.500\ I_o + 0.9924\ I_p$

$\phi = 130° \Rightarrow 5.2\ W/m^2 = \frac{1}{2} I_o + I_p \cos^2(130° - 35°)$

$5.2\ W/m^2 = 0.500\ I_o + 0.0076\ I_p$

Subtract the second equation from the first

$\Rightarrow 19.6\ W/m^2 = 0.9848\ I_p$

$I_p = \underline{19.9\ W/m^2}$

And then $I_o = 2(5.2\ W/m^2 - 0.0076\,(19.9\ W/m^2)) = \underline{10.1\ W/m^2}$

34-45

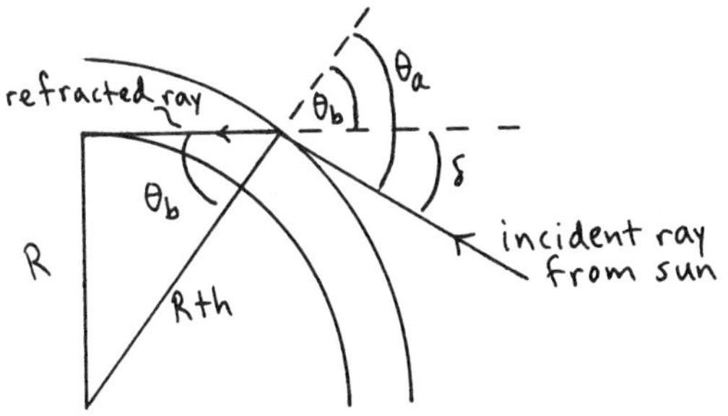

$\delta = \theta_a - \theta_b$

From the diagram $\sin\theta_b = \frac{R}{R+h} \Rightarrow \theta_b = \arcsin\left(\frac{R}{R+h}\right)$

Apply Snell's law to the refraction that occurs at the top of the atmosphere:

$n_a \sin\theta_a = n_b \sin\theta_b$ (a = vacuum of space, refractive index 1.0; b = atmosphere, refractive index n)

$\sin\theta_a = n \sin\theta_b = n\left(\frac{R}{R+h}\right) \Rightarrow \theta_a = \arcsin\left(\frac{nR}{R+h}\right)$

$\delta = \theta_a - \theta_b = \arcsin\left(\frac{nR}{R+h}\right) - \arcsin\left(\frac{R}{R+h}\right)$

b) $\frac{R}{R+h} = \frac{6.38 \times 10^6\ m}{6.38 \times 10^6\ m + 20 \times 10^3\ m} = 0.99688$

$\frac{nR}{R+h} = 1.0003\,(0.99688) = 0.99718$

$\theta_b = \arcsin\left(\frac{R}{R+h}\right) = 85.47°$

$\theta_a = \arcsin\left(\frac{nR}{R+h}\right) = 85.70°$

$\delta = \theta_a - \theta_b = 85.70° - 85.47° = \underline{0.23°}$

This is about the same as the angular radius of the sun.

34-47

a) The light travels a distance $\sqrt{h_1^2+x^2}$ in traveling from point A to the interface. Along this path the speed of the light is v_1, so the time it takes to travel this distance is $t_1 = \frac{\sqrt{h_1^2+x^2}}{v_1}$.

The light travels a distance $\sqrt{h_2^2+(l-x)^2}$ in traveling from the interface to point B. Along this path the speed of the light is v_2, so the time it takes to travel this distance is $t_2 = \frac{\sqrt{h_2^2+(l-x)^2}}{v_2}$.

The total time to go from A to B is $t = t_1 + t_2 = \frac{\sqrt{h_1^2+x^2}}{v_1} + \frac{\sqrt{h_2^2+(l-x)^2}}{v_2}$

b) $\frac{dt}{dx} = 0 = \frac{1}{v_1}\left(\frac{1}{2}\right)(h_1^2+x^2)^{-1/2}(2x) + \frac{1}{v_2}\left(\frac{1}{2}\right)(h_2^2+(l-x)^2)^{-1/2}2(l-x)$

$$\frac{x}{v_1\sqrt{h_1^2+x^2}} = \frac{l-x}{v_2\sqrt{h_2^2+(l-x)^2}}$$

Multiply both sides by $c \Rightarrow \frac{c}{v_1}\frac{x}{\sqrt{h_1^2+x^2}} = \frac{c}{v_2}\frac{l-x}{\sqrt{h_2^2+(l-x)^2}}$

$\frac{c}{v_1} = n_1$ and $\frac{c}{v_2} = n_2$ (Eq. 34-1)

From Fig. 34-40, $\sin\theta_1 = \frac{x}{\sqrt{h_1^2+x^2}}$ and $\sin\theta_2 = \frac{l-x}{\sqrt{h_2^2+(l-x)^2}}$

So $n_1 \sin\theta_1 = n_2 \sin\theta_2$, which is Snell's law.

CHAPTER 35

Exercises 1, 7, 9, 17, 19, 21, 25, 29, 31, 33

Problems 37, 39, 41, 45, 47, 51, 55, 57, 59, 61, 65, 69, 71, 73, 75

Exercises

35-1

Plane mirror: $s = -s'$ (Eq. 35-1) and $m = \frac{y'}{y} = -\frac{s'}{s} = +1$ (Eq. 35-2).

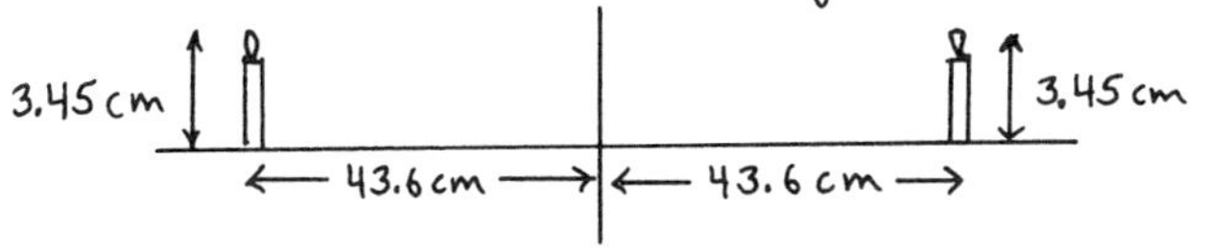

$s' = -s = -43.6$ cm

$|y'| = |m||y| = (+1)(3.45\text{ cm}) = 3.45$ cm

The image is 43.6 cm to the right of the mirror and is 3.45 cm tall.

35-7

concave $\Rightarrow R = +20.0$ cm ; $f = \frac{R}{2} = +10.0$ cm

a)

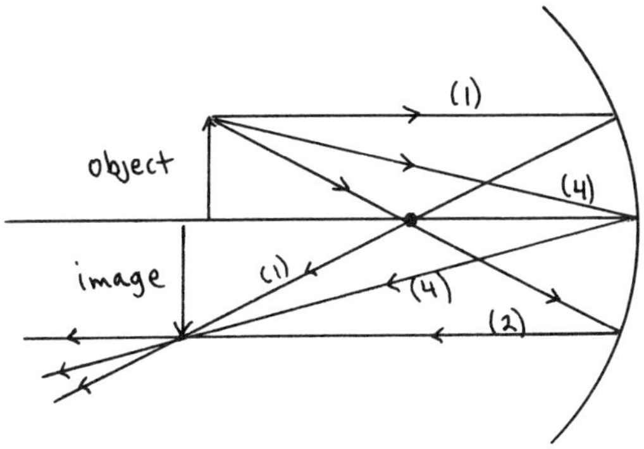

Three principal rays, numbered as in Sect. 35-4 are shown.

The principal ray diagram shows that the image is real, inverted, and enlarged.

b) $\frac{1}{s} + \frac{1}{s'} = \frac{1}{f}$

$\frac{1}{s'} = \frac{1}{f} - \frac{1}{s} = \frac{s-f}{sf} \Rightarrow s' = \frac{sf}{s-f} = \frac{(18.0\text{ cm})(10.0\text{ cm})}{18.0\text{ cm} - 10.0\text{ cm}} = +22.5\text{cm}$

$s' > 0 \Rightarrow$ real image, 22.5 cm to left of mirror vertex

$m = -\frac{s'}{s} = -\frac{22.5\text{cm}}{18.0\text{ cm}} = -1.25$ ($m < 0 \Rightarrow$ inverted image)

$|y'| = |m||y| = 1.25\,(0.400\text{ cm}) = 0.500$ cm

The image is 22.5 cm to the left of the mirror vertex. It is real, inverted, and 0.500 cm tall (enlarged). The calculation agrees with the image characterization from the principal ray diagram.

35-9

convex $\Rightarrow R = -24.0$ cm ; $f = \frac{R}{2} = -12.0$ cm

$s = 9.00$ cm, so $|s| < |f|$

35-9 (cont)

a)

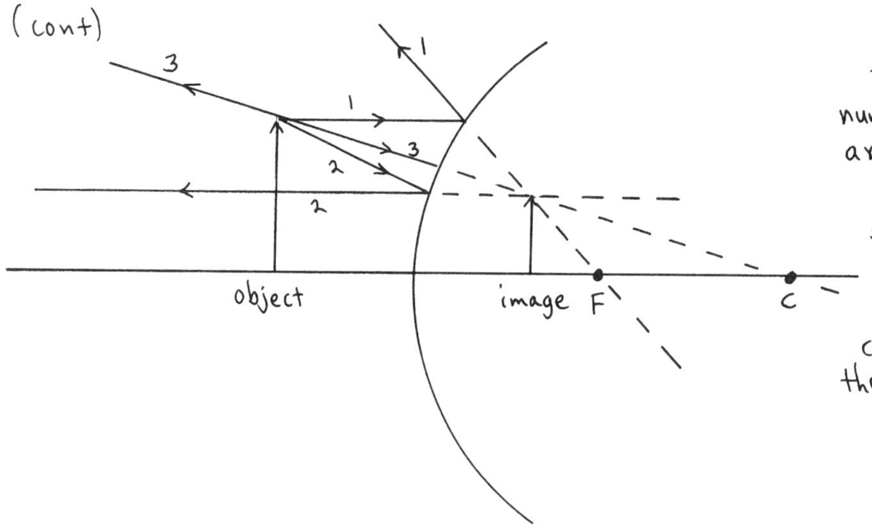

Three principal rays, numbered as in Sect. 35-4, are shown.

The principal ray diagram shows that the image is virtual, erect, reduced in size, and located closer to the mirror than the focal point.

b) $\frac{1}{s} + \frac{1}{s'} = \frac{1}{f}$

$\frac{1}{s'} = \frac{1}{f} - \frac{1}{s} = \frac{s-f}{sf} \Rightarrow s' = \frac{sf}{s-f} = \frac{(9.00\text{cm})(-12.0\text{ cm})}{9.00\text{cm} - (-12.0\text{ cm})} = -5.14\text{ cm}$

$s' < 0 \Rightarrow$ virtual image 5.14 cm to right of mirror vertex

$m = -\frac{s'}{s} = -\frac{-5.14\text{ cm}}{9.00\text{ cm}} = +0.571 \quad (m > 0 \Rightarrow \text{erect image})$

$|y'| = |m|\,|y| = (+0.571)(1.50\text{ cm}) = 0.86\text{ cm}$

The image is 5.14 cm to the right of the mirror vertex. It is 0.86 cm tall (reduced), erect, and virtual. The calculation agrees with the image characterization from the principal ray diagram.

35-17

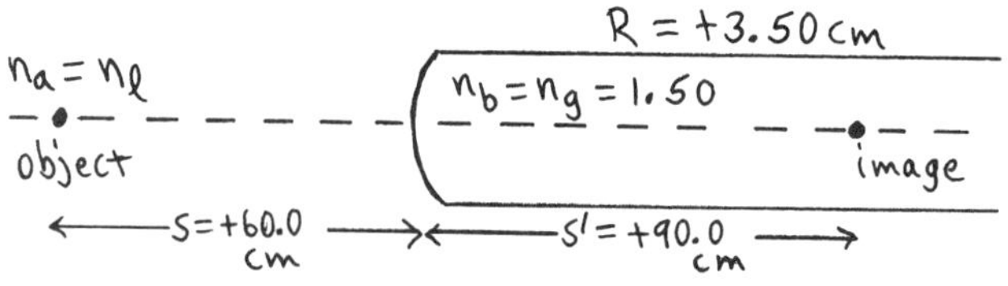

Let n_ℓ be the refractive index of the liquid. s, s', and R are all positive.

$\frac{n_a}{s} + \frac{n_b}{s'} = \frac{n_b - n_a}{R}$ (Eq. 35-11)

$\frac{n_\ell}{60.0\text{cm}} + \frac{1.50}{90.0\text{ cm}} = \frac{1.50 - n_\ell}{3.50\text{cm}}$

Multiply the equation by 1260 cm $\Rightarrow 21.0 n_\ell + 21.0 = 540.0 - 360.0 n_\ell$

$381.0 n_\ell = 519.0 \Rightarrow n_\ell = 1.36$

35-19

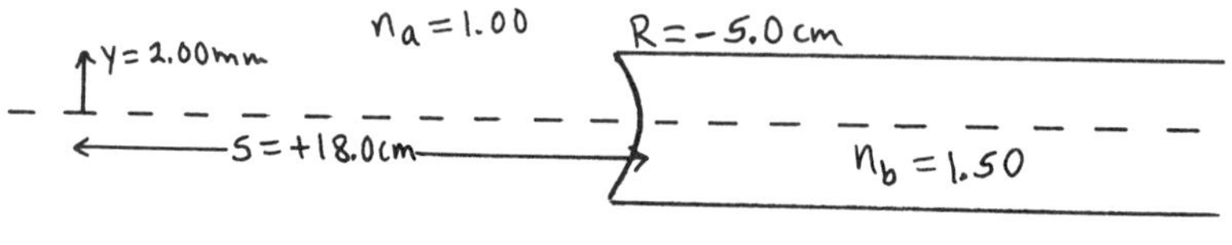

$\frac{n_a}{s} + \frac{n_b}{s'} = \frac{n_b - n_a}{R}$

35-19 (cont)

$$\frac{1.00}{18.0\text{ cm}} + \frac{1.50}{s'} = \frac{1.50-1.00}{-5.0\text{ cm}}$$

Multiply by 18.0 cm $\Rightarrow$ $1.00 + \dfrac{27.0\text{ cm}}{s'} = -1.80$

$$\frac{27.0\text{ cm}}{s'} = -2.80 \Rightarrow s' = -\frac{27.0\text{ cm}}{2.80} = -9.64\text{ cm}$$

Eq. (35-12): $m = -\dfrac{n_a s'}{n_b s} = -\dfrac{(1.00)(-9.64\text{ cm})}{(1.50)(+18.0\text{ cm})} = +0.357$

$|y'| = |m||y| = (0.357)(2.00\text{ mm}) = 0.71\text{ mm}$

The image is virtual ($s'<0$) and is 9.64 cm to the left of the vertex. The image is <u>erect</u> ($m>0$) and is 0.71 mm tall.

35-21

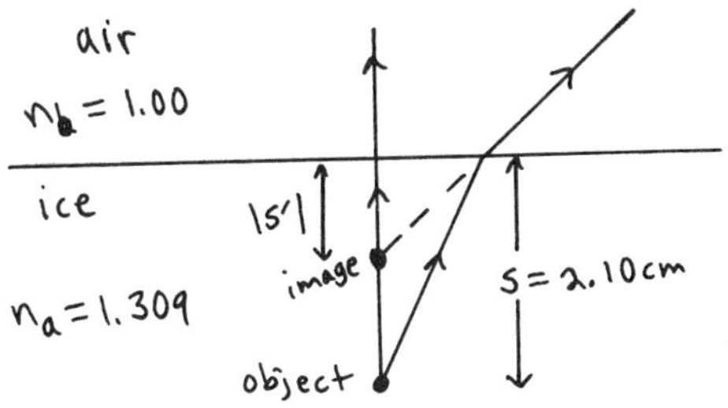

$\dfrac{n_a}{s} + \dfrac{n_b}{s'} = \dfrac{n_b - n_a}{R}$; $R \to \infty$ (flat surface), so

$$\frac{n_a}{s} + \frac{n_b}{s'} = 0$$

$$s' = -\frac{n_b s}{n_a} = -\frac{(1.00)(2.10\text{ cm})}{1.309} = -1.60\text{ cm}$$

The apparent depth is <u>1.60 cm.</u>

35-25

$f = -10.0$ cm

real image 14.0 cm to right of lens $\Rightarrow s' = +14.0$ cm

$$\frac{1}{s} + \frac{1}{s'} = \frac{1}{f} \Rightarrow \frac{1}{s} = \frac{1}{f} - \frac{1}{s'} = \frac{s'-f}{s'f}$$

$$s = \frac{s'f}{s'-f} = \frac{(+14.0\text{ cm})(-10.0\text{ cm})}{+14.0\text{ cm} - (-10.0\text{ cm})} = -\frac{140\text{ cm}}{24} = -5.83\text{ cm}$$

$$m = -\frac{s'}{s} = -\frac{14.0\text{ cm}}{-5.83\text{ cm}} = +2.40$$

$$m = \frac{y'}{y} \Rightarrow |y| = \frac{|y'|}{|m|} = \frac{1.00\text{ cm}}{2.40} = 0.42\text{ cm}$$

$s<0 \Rightarrow$ virtual object 5.83 cm to right of lens

$m>0 \Rightarrow$ image is erect with respect to the object

The height of the object is 0.42 cm.

35-29

$f = +6.00$ cm ($f>0$ since the lens is converging)

$m = -\dfrac{s'}{s}$; image is erect $\Rightarrow m>0$, so must have $s'<0$ (image is virtual)

$$m = \frac{y'}{y} \Rightarrow |m| = \frac{|y|}{|y'|} = \frac{35.0\text{ cm}}{5.00\text{ cm}} = 7.00$$

$s' = -ms = -(7.00)s$

Use this in $\dfrac{1}{s} + \dfrac{1}{s'} = \dfrac{1}{f} \Rightarrow \dfrac{1}{s} + \dfrac{1}{-7.00s} = \dfrac{1}{6.00\text{ cm}}$

35-29 (cont)

$$\frac{6}{7.00\,s} = \frac{1}{6.00\text{cm}} \Rightarrow s = \frac{6(6.00\text{cm})}{7.00} = 5.14\text{ cm}$$

$$s' = -(7.00)s = -(7.00)(5.14\text{cm}) = -36.0\text{cm}$$

The object is 5.14 cm to the left of the lens.
$s' < 0$, so the image is 36.0 cm to the left of the lens and is virtual.

35-31

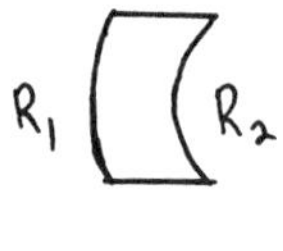

$R_1 = +4.00\text{cm}$
$R_2 = +2.50\text{cm}$
$n = 1.48$

$$\frac{1}{f} = (n-1)\left(\frac{1}{R_1} - \frac{1}{R_2}\right)$$

$$\frac{1}{f} = (0.48)\left(\frac{1}{4.00\text{cm}} - \frac{1}{2.50\text{cm}}\right)$$

$$f = -13.89\text{ cm}$$

$s = 18.0\text{cm}$

$$\frac{1}{s} + \frac{1}{s'} = \frac{1}{f} \Rightarrow \frac{1}{s'} = \frac{1}{f} - \frac{1}{s} = \frac{s-f}{sf}$$

$$s' = \frac{sf}{s-f} = \frac{(18.0\text{cm})(-13.89\text{cm})}{18.0\text{ cm} - (-13.89\text{ cm})} = -7.84\text{cm}$$

$s' < 0$ means that the image is virtual and is 7.84cm to the left of the lens, on the same side of the lens as the object.

35-33

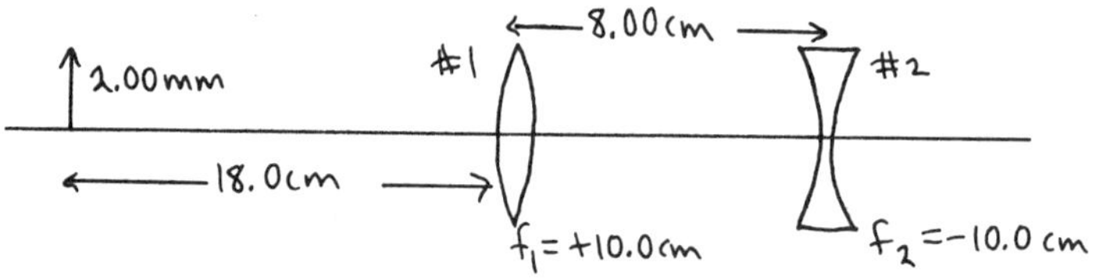

$$\frac{1}{s} + \frac{1}{s'} = \frac{1}{f} \Rightarrow s' = \frac{sf}{s-f}$$

$$m = -\frac{s'}{s}$$

image formed by lens #1

$$s_1' = \frac{s_1 f_1}{s_1 - f_1} = \frac{(18.0\text{cm})(10.0\text{cm})}{18.0\text{cm} - 10.0\text{cm}} = +22.5\text{cm}\,;\quad m_1 = -\frac{s_1'}{s_1} = -\frac{22.5\text{cm}}{18.0\text{cm}} = -1.25$$

This image serves as the object for the second lens. The image is 22.5cm to the right of lens #1 so is 22.5cm − 8.0 cm = 14.5cm to the right of lens #2. So for lens #2, $s_2 = -14.5$ cm.

image formed by lens #2

$$s_2' = \frac{s_2 f_2}{s_2 - f_2} = \frac{(-14.5\text{cm})(-10.0\text{cm})}{-14.5\text{cm} + 10.0\text{cm}} = -32.2\text{cm},\quad m_2 = -\frac{s_2'}{s_2} = -\frac{-32.2\text{cm}}{-14.5\text{cm}} = -2.22$$

a) $s_2' = -32.2$cm, so the final image is 32.2cm to the left of lens #2 and 32.2 cm − 8.0 cm = 24.8cm to the left of lens #1.

b) $s_2' < 0$ so the final image is virtual.

c) $m_{tot} = m_1 m_2 = (-1.25)(-2.22) = +2.78$
$y' = m_{tot}\,y = (2.78)(2.00\text{ mm}) = 5.56\text{ mm}$ (height of final image)

35-33 (cont)

$m_{tot} > 0 \Rightarrow$ final image is erect (with respect to original object)

($m_1 < 0$ so lens #1 inverts the image. But $m_2 < 0$ also, so lens #2 inverts the image, making it erect.)

Problems

35-37

For a plane mirror $s' = -s$. $v = \frac{ds}{dt}$, $v' = \frac{ds'}{dt}$, so $v' = v$. The velocities of the object and image relative to the mirror are equal in magnitude and opposite in direction. Thus both you and your image are approaching the mirror surface at 2.25 m/s, from opposite directions. Your image is therefore moving at 4.50 m/s relative to you.

35-39

a) Image is to be formed on screen $\Rightarrow$ real image; $s' > 0$. Mirror to screen distance is 3.20 m, so $s' = +3.20$ m.

$m = -\frac{s'}{s} < 0$ since both s and s' are positive.

$|m| = \frac{|y'|}{|y|} = \frac{35.0\text{ cm}}{0.500\text{ cm}} = 70.0 \Rightarrow m = -70.0$

Then $m = -\frac{s'}{s} \Rightarrow s = -\frac{s'}{m} = -\frac{3.20\text{ m}}{-70.0} = +0.0457\text{ m} = \underline{+4.57\text{ cm}}$

b) $\frac{1}{s} + \frac{1}{s'} = \frac{2}{R} \Rightarrow \frac{2}{R} = \frac{s+s'}{ss'}$

$R = 2\left(\frac{ss'}{s+s'}\right) = 2\left(\frac{(4.57\text{ cm})(320\text{ cm})}{320\text{ cm} + 4.57\text{ cm}}\right) = \underline{9.01\text{ cm}}$

Note that R is calculated to be positive, which is correct for concave mirror.

35-41

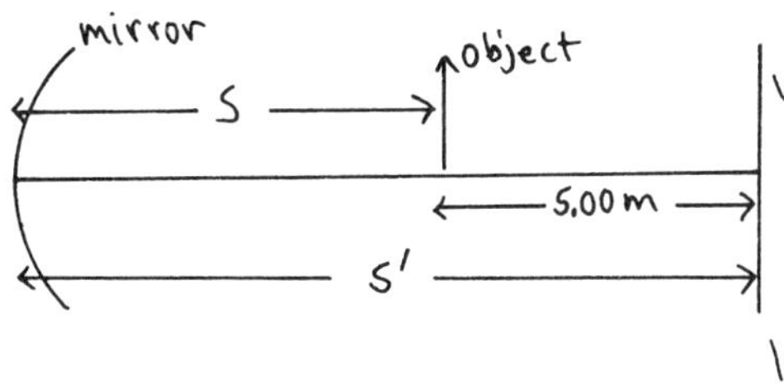

$s > 0$; image projected on wall $\Rightarrow s' > 0$

From the sketch $\boxed{s' = s + 5.00\text{ m.}}$

$m = -\frac{s'}{s}$ and $s > 0$, $s' > 0 \Rightarrow m < 0$

$|m| = \frac{|y'|}{|y|} = 2.75 \Rightarrow m = -2.75$

Then $m = -\frac{s'}{s} \Rightarrow -2.75 = -\frac{s'}{s} \Rightarrow \boxed{s' = 2.75s}$

Use this in the first equation $\Rightarrow$ $2.75s = s + 5.00\text{m}$

$s = \frac{5.00\text{m}}{1.75} = 2.86\text{ m}$

$s' = 2.75s = (2.75)(2.86\text{ m}) = 7.86\text{ m}$

The mirror should be $\underline{7.86\text{m}}$ from the wall.

$\frac{1}{s} + \frac{1}{s'} = \frac{2}{R} \Rightarrow \frac{2}{R} = \frac{s+s'}{ss'}$

$R = 2\left(\frac{ss'}{s+s'}\right) = 2\frac{(2.86\text{ m})(7.86\text{m})}{2.86\text{m} + 7.86\text{ m}} = \underline{4.19\text{ m}}$

35-41 (cont)

Note that R is calculated to be positive, which is the correct sign for a concave mirror.

35-45

a) convex $\Rightarrow R<0$; $R=-14.0\text{ cm}$; $f=\frac{R}{2}=-7.0\text{ cm}$

$$\frac{1}{s}+\frac{1}{s'}=\frac{1}{f} \Rightarrow \frac{1}{s'}=\frac{1}{f}-\frac{1}{s}=\frac{s-f}{sf} \Rightarrow s'=\frac{sf}{s-f}=\frac{(-7.0\text{cm})\,s}{s+7.0\text{cm}}$$

s is negative, so write as $s=-|s| \Rightarrow s'=+\frac{(7.0\text{ cm})|s|}{7.0\text{ cm}-|s|}$

Thus $s'>0$ (real image) for $|s|<7.0\text{cm}$. Since s is negative, this means $-7.0\text{ cm}<s<0$. A real image is formed if the virtual object is closer to the mirror vertex than the focus.

b) $m=-\frac{s'}{s}$; real image $\Rightarrow s'>0$; virtual object $\Rightarrow s<0$

Thus $m>0$ and the image is erect.

c)

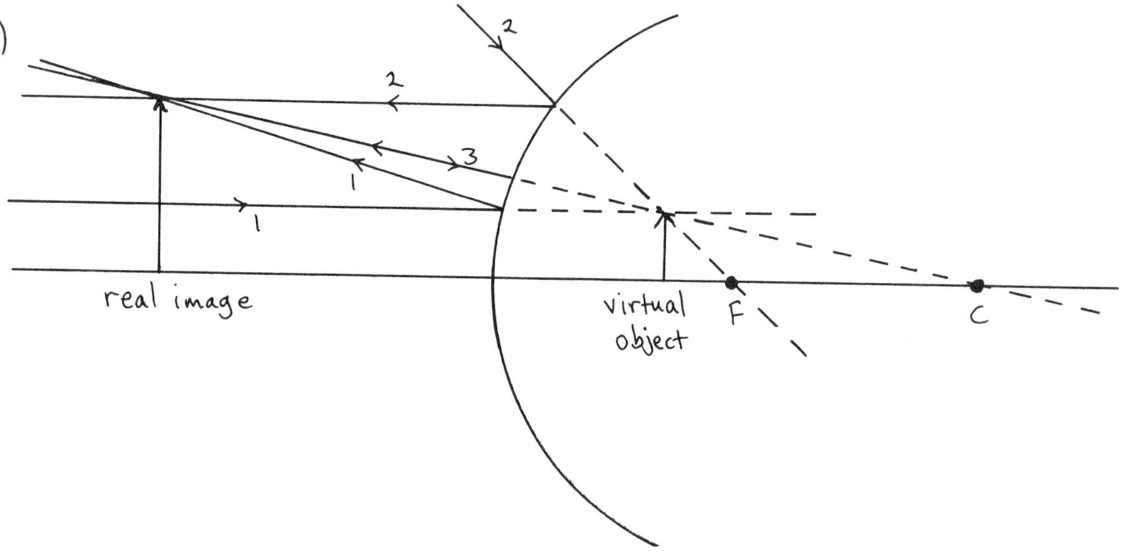

35-47

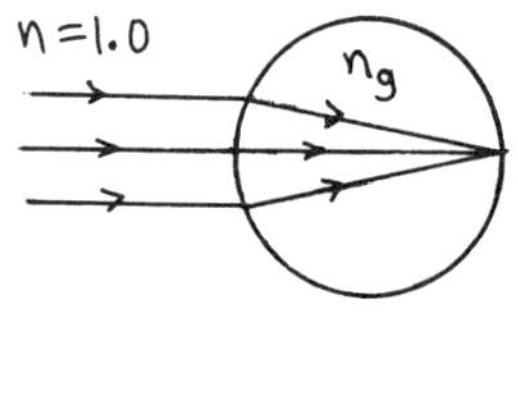

Image is formed by refraction at the front surface of the sphere. Let n_g be the index of refraction of the glass.

$s=\infty$

$s'=+2r$, where r is the radius of the sphere

$n_a=1.00$, $n_b=n_g$, $R=+r$

$$\frac{n_a}{s}+\frac{n_b}{s'}=\frac{n_b-n_a}{R}$$

$$\frac{1}{\infty}+\frac{n_g}{2r}=\frac{n_g-1.00}{r}$$

$$\frac{n_g}{2r}=\frac{n_g}{r}-\frac{1}{r} \Rightarrow \frac{n_g}{2r}=\frac{1}{r} \Rightarrow n_g=\underline{2.00}$$

35-51

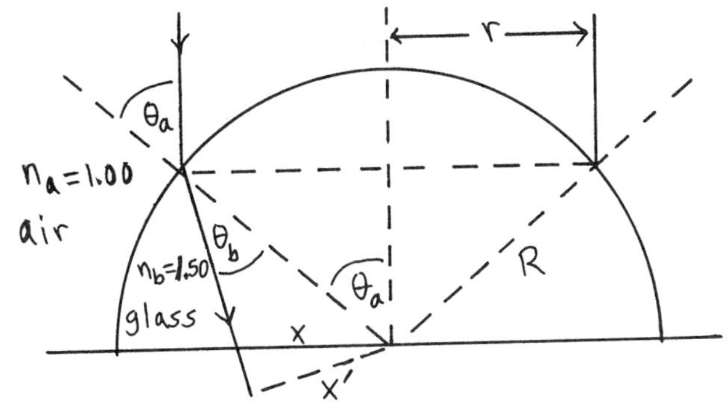

The width of the incident beam is exaggerated in the sketch, to make it easier to draw.

The diameter of the circle of light formed on the table is $2x$. Note the two right triangles containing the angles θ_a and θ_b.

$r = 0.230$ cm is the radius of the incident beam.

$R = 10.0$ cm is the radius of the glass hemisphere

θ_a and θ_b small $\Rightarrow x \simeq x'$; $\sin\theta_a = \frac{r}{R}$, $\sin\theta_b = \frac{x'}{R} \simeq \frac{x}{R}$

Snell's law: $n_a \sin\theta_a = n_b \sin\theta_b$

Use the above expressions for $\sin\theta_a$, $\sin\theta_b \Rightarrow n_a \frac{r}{R} = n_b \frac{x}{R}$

$n_a r = n_b x \Rightarrow x = \frac{n_a r}{n_b} = \frac{1.00\,(0.230\text{ cm})}{1.50} = 0.153$ cm

The diameter of the circle on the table is $2x = 2(0.153\text{ cm}) = \underline{0.306\text{ cm}}$.

b) R divides out of the expression; the result for the diameter of the spot is independent of the radius R of the hemisphere.

35-55

convex mirror $\Rightarrow R = -1.80$ m; $f = \frac{R}{2} = -0.900$ m

Find the relation between s and s':

$$\frac{1}{s} + \frac{1}{s'} = \frac{1}{f}$$

$$\frac{1}{s'} = \frac{1}{f} - \frac{1}{s} = \frac{s-f}{sf} \Rightarrow s' = \frac{sf}{s-f}$$

The speed of the image is $v' = \frac{ds'}{dt} = \frac{ds'}{ds}\frac{ds}{dt}$

$$\frac{ds'}{ds} = \frac{f}{s-f} - \frac{sf}{(s-f)^2} = \frac{(s-f)f - sf}{(s-f)^2} = -\frac{f^2}{(s-f)^2} = -\left(\frac{f}{s-f}\right)^2$$

$\frac{ds}{dt} = v$, the speed of the object

Thus $v' = -\left(\frac{f}{s-f}\right)^2 v$. (The quantity $\left(\frac{f}{s-f}\right)^2$ is always positive, so the minus sign means that the object and image move in opposite directions.)

a) $s = 10.0$ m

$$v' = -\left(\frac{-0.900\text{ m}}{10.0\text{ m} - (-0.900\text{ m})}\right)^2 (3.00\text{ m/s}) = \underline{-0.0205\text{ m/s}}$$

b) $s = 2.0$ m

$$v' = -\left(\frac{-0.900\text{ m}}{2.0\text{ m} + 0.900\text{ m}}\right)^2 (3.00\text{ m/s}) = \underline{-0.289\text{ m/s}}$$

35-57

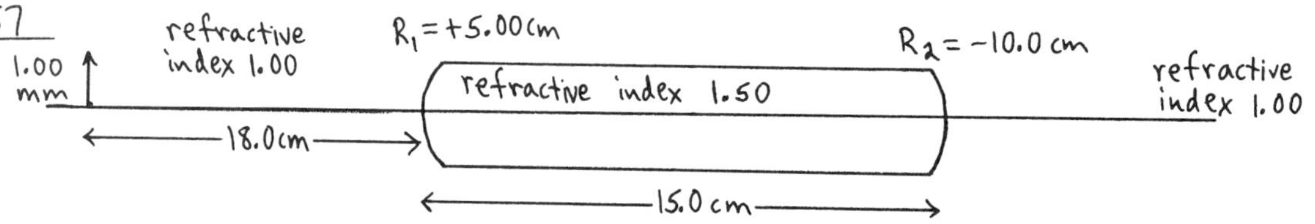

a) image formed by refraction at first surface (left end of rod):

$s = +18.0\text{ cm},\ n_a = 1.0,\ n_b = 1.50,\ R = +5.00\text{ cm}$

$$\frac{n_a}{s} + \frac{n_b}{s'} = \frac{n_b - n_a}{R}$$

$$\frac{1}{18.0\text{ cm}} + \frac{3}{2s'} = \frac{1.50 - 1.00}{5.00\text{ cm}}$$

$$\frac{3}{2s'} = \frac{1}{10.0\text{ cm}} - \frac{1}{18.0\text{ cm}} = \frac{9-5}{90.0\text{ cm}} = \frac{4}{90.0\text{ cm}}$$

$s' = \frac{3}{2}\left(\frac{90.0\text{ cm}}{4}\right) = +33.8\text{ cm}$; image is 33.8 cm to right of first vertex.

This image serves as the object for the refraction at the second surface (right end of rod). It is $33.8\text{ cm} - 15.0\text{ cm} = 18.8\text{ cm}$ to the right of the second vertex. For the second surface $s = \underline{-18.8\text{ cm}}$ (virtual object)

b) Object is on side of outgoing light, so is a <u>virtual</u> object.

c) image formed by refraction at second surface (right end of rod)

$s = -18.8\text{ cm},\ n_a = 1.50,\ n_b = 1.00,\ R = -10.0\text{ cm}$

$$\frac{n_a}{s} + \frac{n_b}{s'} = \frac{n_b - n_a}{R}$$

$$\frac{1.50}{-18.8\text{ cm}} + \frac{1.00}{s'} = \frac{1.00 - 1.50}{-10.0\text{ cm}}$$

$$-0.07979\text{ cm}^{-1} + \frac{1.00}{s'} = +0.05000\text{ cm}^{-1}$$

$s' = +7.70\text{ cm}$; $s' > 0$ so final image is 7.70 cm to right of vertex at right-hand end of rod.

d) $s' > 0 \Rightarrow$ final image is <u>real</u>

magnification for first surface

$$m_1 = -\frac{n_a s'}{n_b s} = -\frac{1.00(+33.8\text{ cm})}{1.50(+18.0\text{ cm})} = -1.25$$

magnification for second surface

$$m_2 = -\frac{n_a s'}{n_b s} = -\frac{1.50(+7.70\text{ cm})}{1.00(-18.8\text{ cm})} = +0.614$$

The overall magnification is $m_{tot} = m_1 m_2 = (-1.25)(+0.614) = -0.768$

$m_{tot} < 0 \Rightarrow$ final image is <u>inverted</u> with respect to the original object

e) $y' = m_{tot}\, y = (-0.768)(1.00\text{ mm}) = -0.768\text{ mm}$

The final image has a height of <u>0.768 mm</u>.

35-59

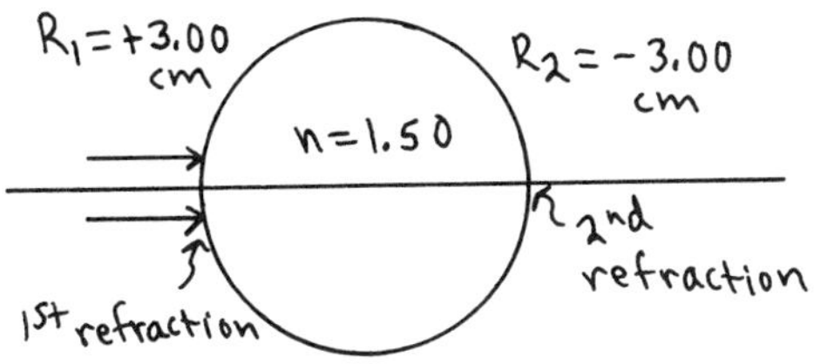

The image formed by the first refraction serves as the object for the second refraction.

The two vertexes are a distance $2R = 6.00$ cm apart.

1st refraction

$n_a = 1.00$
$n_b = 1.50$
$R = +3.00$ cm
$s = \infty$ (parallel rays)

$$\frac{n_a}{s} + \frac{n_b}{s'} = \frac{n_b - n_a}{R}$$

$$\frac{1.0}{\infty} + \frac{1.50}{s'} = \frac{1.50 - 1.00}{+3.00\text{ cm}}$$

$$\frac{1.50}{s'} = \frac{1}{6.00\text{cm}} \Rightarrow s' = (1.50)(6.00\text{cm}) = 9.00\text{ cm}$$

$s' > 0 \Rightarrow$ the first image is 9.00 cm to the right of the first vertex, so is 3.00 cm to the right of the second vertex. This image serves as a virtual object for the second refraction, with $s = -3.00$ cm.

2nd refraction

$n_a = 1.50$
$n_b = 1.00$
$R = -3.00$ cm
$s = -3.00$ cm

$$\frac{n_a}{s} + \frac{n_b}{s'} = \frac{n_b - n_a}{R}$$

$$\frac{1.50}{-3.00\text{ cm}} + \frac{1.00}{s'} = \frac{1.00 - 1.50}{-3.00\text{cm}}$$

$$\frac{1}{s'} = \frac{3}{6.00\text{ cm}} + \frac{1}{6.00\text{cm}} = \frac{4}{6.00\text{cm}} \Rightarrow s' = 1.50\text{ cm}$$

$s' > 0$, so this final image is 1.50 cm to the right of the second vertex, or 4.50 cm from the center of the sphere.

35-61

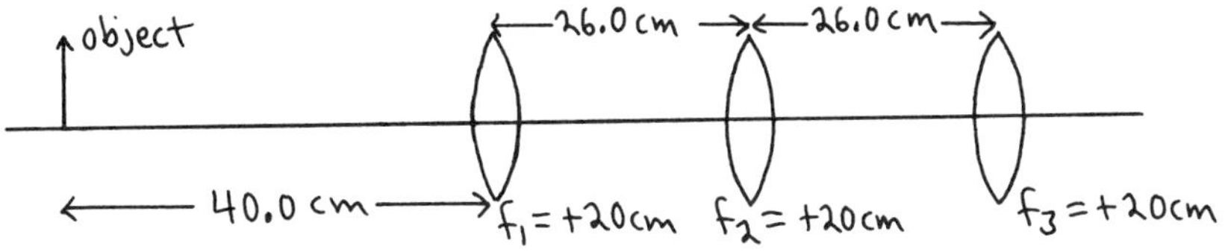

$$\frac{1}{s} + \frac{1}{s'} = \frac{1}{f}$$

$$\frac{1}{s'} = \frac{1}{f} - \frac{1}{s} = \frac{s-f}{sf}$$

$$s' = \frac{sf}{s-f}$$

lens #1

$s = +40.0$cm
$f = +20.0$ cm

$$s' = \frac{sf}{s-f} = \frac{(+40.0\text{cm})(+20.0\text{cm})}{+40.0\text{cm} - 20.0\text{ cm}} = +40.0\text{ cm}$$

The image formed by the first lens is 40.0 cm to the right of the first lens, so it is 40.0 cm − 26.0 cm = 14.0 cm to the right of the second lens.

lens #2

$s = -14.0$cm
$f = +20.0$cm

$$s' = \frac{sf}{s-f} = \frac{(-14.0\text{cm})(+20.0\text{cm})}{-14.0\text{cm} - 20.0\text{cm}} = +8.235\text{ cm}$$

The image formed by the second lens is 8.235 cm to the right of the second lens, so it is 26.0 cm − 8.235 cm = 17.76 cm to the left of the third lens.

35-61 (cont)

lens #3

$s = +17.76$ cm $\quad s' = \frac{sf}{s-f} = \frac{(17.76 \text{ cm})(+20.0 \text{ cm})}{17.76 \text{ cm} - 20.0 \text{ cm}} = -159$ cm

$f = +20.0$ cm

The final image is 159 cm to the left of the third lens, so is 159 cm − 26 cm − 26 cm − 40 cm = 67 cm to the left of the object.

35-65

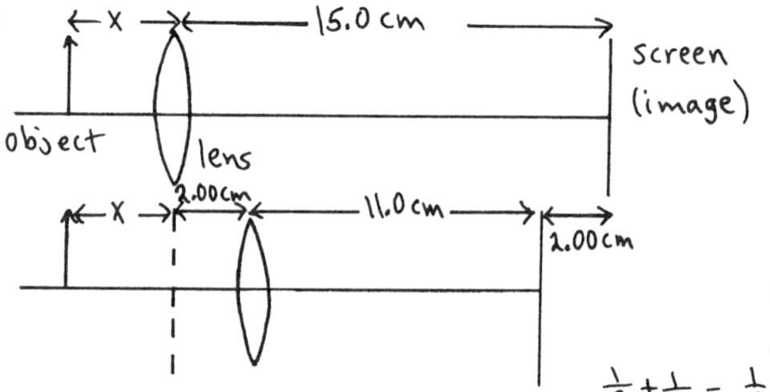

$s = x$

$s' = 15.0$ cm

$\frac{1}{s} + \frac{1}{s'} = \frac{1}{f} \Rightarrow \frac{1}{x} + \frac{1}{15.0 \text{ cm}} = \frac{1}{f}$

$s = x + 2.00$ cm

$s' = 11.0$ cm

$\frac{1}{s} + \frac{1}{s'} = \frac{1}{f} \Rightarrow \frac{1}{x + 2.00 \text{ cm}} + \frac{1}{11.0 \text{ cm}} = \frac{1}{f}$

Equate these two expressions for $\frac{1}{f}$

$$\Rightarrow \frac{1}{x} + \frac{1}{15.0 \text{ cm}} = \frac{1}{x + 2.00 \text{ cm}} + \frac{1}{11.0 \text{ cm}}$$

$$\frac{1}{x} - \frac{1}{x + 2.00 \text{ cm}} = \frac{1}{11.0 \text{ cm}} - \frac{1}{15.0 \text{ cm}}$$

$$\frac{x + 2.00 \text{ cm} - x}{x(x + 2.00 \text{ cm})} = \frac{15 - 11}{165 \text{ cm}} \Rightarrow \frac{2.00 \text{ cm}}{x(x + 2.00 \text{ cm})} = \frac{4}{165 \text{ cm}}$$

$$x^2 + (2.00 \text{ cm})x - 82.5 \text{ cm}^2 = 0$$

$$x = \tfrac{1}{2}\left(-2.00 \pm \sqrt{4.00 + 4(82.5)}\right) \text{ cm}$$

x must be positive $\Rightarrow x = \frac{1}{2}(-2.00 + 18.28)$ cm $= 8.14$ cm

Then $\frac{1}{x} + \frac{1}{15.0 \text{ cm}} = \frac{1}{f} \Rightarrow \frac{1}{f} = \frac{1}{8.14 \text{ cm}} + \frac{1}{15.0 \text{ cm}} \Rightarrow f = \underline{+5.28 \text{ cm}}$.

($f > 0$; the lens is converging)

35-69

a)

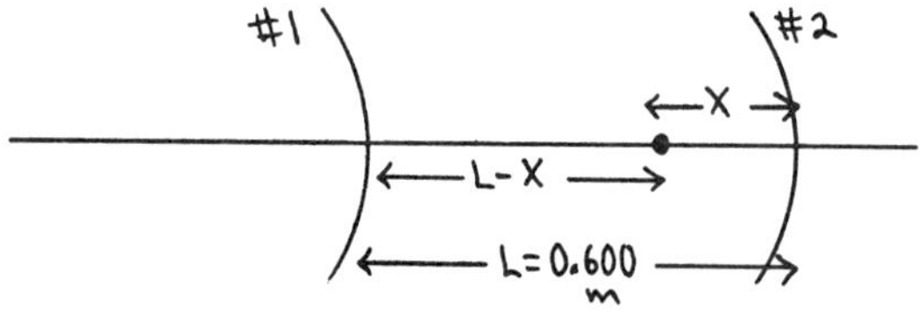

$|R| = 0.360$ m

$|f| = \frac{|R|}{2} = 0.180$ m

Image formed by convex mirror (mirror #1):

convex $\Rightarrow f_1 = -0.180$ m

$s_1 = L - x$

$$s_1' = \frac{s_1 f_1}{s_1 - f_1} = \frac{(L-x)(-0.180 \text{ m})}{L - x + 0.180 \text{ m}} = -(0.180 \text{ m})\left(\frac{0.600 \text{ m} - x}{0.780 \text{ m} - x}\right) < 0$$

35-69 (cont)

The image is $(0.180\text{ m})\left(\frac{0.600\text{ m}-x}{0.780\text{ m}-x}\right)$ to the left of mirror #1, so is

$$0.600\text{m}+(0.180\text{m})\left(\frac{0.600\text{m}-x}{0.780\text{ m}-x}\right)=\frac{0.468\text{ m}^2-0.600\text{m }x+0.108\text{ m}^2-0.180\text{m }x}{0.780\text{ m}-x}=\frac{0.576\text{ m}^2-0.780\text{m }x}{0.780\text{ m}-x}$$

to the left of mirror #2

Image formed by concave mirror (mirror #2):

concave $\Rightarrow f_2=+0.180\text{m}$

$$s_2=\frac{0.576\text{m}^2-0.780\text{m }x}{0.780\text{ m}-x}$$

Rays return to the source $\Rightarrow s_2'=x$

Use these expressions in $s_2=\frac{s_2'f_2}{s_2'-f_2}\Rightarrow\frac{0.576\text{m}^2-(0.780\text{m})x}{0.780\text{m}-x}=\frac{(0.180\text{m})x}{x-0.180\text{ m}}$

$$0.600x^2-(0.576\text{m})x+0.10368\text{ m}^2=0$$

$$x=\frac{1}{1.20}\left(0.576\pm\sqrt{(0.576)^2-4(0.600)(0.10368)}\right)\text{m}=\frac{1}{1.20}(0.576\pm0.288)\text{ m}$$

$x=0.72\text{m}$ (impossible; can't have $x>L=0.600\text{ m}$) or $\underline{x=0.24\text{ m}}$

b)

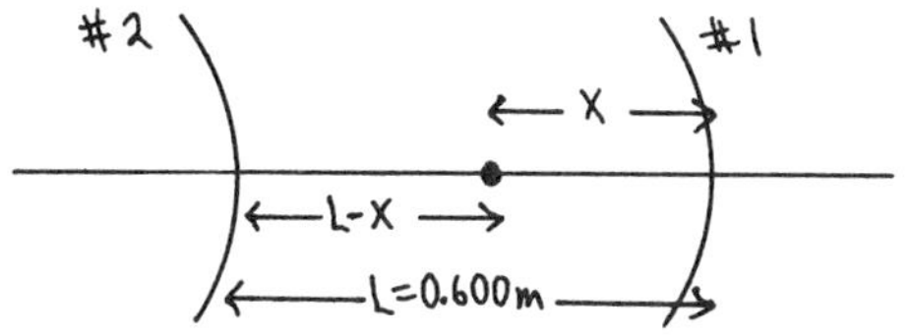

Image formed by concave mirror (mirror #1):

concave $\Rightarrow f_1=+0.180\text{ m}$

$s_1=x$

$$s_1'=\frac{s_1f_1}{s_1-f_1}=\frac{(0.180\text{m})x}{x-0.180\text{m}}$$

The image is $\frac{(0.180\text{m})x}{x-0.180\text{ m}}$ to left of mirror #1, so $s_2=0.600\text{m}-\frac{(0.180\text{m})x}{x-0.180\text{ m}}=\frac{(0.420\text{m})x-0.108\text{m}^2}{x-0.180\text{ m}}$

Image formed by convex mirror (mirror #2):

convex $\Rightarrow f_2=-0.180\text{ m}$

rays return to the source $\Rightarrow s_2'=L-x=0.600\text{m}-x$

$$\frac{1}{s}+\frac{1}{s'}=\frac{1}{f}\Rightarrow\frac{x-0.180\text{ m}}{(0.420\text{m})x-0.108\text{m}^2}+\frac{1}{0.600\text{m}-x}=-\frac{1}{0.180\text{m}}$$

$$\frac{x-0.180\text{ m}}{(0.420\text{m})x-0.108\text{m}^2}=-\left(\frac{0.780\text{ m}-x}{0.108\text{m}^2-0.180\text{m }x}\right)$$

$$0.600x^2-(0.576\text{m})x+0.10368\text{ m}^2=0$$

This is the same quadratic equation as obtained in part (a), so again $\underline{x=0.24\text{ m}}$.

35-71

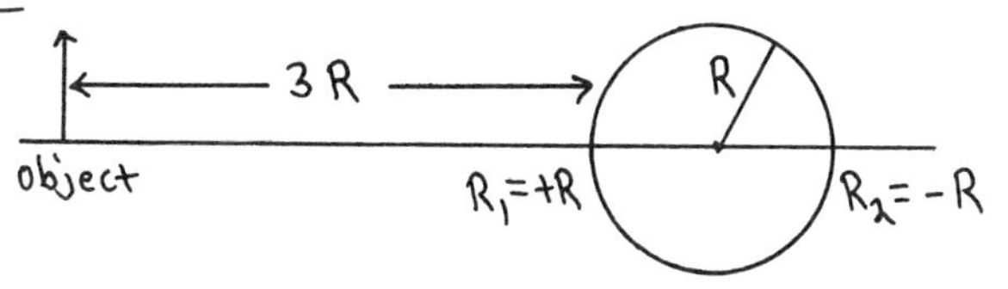

Thin-walled glass ⇒ the glass has no effect on the light rays. The problem is that of refraction by a sphere of water surrounded by air.

35-71 (cont)

First refraction (air→water):

$n_a = 1.00$ (air)

$n_b = 1.333$ (water)

$s_1 = 3R$

$R_1 = +R$

$$\frac{n_a}{s_1} + \frac{n_b}{s_1'} = \frac{n_b - n_a}{R_1}$$

$$\frac{1}{3R} + \frac{1.333}{s_1'} = \frac{1.333 - 1}{R}$$

$$\frac{1.333}{s_1'} = \frac{0.333}{R} - \frac{0.333}{R} = 0 \Rightarrow s_1' = \infty \text{ (parallel rays)}$$

Second refraction (water → air):

$n_a = 1.333$ (water)

$n_b = 1.00$

$s_2 = -\infty$

$R_2 = -R$

$$\frac{n_a}{s_2} + \frac{n_b}{s_2'} = \frac{n_b - n_a}{R_2}$$

$$\frac{1.333}{-\infty} + \frac{1.00}{s_2'} = \frac{1.00 - 1.333}{-R}$$

$$\frac{1}{s_2'} = +\frac{1}{3R} \Rightarrow s_2' = +3R$$

The final image is 3R to the right of the second surface ⇒ 4R from the center of the sphere, on the opposite side from the object.

35-73

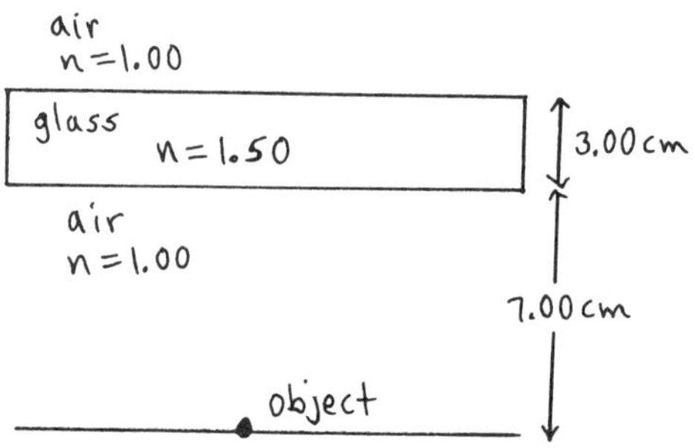

For a plane (flat) surface $R \to \infty$,

so $\frac{n_a}{s} + \frac{n_b}{s'} = 0$

$$s' = -\left(\frac{n_b}{n_a}\right)s$$

First refraction (air→glass):

$n_a = 1.00$

$n_b = 1.50$

$s = 7.00\text{ cm}$

$$s' = -\left(\frac{n_b}{n_a}\right)s = -\frac{1.50}{1.00}(7.00\text{ cm}) = -10.5\text{ cm}$$

The image is 10.5 cm below the lower surface of the glass, so is 10.5 cm + 3.00 cm = 13.5 cm below the upper surface.

Second refraction (glass → air):

$n_a = 1.50$

$n_b = 1.00$

$s = +13.5\text{ cm}$

$$s' = -\left(\frac{n_b}{n_a}\right)s = -\frac{1.00}{1.50}(13.5\text{ cm}) = -9.00\text{ cm}$$

The image of the page is 9.00 cm below the top surface of the glass plate and therefore 1.00 cm above the page.

35-75

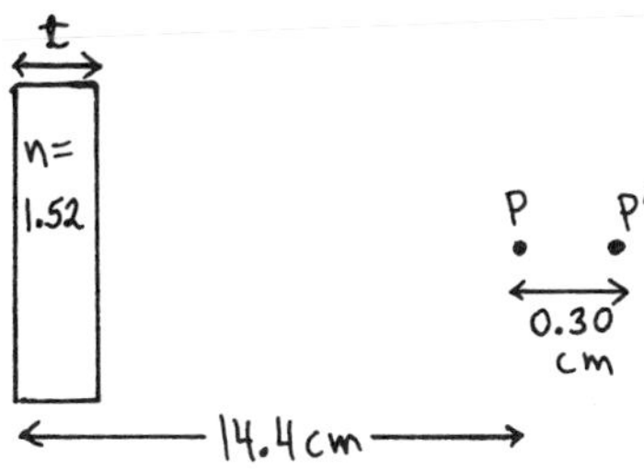

plane faces $\Rightarrow R \to \infty$ and $\frac{n_a}{s} + \frac{n_b}{s'} = 0$

$$s' = -\frac{n_b}{n_a} s$$

refraction at first (left-hand) surface of the piece of glass:

The rays converging toward point P constitute a virtual object for this surface, so $s = -14.4$ cm.

$n_a = 1.00$, $n_b = 1.52$

$$s' = -\frac{1.52}{1.00}(-14.4\text{ cm}) = +21.9\text{ cm}$$

This image is 21.9 cm to the right of the first surface so is a distance 21.9 cm $- t$ to the right of the second surface. This image serves as a virtual object for the second surface.

refraction at second (right-hand) surface of the piece of glass:

The image is at P′ $\Rightarrow s' = 14.4\text{ cm} + 0.30\text{ cm} - t = 14.7\text{ cm} - t$

$$s = -(21.9\text{ cm} - t)$$

$$n_a = 1.52,\quad n_b = 1.00$$

$s' = -\frac{n_b}{n_a} s$ gives

$$14.7\text{ cm} - t = -\frac{1.00}{1.52}\left(-[21.9\text{ cm} - t]\right)$$

$$14.7\text{ cm} - t = +14.4\text{ cm} - 0.658t$$

$$0.342t = 0.030\text{ cm}$$

$$t = \underline{0.88\text{ cm}}$$

CHAPTER 36

Exercises 3, 5, 9, 11, 13, 17, 19, 23, 27

Problems 31, 33, 37, 39, 41

Exercises

36-3

a) We need $m = -\frac{24\times10^{-3}\text{ m}}{100\text{ m}} = -2.4\times10^{-4}$, or $m = -\frac{36\times10^{-3}\text{ m}}{150\text{ m}} = -2.4\times10^{-4}$.

$s \gg f$ so $s' \simeq f$

Then $m = -\frac{s'}{s} = -\frac{f}{s} \Rightarrow \frac{f}{s} = 2.4\times10^{-4} \Rightarrow f = 2.4\times10^{-4}(120\text{ m}) = 0.0288\text{ m} = 28.8\text{ mm}$

A smaller f means a smaller s' and a smaller m, so with $f = 28$ mm the object's image nearly fills the picture area.

b) We need $m = -\frac{36\times10^{-3}\text{ m}}{0.9\text{ m}} = -4.0\times10^{-2}$.

Then, as in part (a), $\frac{f}{s} = 4.0\times10^{-2} \Rightarrow f = (5.3\text{ m})(4.0\times10^{-2}) = 0.212\text{ m} = 212\text{ mm}$.

Therefore use the <u>200 mm</u> lens.

36-5

a) $f/2.8$ lens $\Rightarrow$ f-number $= 2.8$

f-number $= \frac{f}{D}$ (Eq. 36-1) $\Rightarrow D = \frac{f}{\text{f-number}} = \frac{85.0\text{ mm}}{2.8} = \underline{30.4\text{ mm}}$

b) $f/5.6$ lens $\Rightarrow$ f-number $= 5.6$; $D = \frac{f}{\text{f-number}} = \frac{85.0\text{ mm}}{5.6} = 15.2\text{ mm}$

D is smaller by a factor of 2. The aperture area is smaller by a factor of $2^2 = 4$. So need an exposure time larger by a factor of $4 \Rightarrow 4\left(\frac{1}{120\text{ s}}\right) = \frac{1}{30\text{ s}}$.

36-9

a) $f = 0.100$ m
$s' = 5.00$ m
$s = ?$

$\frac{1}{s} + \frac{1}{s'} = \frac{1}{f} \Rightarrow \frac{1}{s} = \frac{1}{f} - \frac{1}{s'} = \frac{s'-f}{s'f}$

$s = \frac{s'f}{s'-f} = \frac{(5.00\text{ m})(0.100\text{ m})}{5.00\text{ m} - 0.100\text{ m}} = 0.1020\text{ m} = \underline{10.2\text{ cm}}$

b) $f = 0.100$ m
$s = 0.100$ m

$\frac{1}{s} + \frac{1}{s'} = \frac{1}{f}$

$s' = \frac{sf}{s-f} = \frac{(0.100\text{ m})(0.100\text{ m})}{0.100\text{ m} - 0.100\text{ m}} \rightarrow \infty$; no, the screen cannot be moved to achieve focus

36-11

a) $\frac{1}{f} = +2.75$ diopters $\Rightarrow f = +\frac{1}{2.75}\text{ m} = +0.3636\text{ m}$ (converging lens)

The purpose of the corrective lens is to take an object at 25 cm from the eye and form a virtual image at the eye's near point.

36-11 (cont)

$s = 25$ cm

$f = 36.36$ cm

$$\frac{1}{s}+\frac{1}{s'}=\frac{1}{f} \Rightarrow s'=\frac{sf}{s-f} = \frac{(25\text{ cm})(36.36\text{ cm})}{25\text{ cm}-36.36\text{ cm}} = -80\text{ cm}$$

The eye's near point is 80 cm from the eye.

b) $\frac{1}{f} = -0.830$ diopters $\Rightarrow f = -\frac{1}{0.830}\text{ m} = -1.205\text{ m}$ (diverging lens)

The purpose of this corrective lens is to take an object at infinity and form a virtual image of it at the eye's far point.

$f = -120.5$ cm

$s = \infty$

$$\frac{1}{s}+\frac{1}{s'}=\frac{1}{f}$$

$$s=\infty \Rightarrow \frac{1}{s'}=\frac{1}{f} \Rightarrow s=f=-120\text{ cm}$$

The eye's far point is 120 cm from the eye.

36-13

Consider the refraction that occurs at the front surface of the cornea.

$$\frac{n_a}{s}+\frac{n_b}{s'}=\frac{n_b-n_a}{R}$$

$s' = +2.60$ cm, $s = +25.0$ cm, $n_a = 1.00$ (air), $n_b = 1.40$

$$\frac{1.00}{+25.0\text{ cm}}+\frac{1.40}{+2.60\text{ cm}}=\frac{1.40-1.00}{R}$$

$$0.0400\text{ cm}^{-1}+0.5385\text{ cm}^{-1}=\frac{0.40}{R}$$

$R = \underline{0.691\text{ cm}}$

36-17

a) $s' = -25.0$ cm

$f = +9.00$ cm

$s = ?$

$$\frac{1}{s}+\frac{1}{s'}=\frac{1}{f} \Rightarrow \frac{1}{s}=\frac{1}{f}-\frac{1}{s'}=\frac{s'-f}{s'f}$$

$$s=\frac{s'f}{s'-f} = \frac{(-25.0\text{ cm})(+9.00\text{ cm})}{-25.0\text{ cm}-9.00\text{ cm}} = \underline{+6.62\text{ cm}}$$

b) $m = -\frac{s'}{s} = -\frac{-25.0\text{ cm}}{6.62\text{ cm}} = +3.776$

$|m| = \frac{|y'|}{|y|} \Rightarrow |y'| = |m||y| = (3.776)(1.00\text{ mm}) = \underline{3.78\text{ mm}}$

36-19

a)

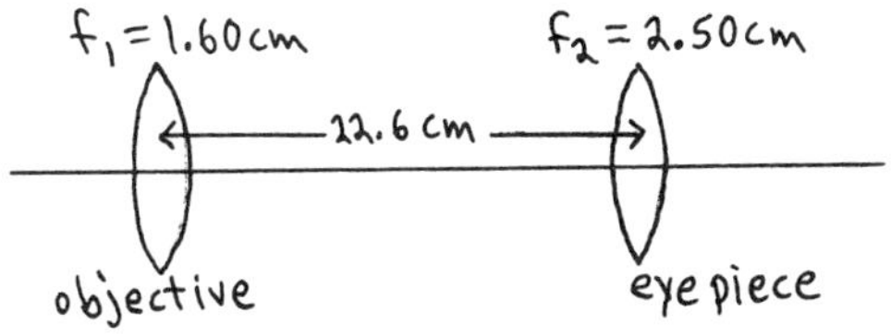

Final image at $\infty \Rightarrow$ object for eyepiece is at its focal point. But the object for the eyepiece is the image of the objective, so the image formed by the objective is 22.6 cm − 2.50 cm = 20.1 cm to the right of the lens.

36-19 (cont)

Image formation by the objective:

$s' = +20.1\text{ cm}$ $\qquad \frac{1}{s}+\frac{1}{s'}=\frac{1}{f} \Rightarrow \frac{1}{s}=\frac{1}{f}-\frac{1}{s'}=\frac{s'-f}{s'f}$

$f = +1.60\text{ cm}$

$s = ?$ $\qquad s=\frac{s'f}{s'-f}=\frac{(20.1\text{ cm})(1.60\text{ cm})}{20.1\text{ cm}-1.60\text{ cm}}=1.738\text{ cm}=\underline{17.4\text{ mm}}$

b) $m_1=-\frac{s'}{s}=-\frac{20.1\text{ cm}}{1.738\text{ cm}}=-11.6$

The magnitude of this linear magnification is $\underline{11.6}$.

c) $M=m_1M_2 \qquad M_2=\frac{25\text{ cm}}{f_2}=\frac{25\text{ cm}}{2.50\text{ cm}}=10.0$

$M=m_1M_2=-(11.6)(10.0)=\underline{-116}$

36-23

$f_1=90.0\text{ cm}$ (objective); $f_2=30.0\text{ cm}$ (eyepiece)

a) Eq. (36-5) $M=-\frac{f_1}{f_2}=-\frac{90.0\text{ cm}}{30.0\text{ cm}}=\underline{-3.00}$

b) $s=2.00\times10^{3}\text{ m}$

$s'=f_1=90.0\text{ cm}$ (since s is very large, $s'\simeq f$)

$m=-\frac{s'}{s}=-\frac{0.900\text{ m}}{2.00\times10^3\text{ m}}=-4.50\times10^{-4}$

$|y'|=|m|\,|y|=(4.50\times10^{-4})(80.0\text{ m})=0.0360\text{ m}=\underline{3.60\text{ cm}}$

c) The angular size of the object is $\theta=\frac{80.0\text{ m}}{2.00\times10^3\text{ m}}=0.0400\text{ rad}$

$M=\frac{\theta'}{\theta}$ (Eq. 36-5) $\Rightarrow$ angular size of image is $\theta'=M\theta=-(3.00)(0.0400\text{ rad})=\underline{-0.120\text{ rad}}$

(the minus sign shows that the final image is inverted.)

36-27

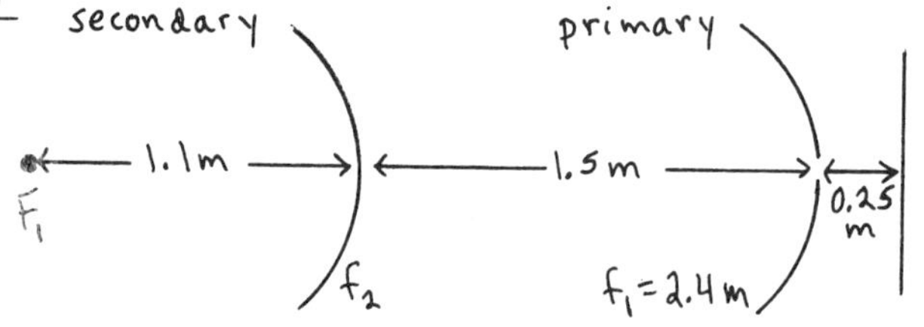

Far distant object ⇒ image of primary mirror is formed at its focus, which is 2.6 m − 1.5 m = 1.1 m behind the secondary mirror. The image formed by the secondary mirror is located at the detector.

Image formation by the secondary mirror:

$s=-1.1\text{ m}$ $\qquad \frac{1}{s}+\frac{1}{s'}=\frac{1}{f} \Rightarrow \frac{1}{f}=\frac{s+s'}{ss'}$

$s'=+1.75\text{ m}$

$f=?$ $\qquad f=\frac{ss'}{s+s'}=\frac{(-1.1\text{ m})(1.75\text{ m})}{-1.1\text{ m}+1.75\text{ m}}=-2.96\text{ m}$

$R=2f=2(-2.96\text{ m})=-5.9\text{ m}$

The secondary mirror should be convex ($R<0$) with radius of curvature 5.9 m.

Problems

36-31

a) $f = 50.0 \times 10^{-3}$ m

$$|m| = \frac{|y'|}{|y|} = \frac{\frac{2}{3}(36.0 \times 10^{-3}\text{ m})}{3.60\text{ m}} = 6.667 \times 10^{-3}$$

Image on film ⇒ image is real. $s>0,\ s'>0 \Rightarrow m = -\frac{s'}{s} < 0 \Rightarrow m = -6.667 \times 10^{-3}$

$s' = -ms = -(-6.667 \times 10^{-3})s = +(6.667 \times 10^{-3})s$

Use this in $\frac{1}{s} + \frac{1}{s'} = \frac{1}{f} \Rightarrow \frac{1}{s} + \frac{1}{(6.667 \times 10^{-3})s} = \frac{1}{f}$

$\frac{151}{s} = \frac{1}{f} \Rightarrow s = 151f = 151(50.0 \times 10^{-3}\text{ m}) = \underline{7.55\text{ m}}$

b) Fill the viewfinder frame ⇒ $|m| = \frac{36.0 \times 10^{-3}\text{ m}}{3.60\text{ m}} = 10.0 \times 10^{-3}$.

Thus $s' = (1.00 \times 10^{-2})s$.

$\frac{1}{s} + \frac{1}{s'} = \frac{1}{f} \Rightarrow \frac{1}{s} + \frac{1}{(1.00 \times 10^{-2})s} = \frac{1}{f}$

$\frac{101}{s} = \frac{1}{f} \Rightarrow s = 101f = 101(50.0 \times 10^{-3}\text{ m}) = \underline{5.05\text{ m}}$

36-33

a)

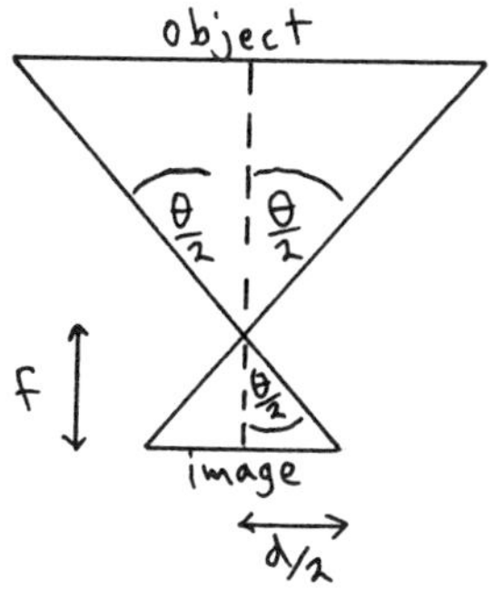

From the sketch, $\tan\frac{\theta}{2} = \frac{d/2}{f} = \frac{d}{2f}$.

$\frac{\theta}{2} = \arctan\left(\frac{d}{2f}\right)$

$\theta = 2\arctan\left(\frac{d}{2f}\right)$

b) $d = \sqrt{(24\text{mm})^2 + (36\text{mm})^2} = 43.3\text{ mm}$

$f = 28\text{ mm} \Rightarrow \theta = 2\arctan\left(\frac{43.3\text{ mm}}{2(28\text{ mm})}\right) = 75°$ (Fig. 36-2d says 75°)

$f = 105\text{ mm} \Rightarrow \theta = 2\arctan\left(\frac{43.3\text{ mm}}{2(105\text{ mm})}\right) = 23°$ (Fig. 36-2d says 25°)

$f = 300\text{ mm} \Rightarrow \theta = 2\arctan\left(\frac{43.3\text{ mm}}{2(300\text{ mm})}\right) = 8.3°$ (Fig. 26-2d says 8°)

36-37

The generalization of Eq. (36-3) is $M = \frac{\text{near point}}{f}$, so $f = \frac{\text{near point}}{M}$

a) age 10, near point = 7 cm

$f = \frac{7\text{ cm}}{2.5} = 2.8\text{ cm}$

36-37 (cont)

b) age 30, near point = 14 cm

$$f = \frac{14\text{ cm}}{2.5} = 5.6\text{ cm}$$

c) age 60; near point = 200 cm

$$f = \frac{200\text{ cm}}{2.5} = 80\text{ cm}$$

d) $f = 2.8$ cm (from part (a)) and near point = 200 cm (for 60-year-old)

$$M = \frac{200\text{ cm}}{2.8\text{ cm}} = 71$$

e) No. The reason $f = 2.8$ cm gives a larger M for a 60-year-old than for a 10-year-old is that the eye of the older person can't focus on as close of an object as the younger person can. The unaided eye of the 60-year-old must view a much smaller angular size, and that is why the same f gives a much larger M. The angular size of the image depends only on f, so it is the same for the two ages.

36-39

$f_1 = 9.00$ mm (objective), $f_2 = 6.00$ cm (eyepiece)

a) The total magnification of the final image is $m_{tot} = m_1 m_2$.

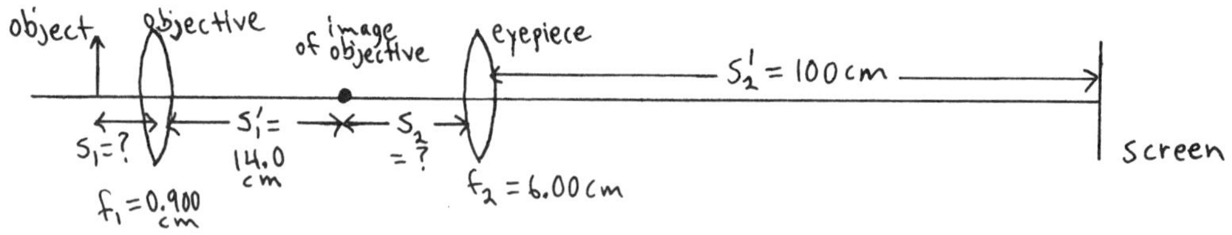

Find the object distance s_1 for the objective:

$s_1' = +14.0$ cm
$f_1 = 0.900$ cm
$s_1 = ?$

$$\frac{1}{s_1} + \frac{1}{s_1'} = \frac{1}{f_1} \Rightarrow \frac{1}{s_1} = \frac{1}{f} - \frac{1}{s_1'} = \frac{s_1' - f_1}{s_1' f_1}$$

$$s_1 = \frac{s_1' f_1}{s_1' - f_1} = \frac{(14.0\text{ cm})(0.900\text{ cm})}{14.0\text{ cm} - 0.900\text{ cm}} = 0.9618\text{ cm}$$

Find the object distance s_2 for the eyepiece:

$s_2' = 100$ cm
$f_2 = 6.00$ cm
$s_2 = ?$

$$\frac{1}{s_2} + \frac{1}{s_2'} = \frac{1}{f_2}$$

$$s_2 = \frac{s_2' f_2}{s_2' - f_2} = \frac{(100\text{ cm})(6.00\text{ cm})}{100\text{ cm} - 6.00\text{ cm}} = 6.383\text{ cm}$$

Now we can calculate the magnification for each lens:

$$m_1 = -\frac{s_1'}{s_1} = -\frac{14.0\text{ cm}}{0.9618\text{ cm}} = -14.56$$

$$m_2 = -\frac{s_2'}{s_2} = -\frac{100\text{ cm}}{6.383\text{ cm}} = -15.67$$

$$m_{tot} = m_1 m_2 = (-14.56)(-15.67) = \underline{228}$$

36-39 (cont)

b) From the sketch we see that the distance between the two lenses is $s_1' + s_2 = 14.0\text{ cm} + 6.383\text{ cm} = \underline{20.4\text{ cm}}$.

36-41

a)

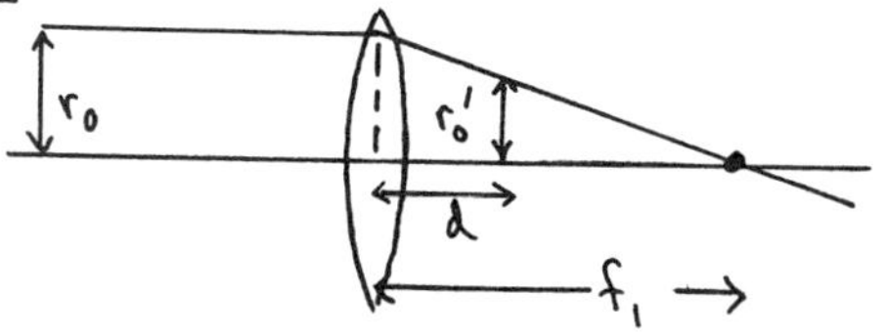

From similar triangles in the sketch

$$\frac{r_0}{f_1} = \frac{r_0'}{f_1 - d}$$

$$r_0' = \left(\frac{f_1 - d}{f_1}\right) r_0,$$ as was to be shown.

b) The image at the focal point of the first lens, a distance f_1 to the right of the first lens, serves as the object for the second lens. The image is a distance $f_1 - d$ to the right of the second lens, so $s_2 = -(f_1 - d) = d - f_1$.

$$s_2' = \frac{s_2 f_2}{s_2 - f_2} = \frac{(d - f_1) f_2}{d - f_1 - f_2}$$

$f_2 < 0$ so $|f_2| = -f_2$ and $s_2' = \dfrac{(f_1 - d)\,|f_2|}{|f_2| - f_1 + d}$, as was to be shown.

c)

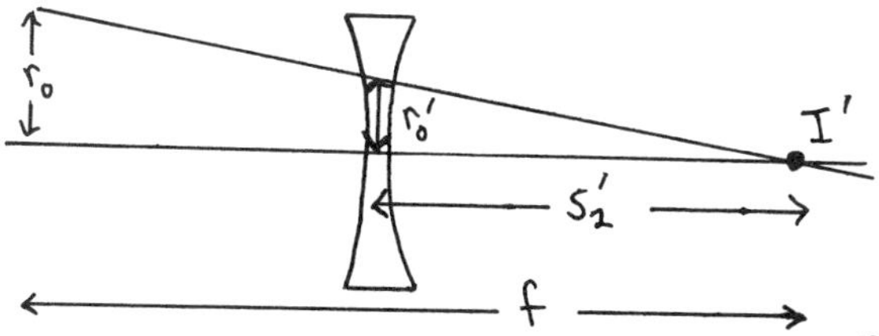

From similar triangles in the sketch, $\frac{r_0}{f} = \frac{r_0'}{s_2'}$

$$\frac{r_0}{r_0'} = \frac{f}{s_2'}$$

From the results of part (a), $\dfrac{r_0}{r_0'} = \dfrac{f_1}{f_1 - d}$

Combine the two results $\Rightarrow \dfrac{f_1}{f_1 - d} = \dfrac{f}{s_2'}$

$$f = s_2'\left(\frac{f_1}{f_1 - d}\right) = \frac{(f_1 - d)\,|f_2|\,f_1}{(|f_2| - f_1 + d)(f_1 - d)} = \frac{f_1\,|f_2|}{|f_2| - f_1 + d},$$ as was to be shown.

d) $$f = \frac{f_1\,|f_2|}{|f_2| - f_1 + d}$$

$f_1 = 10.0\text{ cm}$, $|f_2| = 15.0\text{ cm}$, $f = \dfrac{150.0\text{ cm}^2}{5.0\text{ cm} + d}$

$d = 0$ gives $f = 30.0\text{ cm}$; maximum f

$d = 7.0\text{ cm}$ gives $f = 12.5\text{ cm}$; minimum f

$f = 15.0\text{ cm} \Rightarrow 15.0\text{ cm} = \dfrac{150.0\text{ cm}^2}{5.0\text{ cm} + d}$

$5.0\text{ cm} + d = 10.0\text{ cm}$

$d = \underline{5.0\text{ cm}}$

CHAPTER 37

Exercises 1, 7, 9, 11, 13, 15, 17, 21, 27, 33

Problems 37, 39, 41, 45, 47, 49, 51

Exercises

37-1

For constructive interference the path difference $r_2 - r_1 = 1.80 \times 10^{-6}$ m is an integer number of wavelengths : $r_2 - r_1 = m\lambda$, $m = 1, 2, 3, \ldots$ (Since $r_2 - r_1$ is specified as nonzero and positive, $m = 0$ and negative m values are excluded.)

$$\lambda = \frac{r_2 - r_1}{m} = \frac{1.80 \times 10^{-6}\,\text{m}}{m}, \quad m = 1, 2, 3, \ldots$$

$m = 3 \Rightarrow 0.600 \times 10^{-6}$ m $=$ 600 nm

$m = 4 \Rightarrow 0.450 \times 10^{-6}$ m $=$ 450 nm

Other values of λ are outside the range of visible wavelengths.

37-7

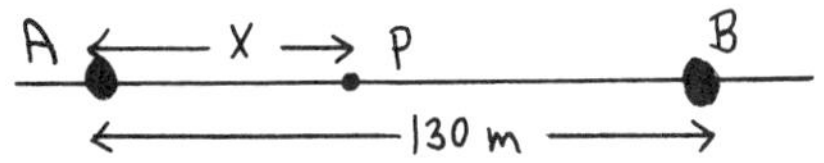

The distance of point P from each coherent source is $r_A = x$ and $r_B = 130\,\text{m} - x$.

The path difference is $r_B - r_A = 130\,\text{m} - 2x$. For constructive interference this path difference is an integer multiple of the wavelength:

$$r_B - r_A = m\lambda, \quad m = 0, \pm 1, \pm 2, \ldots$$

$$\lambda = \frac{c}{f} = \frac{2.998 \times 10^8\,\text{m/s}}{7.50 \times 10^6\,\text{Hz}} = 40.0\,\text{m}$$

Thus $130\,\text{m} - 2x = m(40.0\,\text{m}) \Rightarrow x = \dfrac{130\,\text{m} - m(40.0\,\text{m})}{2} = 65.0\,\text{m} - (20.0\,\text{m})\,m$

x must lie in the range 0 to 130 m, since P is said to be between the two antennas.

$m = 0 \Rightarrow x = 65.0$ m

$m = +1 \Rightarrow x = 65.0\,\text{m} - 20.0\,\text{m} = 45.0$ m

$m = +2 \Rightarrow x = 65.0\,\text{m} - 40.0\,\text{m} = 25.0$ m

$m = +3 \Rightarrow x = 65.0\,\text{m} - 60.0\,\text{m} = 5.0$ m

$m = -1 \Rightarrow x = 65.0\,\text{m} + 20.0\,\text{m} = 85.0$ m

$m = -2 \Rightarrow x = 65.0\,\text{m} + 40.0\,\text{m} = 105.0$ m

$m = -3 \Rightarrow x = 65.0\,\text{m} + 60.0\,\text{m} = 125.0$ m

All other values of m give values of x out of the allowed range.

Constructive interference will occur for x = 5 m, 25 m, 45 m, 65 m, 85 m, 105 m, and 125 m.

37-9

The dark lines correspond to destructive interference and hence are located by Eq. (37-5): $d \sin\theta = (m + \frac{1}{2})\lambda \Rightarrow \sin\theta = \dfrac{(m + \frac{1}{2})\lambda}{d}$, $m = 0, \pm 1, \pm 2, \ldots$

37-9 (cont)

1st dark line $\Rightarrow m=0$

2nd dark line $\Rightarrow m=1 \Rightarrow \sin\theta_1 = \frac{3\lambda}{2d} = \frac{3(600\times10^{-9}\text{m})}{2(0.300\times10^{-3}\text{m})} = 3.00\times10^{-3} \Rightarrow \theta_1 = 3.00\times10^{-3}$ rad

3rd dark line $\Rightarrow m=2 \Rightarrow \sin\theta_2 = \frac{5\lambda}{2d} = \frac{5(600\times10^{-9}\text{m})}{2(0.300\times10^{-3}\text{m})} = 5.00\times10^{-3} \Rightarrow \theta_2 = 5.00\times10^{-3}$ rad

(Note that θ_1 and θ_2 are small so that the approximation $\theta \simeq \sin\theta \simeq \tan\theta$ is valid.)

The distance of each dark line from the center of the central bright band is given by $y_m = R\tan\theta$, where $R = 0.850$ m is the distance to the screen.

$\tan\theta \simeq \theta \Rightarrow y_m = R\theta_m$

$y_1 = R\theta_1 = (0.850\text{m})(3.00\times10^{-3}\text{rad}) = 2.55\times10^{-3}\text{m}$

$y_2 = R\theta_2 = (0.850\text{m})(5.00\times10^{-3}\text{rad}) = 4.25\times10^{-3}\text{m}$

$\Delta y = y_2 - y_1 = 4.25\times10^{-3}\text{m} - 2.55\times10^{-3}\text{m} = 1.70\times10^{-3}\text{m} = \underline{1.70\text{ mm}}$

37-11

The positions of the bright fringes are given by Eq. (37-6): $y_m = R\frac{m\lambda}{d}$. For each one the adjacent bright fringe is located at $y_{m+1} = R\frac{(m+1)\lambda}{d}$. The separation between adjacent fringes is $\Delta y = y_{m+1} - y_m = \frac{R\lambda}{d}$.

$\lambda = \frac{d\,\Delta y}{R} = \frac{(0.380\times10^{-3}\text{m})(2.27\times10^{-3}\text{m})}{2.50\text{ m}} = 3.45\times10^{-7}\text{m} = \underline{345\text{ nm}}$

37-13

a) The minima are located by $d\sin\theta = (m+\frac{1}{2})\lambda$. The first minimum is for $m=0 \Rightarrow \sin\theta = \frac{\lambda}{2d} = \frac{500\times10^{-9}\text{m}}{2(0.150\times10^{-3}\text{m})} = 1.667\times10^{-3} \Rightarrow \theta = 1.667\times10^{-3}$ rad

(θ is small so $\theta \simeq \sin\theta \simeq \tan\theta$)

The distance on the screen is $y_1 = R\tan\theta \simeq R\theta$, where R is the distance to the screen. $y_1 = R\theta_1 = (0.800\text{m})(1.667\times10^{-3}\text{rad}) = 1.33\times10^{-3}\text{m} = \underline{1.33\text{ mm}}$.

b) Since θ is small we can use Eq. (37-15): $I = I_0\cos^2\left(\frac{\pi d y}{\lambda R}\right)$.

We want I for $y = \frac{1}{2}y_1 = 0.666\times10^{-3}$ m.

$I = (6.00\times10^{-6}\text{ W/m}^2)\left(\cos\left[\frac{\pi(0.150\times10^{-3}\text{m})(0.666\times10^{-3}\text{m})}{(500\times10^{-9}\text{m})(0.800\text{ m})}\right]\right)^2 = \underline{3.00\times10^{-6}\text{ W/m}^2}$.

37-15

a) The minima are located at angles θ given by $d\sin\theta = (m+\frac{1}{2})\lambda$. The first minimum corresponds to $m=0$, so $\sin\theta = \frac{\lambda}{2d} = \frac{620\times10^{-9}\text{m}}{2(0.300\times10^{-3}\text{m})} = 1.033\times10^{-3}$ and $\theta = 1.033\times10^{-3}$ rad.

The distance on the screen is $y = R\tan\theta = (0.600\text{ m})\tan(1.033\times10^{-3}\text{rad}) = \underline{0.620\text{ mm}}$.

37-15 (cont)

b) Eq. (37-15) gives the intensity I as a function of the position y on the screen:

$$I = I_0 \cos^2\left(\frac{\pi d y}{\lambda R}\right)$$

$$I = \tfrac{1}{2} I_0 \Rightarrow \cos^2\left(\frac{\pi d y}{\lambda R}\right) = \tfrac{1}{2}$$

$$\cos\left(\frac{\pi d y}{\lambda R}\right) = \frac{1}{\sqrt{2}} \text{ so } \frac{\pi d y}{\lambda R} = \frac{\pi}{4} \text{ rad}$$

$$y = \frac{\lambda R}{4d} = \frac{(620\times10^{-9}\text{m})(0.600\text{m})}{4(0.300\times10^{-3}\text{ m})} = \underline{0.310 \text{ mm}}$$

37-17

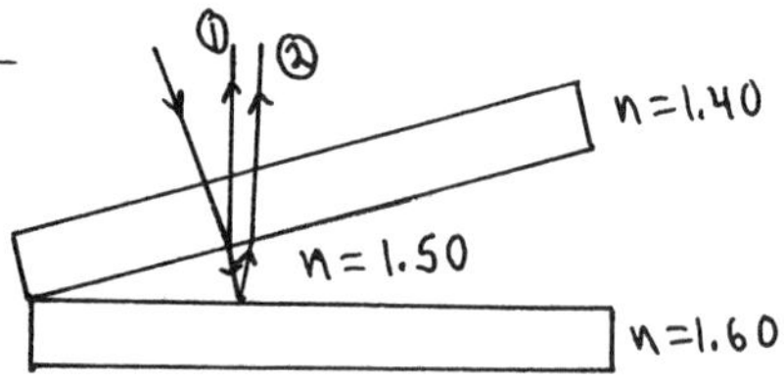

When ray ① reflects off the top of the wedge of silicone grease it undergoes a 180° phase change ($n=1.40 < n=1.50$).

When ray ② reflects off the top of the lower plate it undergoes a 180° phase change ($n=1.50 < n=1.60$).

There is no net phase difference introduced by the phase changes on reflection, and the condition for an interference minimum is $2t = (m+\frac{1}{2})\lambda$, where λ is the wavelength in the silicone grease, $\lambda = \frac{\lambda_0}{n} = \frac{500\times10^{-9}\text{m}}{1.50} = 333\times10^{-9}\text{m}$

As in Example 37-4 (Fig. 37-11), $\frac{t}{x} = \frac{h}{l} \Rightarrow t = \frac{hx}{l}$. Then $\frac{2hx}{l} = (m+\frac{1}{2})\lambda$.

$$x_m = (m+\tfrac{1}{2})\frac{\lambda l}{2h} \; ; \; x_{m+1} = (m+\tfrac{3}{2})\frac{\lambda l}{2h}$$

The spacing between adjacent dark fringes is $\Delta x = x_{m+1} - x_m = \frac{\lambda l}{2h} = \frac{(333\times10^{-9}\text{m})(0.10\text{m})}{2(0.020\times10^{-3}\text{m})}$

$$\Delta x = 0.83\times10^{-3}\text{m} = \underline{0.83 \text{ mm}}.$$

37-21

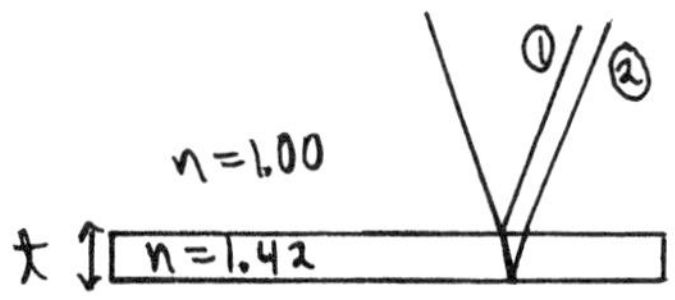

Both rays ① and ② undergo a 180° phase change on reflection, so there is no net phase difference introduced and the condition for destructive interference is $2t = (m+\frac{1}{2})\lambda$.

$$t = \frac{(m+\frac{1}{2})\lambda}{2} \; ; \text{ thinnest film} \Rightarrow m=0 \Rightarrow t = \frac{\lambda}{4}.$$

$$\lambda = \frac{\lambda_0}{1.50} \Rightarrow t = \frac{\lambda_0}{4(1.42)} = \frac{500\times10^{-9}\text{m}}{4(1.42)} = 8.80\times10^{-8}\text{m} = \underline{88.0 \text{ nm}}$$

37-27

$$\text{Eq. (37-19)} \Rightarrow x = m\frac{\lambda}{2} = 1600\left(\frac{606\times10^{-9}\text{ m}}{2}\right) = 485\times10^{-6}\text{m} = \underline{0.485 \text{ mm}}$$

37-33

By conservation of linear momentum, since the initial momentum is zero the nucleus and the photon must have equal and opposite momenta.

37-33 (cont)

Find the recoil speed of the nucleus:

$$K = \frac{1}{2}mv^2 \Rightarrow v = \sqrt{\frac{2K}{m}} = \sqrt{\frac{2(3.786\times10^{-22}\text{ J})}{9.454\times10^{-26}\text{ kg}}} = 89.49\text{ m/s}$$

The momentum of the recoiling nucleus is

$$p = mv = (9.454\times10^{-26}\text{ kg})(89.49\text{ m/s}) = \underline{8.46\times10^{-24}\text{ kg}\cdot\text{m/s}}$$

The gamma ray photon must have $p = 8.46\times10^{-24}$ kg·m/s.

$$E = pc = (8.46\times10^{-24}\text{ kg}\cdot\text{m/s})(2.998\times10^{8}\text{ m/s}) = \underline{2.54\times10^{-15}\text{ J}}$$

$$p = \frac{h}{\lambda} \Rightarrow \lambda = \frac{h}{p} = \frac{6.626\times10^{-34}\text{ J}\cdot\text{s}}{8.46\times10^{-24}\text{ kg}\cdot\text{m/s}} = 7.83\times10^{-11}\text{ m} = \underline{0.0783\text{ nm}}$$

$$c = f\lambda \Rightarrow f = \frac{c}{\lambda} = \frac{2.998\times10^{8}\text{ m/s}}{7.83\times10^{-11}\text{ m}} = \underline{3.83\times10^{18}\text{ Hz}}$$

Problems

37-37

The only effect of the water is to change the wavelength to $\lambda = \frac{\lambda_0}{n} = \frac{600\times10^{-9}\text{ m}}{1.333} = 450\times10^{-9}$ m.

$$\theta_1 \simeq \sin\theta_1 = \frac{3\lambda}{2d} = \frac{3(450\times10^{-9}\text{ m})}{2(0.300\times10^{-3}\text{ m})} = 2.25\times10^{-3}\text{ rad}$$

$$\theta_2 \simeq \sin\theta_2 = \frac{5\lambda}{2d} = \frac{5(450\times10^{-9}\text{ m})}{2(0.300\times10^{-3}\text{ m})} = 3.75\times10^{-3}\text{ rad}$$

$$y_1 \simeq R\theta_1 = (0.850\text{ m})(2.25\times10^{-3}\text{ rad}) = 1.91\times10^{-3}\text{ m}$$

$$y_2 \simeq R\theta_2 = (0.850\text{ m})(3.75\times10^{-3}\text{ rad}) = 3.17\times10^{-3}\text{ m}$$

$$\Delta y = y_2 - y_1 = 3.17\times10^{-3}\text{ m} - 1.91\times10^{-3}\text{ m} = 1.26\times10^{-3}\text{ m} = \underline{1.26\text{ mm}}$$

37-39

a) There must be destructive interference between the sound waves from the two speakers.

b) The change in path length must be $\frac{\lambda}{2}$, so $\frac{\lambda}{2} = 0.32$ m and $\lambda = 0.64$ m.

$$v = f\lambda \Rightarrow f = \frac{v}{\lambda} = \frac{340\text{ m/s}}{0.64\text{ m}} = 531\text{ Hz}$$

c) The change in path length must equal λ to go from one point of constructive interference to the next, so the speakers must be moved 0.64 m.

37-41

The dark lines are located by $d\sin\theta = (m+\frac{1}{2})\lambda$.

first dark line $\Rightarrow m = 0 \Rightarrow d\sin\theta_1 = \frac{\lambda}{2}$

$$\sin\theta_1 = \frac{\lambda}{2d} = \frac{600\times10^{-9}\text{ m}}{2(1.50\times10^{-6}\text{ m})} = 0.200 \Rightarrow \theta_1 = 11.54°.$$

37-41 (cont)

second dark line $\Rightarrow m=1 \Rightarrow d\sin\theta_2 = \frac{3\lambda}{2}$

$$\sin\theta_2 = \frac{3\lambda}{2d} = \frac{3(600\times10^{-9}\text{m})}{2(1.50\times10^{-6}\text{m})} = 0.600 \Rightarrow \theta_2 = 36.87°$$

$$y_1 = R\tan\theta_1 = (0.450\text{m})\tan 11.54° = 0.0919\text{m}$$

$$y_2 = R\tan\theta_2 = (0.450\text{m})\tan 36.87° = 0.3375\text{m}$$

The distance between the lines is $\Delta y = y_2 - y_1 = 0.3375\text{m} - 0.0919\text{m} = 0.246\text{m} = \underline{24.6\text{cm}}$.

37-45

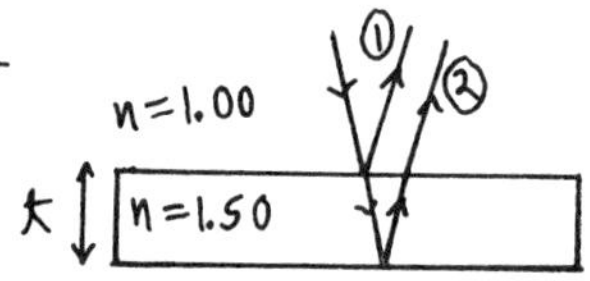

Ray ① undergoes a 180° phase change on reflection at the top surface of the glass.

Ray ② has no phase change on reflection from the lower surface of the glass.

a) Intensified in the reflected beam $\Rightarrow$ constructive interference for the reflected light. The reflections produce a net phase difference of 180°, so the condition for constructive interference is $2t = (m+\frac{1}{2})\lambda$, $m = 0, 1, 2, \ldots \Rightarrow \lambda = \frac{2t}{m+\frac{1}{2}}$.

λ is the wavelength in the glass plate; $\lambda = \lambda_0/n$.

$$\frac{\lambda_0}{n} = \frac{2t}{m+\frac{1}{2}} \Rightarrow \lambda_0 = \frac{2tn}{m+\frac{1}{2}} = \frac{2(0.425\times10^{-6}\text{m})(1.50)}{m+\frac{1}{2}} = \frac{1275\text{nm}}{m+\frac{1}{2}}$$

m	λ
0	2550 nm
1	850 nm
2	510 nm
3	364 nm
⋮	

Only $\underline{\lambda = 510\text{nm}}$ is within the limits of the visible spectrum.

b) Intensified in the transmitted light $\Rightarrow$ destructive interference in the reflected light. The condition for destructive interference between the light reflected at the top and at the bottom of the oil film is $2t = m\frac{\lambda_0}{n}$.

$$\lambda_0 = \frac{2tn}{m} = \frac{2(0.425\times10^{-6}\text{m})(1.50)}{m} = \frac{1275\text{nm}}{m}$$

m	λ_0
1	1275 nm
2	638 nm
3	425 nm
4	318 nm
⋮	

The wavelengths within the visible spectrum are 638 nm and 425 nm.

37-47

This problem deals with Newton's rings (Sect. 37-5). The interference is between rays reflecting from the top and bottom edges of the air that is between the lens and the plate.

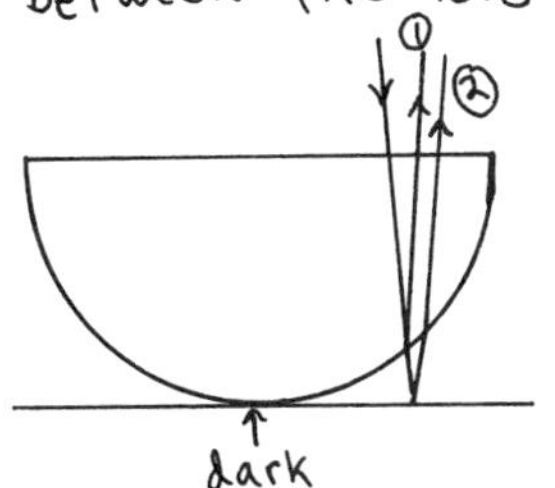

Ray ① does not undergo any phase change on reflection.
Ray ② does undergo a 180° phase change on reflection.

Hence the path difference $2t$, where t is the thickness of the air wedge, must satisfy

$$2t = (m + \tfrac{1}{2})\lambda, \quad m = 0, 1, 2, \ldots \text{ for constructive interference.}$$

Third bright ring $\Rightarrow m = 2$ and $t = \frac{5\lambda}{4} = \frac{5(650 \times 10^{-9}\,\text{m})}{4} = 8.125 \times 10^{-7}\,\text{m}$.

Now must relate this to the diameter of the ring:

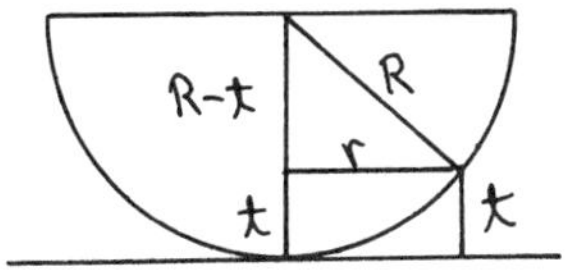

The radius of the ring is r.

$$r^2 + (R-t)^2 = R^2$$

$$r = \sqrt{R^2 - (R-t)^2} = \sqrt{R^2 - R^2 + 2Rt - t^2}$$

$$r = \sqrt{2Rt - t^2}$$

But $R = 1.15\,\text{m} \gg t$, so can neglect t^2 relative to $2Rt$.

$$r = \sqrt{2Rt} = \sqrt{2(1.15\,\text{m})(8.125 \times 10^{-7}\,\text{m})} = 1.37 \times 10^{-3}\,\text{m}$$

This is the radius. The diameter of the ring is $2r = 2.74 \times 10^{-3}\,\text{m} = \underline{2.74\,\text{mm}}$.

37-49

a)

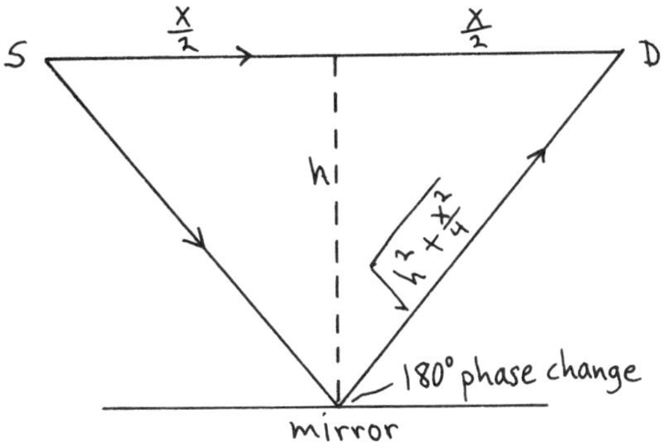

$$\text{path difference} = 2\sqrt{h^2 + \frac{x^2}{4}} - x = \sqrt{4h^2 + x^2} - x$$

Since there is a 180° phase change for the reflected ray, the condition for constructive interference is

$$\text{path difference} = (m + \tfrac{1}{2})\lambda$$

and the condition for destructive interference is

$$\text{path difference} = m\lambda.$$

Thus
$$\sqrt{4h^2 + x^2} - x = (m + \tfrac{1}{2})\lambda \quad \text{(constructive interference)}$$
$$\sqrt{4h^2 + x^2} - x = m\lambda \quad \text{(destructive interference)}$$

b) Constructive interference: $(m + \tfrac{1}{2})\lambda = \sqrt{4h^2 + x^2} - x$

$$\lambda = \frac{\sqrt{4h^2 + x^2} - x}{m + 1/2}$$

longest $\lambda \Rightarrow m = 0$ and $\lambda = 2\left(\sqrt{4h^2 + x^2} - x\right) = 2\left(\sqrt{4(0.300\,\text{m})^2 + (0.080\,\text{m})^2} - 0.080\,\text{m}\right)$

$$\lambda = \underline{1.05\,\text{m}}$$

37-51

a) The wavelength in the glass is decreased by a factor of $\frac{1}{n}$, so for the light through the upper slit a shorter path is needed to produce the same phase at the screen. Therefore, the interference pattern is shifted downward on the screen.

b) At a point on the screen located by the angle θ the difference in path length is $d\sin\theta$. This introduces a phase difference of $\phi = \frac{2\pi}{\lambda_0}(d\sin\theta)$, where λ_0 is the wavelength of the light in air or vacuum.

In the thickness L of glass the number of wavelengths is $\frac{L}{\lambda} = \frac{nL}{\lambda_0}$. A corresponding length L of the path of the ray through the lower slit, in air, contains $\frac{L}{\lambda_0}$ wavelengths. The phase difference this introduces is $\phi = 2\pi\left(\frac{nL}{\lambda_0} - \frac{L}{\lambda_0}\right)$
$\Rightarrow \phi = 2\pi(n-1)\frac{L}{\lambda_0}$.

The total phase difference is the sum of these two,
$$\phi = \frac{2\pi}{\lambda_0}(d\sin\theta) + 2\pi(n-1)\frac{L}{\lambda_0} = \frac{2\pi}{\lambda_0}(d\sin\theta + L(n-1)).$$

Eq. (37-10) then gives $I = I_0\cos^2\left(\frac{\pi}{\lambda_0}(d\sin\theta + L(n-1)\right)$.

c) Maxima $\Rightarrow \cos\frac{\phi}{2} = \pm 1 \Rightarrow \frac{\phi}{2} = m\pi, \quad m = 0, \pm 1, \pm 2, \ldots$

$$\frac{\pi}{\lambda_0}(d\sin\theta + L(n-1)) = m\pi$$
$$d\sin\theta + L(n-1) = m\lambda_0$$
$$\sin\theta = \frac{m\lambda_0 - L(n-1)}{d}$$

CHAPTER 38

Exercises 3, 5, 9, 11, 13, 15, 17, 19, 23, 27, 31, 33

Problems 37, 39, 41, 47, 51, 53

Exercises

38-3

The minima are located by Eq. (38-2): $\sin\theta = \frac{m\lambda}{a}$, $m = \pm 1, \pm 2, \ldots$

First minimum $\Rightarrow m = 1 \Rightarrow \sin\theta_1 = \frac{\lambda}{a} \Rightarrow \lambda = a\sin\theta_1$.

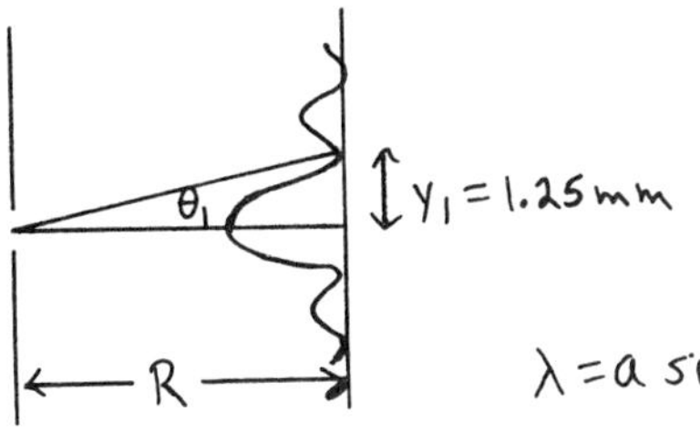

$$y_1 = R\tan\theta_1$$

$$\tan\theta_1 = \frac{y_1}{R} = \frac{1.25\times10^{-3}\text{m}}{3.00\text{ m}} = 0.4167\times10^{-3}$$

$$\theta_1 = 0.4167\times10^{-3}\text{ rad}$$

$$\lambda = a\sin\theta_1 = (0.800\times10^{-3}\text{ m})\sin(0.4167\times10^{-3}\text{ rad}) = \underline{333\text{ nm}}$$

38-5

a)

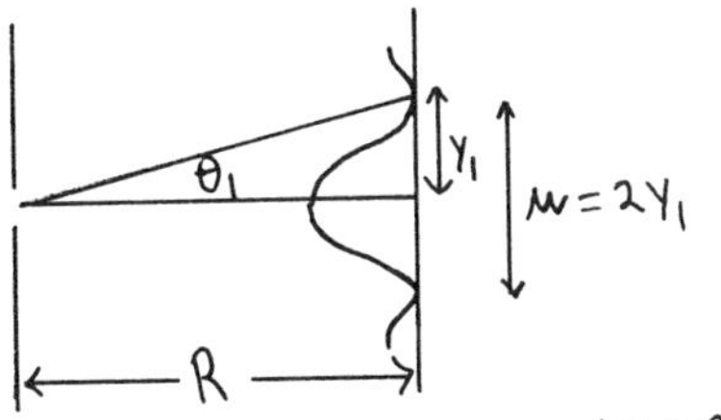

The first minimum is located by

$$\sin\theta_1 = \frac{\lambda}{a} = \frac{633\times10^{-9}\text{m}}{0.250\times10^{-3}\text{m}} = 2.532\times10^{-3}$$

$$\theta_1 = 2.532\times10^{-3}\text{ rad}$$

$$y_1 = R\tan\theta_1 = (4.0\text{m})\tan(2.532\times10^{-3}\text{rad}) = 1.0128\times10^{-2}\text{ m}$$

$$w = 2y_1 = 2(1.0128\times10^{-2}\text{ m}) = 2.03\times10^{-2}\text{m} = \underline{2.03\text{ cm}}$$

b)

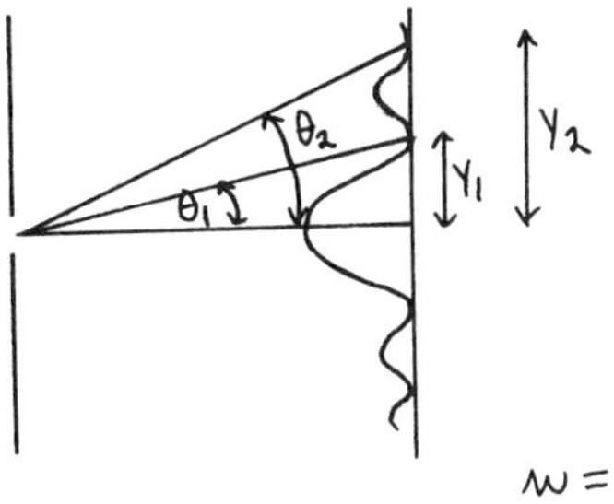

$$w = y_2 - y_1$$

$$y_1 = 1.0128\times10^{-2}\text{ m (part (a))}$$

$$\sin\theta_2 = \frac{2\lambda}{a} = 5.064\times10^{-3} \Rightarrow \theta_2 = 5.064\times10^{-3}\text{rad}$$

$$y_2 = R\tan\theta_2 = 2.026\times10^{-2}\text{m}$$

$$w = y_2 - y_1 = 2.026\times10^{-2}\text{m} - 1.013\times10^{-2}\text{ m} = 1.013\times10^{-2}\text{m} = \underline{1.01\text{cm}}$$

38-9

a) $\theta = 5.35°$, $\beta = 48.0$ rad, $a = 0.125\times10^{-3}$ m

Eq. (38-6): $\beta = \frac{2\pi}{\lambda}a\sin\theta$

$$\lambda = \frac{2\pi a\sin\theta}{\beta} = \frac{2\pi\,(0.125\times10^{-3}\text{m})\sin 5.35°}{48.0\text{ rad}} = 1.53\,\mu\text{m}.$$

b) $$I = I_0\left[\frac{\sin\beta/2}{\beta/2}\right]^2 = I_0\frac{4}{\beta^2}(\sin\beta/2)^2 = I_0\frac{4}{(48.0\text{rad})^2}\left[\sin(24.0\text{ rad})\right]^2 = \underline{1.42\times10^{-3}\,I_0}$$

38-11

a)

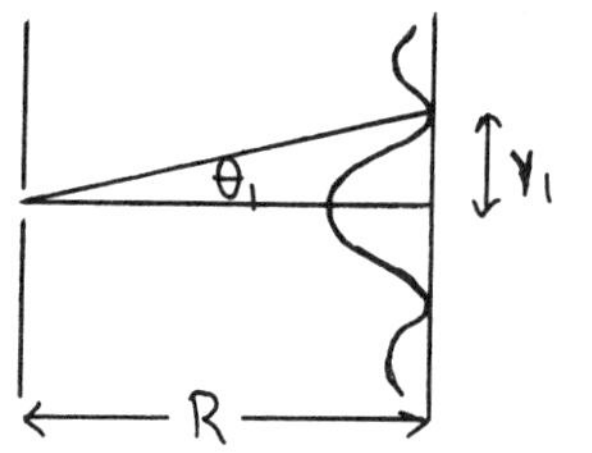

$\sin\theta_1 = \frac{\lambda}{a} = \frac{590\times10^{-9}\text{ m}}{0.200\times10^{-3}\text{ m}} = 2.95\times10^{-3}$; $\theta_1 = 2.95\times10^{-3}$ rad

$y_1 = R\tan\theta_1 = (4.00\text{ m})\tan(2.95\times10^{-3}\text{ rad}) = 1.18\times10^{-2}\text{ m} = \underline{1.18\text{ cm}}$

b) Midway between the center of the central maximum and the first minimum $\Rightarrow y = \frac{1}{2}(1.18\text{ cm}) = 5.90\times10^{-3}\text{ m}$

$\tan\theta = \frac{y}{R} = \frac{5.90\times10^{-3}\text{ m}}{4.00\text{ m}} = 1.475\times10^{-3}$; $\theta = 1.475\times10^{-3}$ rad

The phase angle β at this point on the screen is

$\beta = \frac{2\pi}{\lambda}a\sin\theta = \frac{2\pi}{590\times10^{-9}\text{ m}}(0.200\times10^{-3}\text{ m})\sin(1.475\times10^{-3}\text{ rad}) = \pi$

Then $I = I_0\left[\frac{\sin\beta/2}{\beta/2}\right]^2 = (5.00\times10^{-6}\text{ W/m}^2)\left[\frac{\sin\frac{\pi}{2}}{\pi/2}\right]^2 = \frac{4}{\pi^2}(5.00\times10^{-6}\text{ W/m}^2)$

$I = \underline{2.03\times10^{-6}\text{ W/m}^2}$

38-13

Eq. (38-6): $\beta = \frac{2\pi}{\lambda}a\sin\theta$

$\lambda = \frac{2\pi}{\beta}a\sin\theta = \frac{2\pi}{\frac{\pi}{2}\text{ rad}}(0.400\times10^{-3}\text{ m})\sin 3.2° = 8.931\times10^{-5}\text{ m} = \underline{89.3\ \mu\text{m}}$

38-15

(i) $\phi = \frac{\pi}{2}$

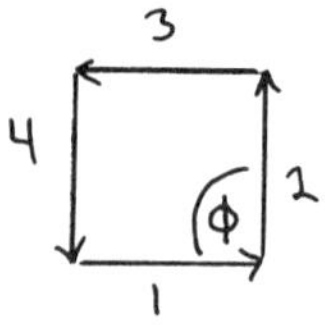

There is destructive interference between the light through slits 1 and 3 and between 2 and 4.

(ii) $\phi = \pi$

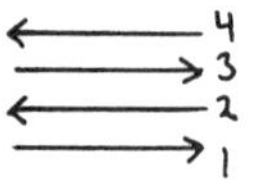

There is destructive interference between the light through slits 1 and 2 and between 3 and 4.

(iii) $\phi = \frac{3\pi}{2}$

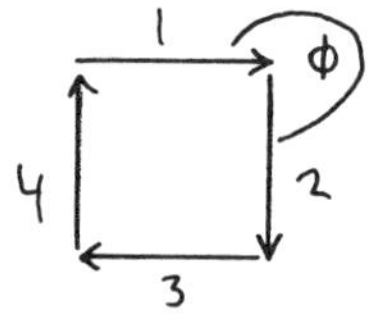

There is destructive interference between the light through slits 1 and 3 and between 2 and 4.

38-17

a) The interference fringes (maxima) are located by $d\sin\theta = m\lambda$, with $m = 0, \pm1, \pm2, \ldots$ The intensity I in the diffraction pattern is given by $I = I_0\left(\frac{\sin\beta/2}{\beta/2}\right)^2$, with $\beta = \frac{2\pi}{\lambda}a\sin\theta$.

We want $m = \pm3$ in the first equation to give θ that makes $I = 0$ in the second equation. $d\sin\theta = 3\lambda \Rightarrow \beta = \frac{2\pi}{\lambda}a\left(\frac{3\lambda}{d}\right) = 2\pi\left(3\frac{a}{d}\right)$

38-17 (cont)

$I = 0 \Rightarrow \dfrac{\sin(\beta/2)}{\beta/2} = 0 \Rightarrow \beta = 2\pi \Rightarrow 2\pi = 2\pi\left(3\dfrac{a}{d}\right) \Rightarrow \dfrac{d}{a} = 3.$

b) Fringes $m = 0, \pm 1, \pm 2$ are within the central diffraction maximum and the $m = \pm 3$ fringes coincide with the first diffraction minimum.

Find the value of m for the fringe that coincides with the second diffraction minimum:

second minimum $\Rightarrow \beta = 4\pi$

$\beta = \dfrac{2\pi}{\lambda} a \sin\theta = \dfrac{2\pi}{\lambda} a \dfrac{m\lambda}{d} = 2\pi m\left(\dfrac{a}{d}\right) = 2\pi \dfrac{m}{3}$

$\beta = 4\pi \Rightarrow 4\pi = 2\pi \dfrac{m}{3} \Rightarrow m = 6$

Therefore; the $m = +4$ and $m = +5$ fringes are contained within the first diffraction maximum on one side of the central maximum; two fringes.

38-19

a) If the slits are very narrow then the central maximum of the diffraction pattern for each slit completely fills the screen and the intensity distribution is given solely by the two-slit interference. The maxima are given by

$d \sin\theta = m\lambda \Rightarrow \sin\theta = \dfrac{m\lambda}{d}$

1st order maximum: $m = 1$, so $\sin\theta = \dfrac{\lambda}{d} = \dfrac{490 \times 10^{-9}\ m}{0.630 \times 10^{-3}\ m} = 7.778 \times 10^{-4}$; $\underline{\theta = 0.0446^\circ}$

2nd order maximum: $m = 2$, so $\sin\theta = \dfrac{2\lambda}{d} = 1.556 \times 10^{-3}$; $\underline{\theta = 0.0891^\circ}$

b) The intensity is given by Eq. (38-12): $I = I_0 \cos^2\left(\dfrac{\phi}{2}\right)\left[\dfrac{\sin\beta/2}{\beta/2}\right]^2$

$\phi = \dfrac{2\pi d}{\lambda}\sin\theta = \dfrac{2\pi d}{\lambda}\left(\dfrac{m\lambda}{d}\right) = 2\pi m$, so $\cos^2\left(\dfrac{\phi}{2}\right) = \cos^2(m\pi) = 1$

(Since the angular positions in part (a) correspond to interference maxima.)

$\beta = \dfrac{2\pi a}{\lambda}\sin\theta = \left(\dfrac{2\pi a}{\lambda}\right)\left(\dfrac{m\lambda}{d}\right) = 2\pi m\left(\dfrac{a}{d}\right) = m\, 2\pi\left(\dfrac{0.420\ mm}{0.630\ mm}\right) = m(4.189\ rad)$

1st order maximum: $m = 1$, so $I = I_0 (1)\left[\dfrac{\sin(4.189/2)\ rad}{(4.189/2)\ rad}\right]^2 = \underline{0.171\ I_0}$

2nd order maximum: $m = 2$, so $I = I_0 (1)\left[\dfrac{\sin 4.189\ rad}{4.189\ rad}\right]^2 = \underline{0.0428\ I_0}$

38-23

4000 lines/cm $\Rightarrow 4.00 \times 10^5$ lines/m

The slit spacing is $d = \dfrac{1}{4.00 \times 10^5}\ m = 2.50 \times 10^{-6}\ m$

a) The line positions are given by $\sin\theta = m\dfrac{\lambda}{d}$.

first order $\Rightarrow m = 1$

$\sin\theta_\alpha = \dfrac{\lambda_\alpha}{d} = \dfrac{656 \times 10^{-9}\ m}{2.50 \times 10^{-6}\ m} = 0.2624 \Rightarrow \theta_\alpha = 15.21^\circ$

38-23 (cont)

$$\sin\theta_\delta = \frac{\lambda_\delta}{d} = \frac{410\times10^{-9}\text{ m}}{2.50\times10^{-6}\text{ m}} = 0.1640 \Rightarrow \theta_\delta = 9.44°$$

The angular separation is $\theta_\alpha - \theta_\delta = 15.21° - 9.44° = \underline{5.77°}$.

b) second order $\Rightarrow m = 2$

$$\sin\theta_\alpha = \frac{2\lambda_\alpha}{d} = 0.5248 \Rightarrow \theta_\alpha = 31.65°$$

$$\sin\theta_\delta = \frac{2\lambda_\delta}{d} = 0.3280 \Rightarrow \theta_\delta = 19.15°$$

The angular separation is $\theta_\alpha - \theta_\delta = 31.65° - 19.15° = \underline{12.5°}$.

Note that the angular separation is larger in higher orders.

38-27

The maxima occur at angles θ given by Eq. (38-16):
$2d\sin\theta = m\lambda$, where d is the spacing between adjacent atomic planes.
Second order $\Rightarrow m = 2$

$$d = \frac{m\lambda}{2\sin\theta} = \frac{2(0.0820\times10^{-9}\text{ m})}{2\sin 23.5°} = 2.056\times10^{-10}\text{ m} = \underline{0.206\text{ nm}}$$

38-31

The angular radius of the first dark ring is given by $\sin\theta_1 = 1.22\frac{\lambda}{D}$ (Eq. 38-17).

$$\sin\theta_1 = 1.22\left(\frac{480\times10^{-9}\text{ m}}{8.4\times10^{-6}\text{ m}}\right) = 0.06971;\ \theta_1 = 0.06977\text{ rad}$$

The radius of the Airy disk (central bright spot) is $r = (4.0\text{ m})\tan\theta_1 = 0.2795\text{ m}$.
The diameter is $2r = 0.56\text{ m} = \underline{56\text{ cm}}$.

38-33

Resolved by Rayleigh's criterion $\Rightarrow$ angular separation θ of the objects equals $1.22\frac{\lambda}{D}$. The angular separation θ of the objects is

$$\theta = \frac{30.0\times10^{3}\text{ m}}{1.50\times10^{11}\text{ m}},$$ where 1.50×10^{11} m is the distance from the earth to the sun.

Thus $\theta = 2.00\times10^{-7}$.

$$\theta = 1.22\frac{\lambda}{D} \Rightarrow D = \frac{1.22\lambda}{\theta} = \frac{1.22(500\times10^{-9}\text{ m})}{2.00\times10^{-7}} = \underline{3.05\text{ m}}$$

Problems

38-37

a) $I = I_0\left(\frac{\sin\beta/2}{\beta/2}\right)^2$, where $\beta = \frac{2\pi}{\lambda}a\sin\theta$

38-37 (cont)

$$I = \tfrac{1}{2} I_0 \Rightarrow \frac{\sin \beta/2}{\beta/2} = \frac{1}{\sqrt{2}}$$

Let $x = \frac{\beta}{2}$; the equation for x is $\frac{\sin x}{x} = \frac{1}{\sqrt{2}} = 0.7071$.

Use trial and error to find the value of x that is a solution to this equation.

x	$\frac{\sin x}{x}$	x	$\frac{\sin x}{x}$
1.0 rad	0.841	1.4 rad	0.7039
1.5 rad	0.665	1.39 rad	0.7077 $\Rightarrow x = 1.39$ rad; $\beta = 2x = 2.78$ rad
1.2 rad	0.777		

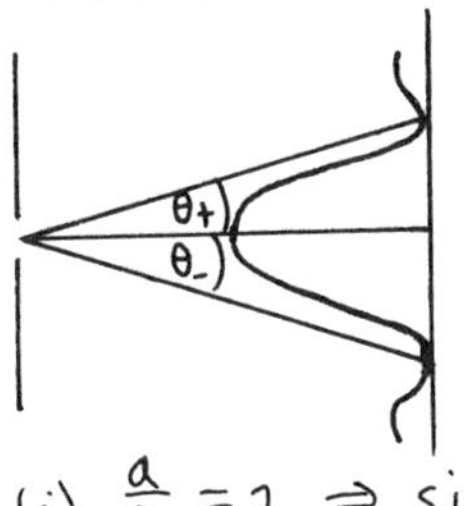

$$\Delta\theta = |\theta_+ - \theta_-| = 2\theta_+$$

$$\sin\theta_+ = \frac{\lambda\beta}{2\pi a} = \frac{\lambda}{a}\left(\frac{2.78 \text{ rad}}{2\pi \text{ rad}}\right) = 0.4425\frac{\lambda}{a}$$

(i) $\frac{a}{\lambda} = 2 \Rightarrow \sin\theta_+ = 0.4425\left(\frac{1}{2}\right) = 0.2212 \Rightarrow \theta_+ = 12.78°$; $\Delta\theta = 2\theta_+ = \underline{25.6°}$

(ii) $\frac{a}{\lambda} = 5 \Rightarrow \sin\theta_+ = 0.4425\left(\frac{1}{5}\right) = 0.0885 \Rightarrow \theta_+ = 5.077°$; $\Delta\theta = 2\theta_+ = \underline{10.2°}$

(iii) $\frac{a}{\lambda} = 10 \Rightarrow \sin\theta_+ = 0.4425\left(\frac{1}{10}\right) = 0.04425 \Rightarrow \theta_+ = 2.536°$; $\Delta\theta = \underline{5.1°}$

b) $\sin\theta_0 = \frac{\lambda}{a}$ locates the first minimum

(i) $\frac{a}{\lambda} = 2 \Rightarrow \sin\theta_0 = \frac{1}{2} \Rightarrow \theta_0 = 30.0°$; $2\theta_0 = \underline{60.0°}$

(ii) $\frac{a}{\lambda} = 5 \Rightarrow \sin\theta_0 = \frac{1}{5} \Rightarrow \theta_0 = 11.54°$; $2\theta_0 = \underline{23.1°}$

(iii) $\frac{a}{\lambda} = 10 \Rightarrow \sin\theta_0 = \frac{1}{10} \Rightarrow \theta_0 = 5.74°$; $2\theta_0 = \underline{11.5°}$

Note: Either definition of the width shows that the central maximum gets narrower as the slit gets wider.

38-39

a) The angular position of the first minimum is given by $a\sin\theta = m\lambda$ (Eq. 38-2), with $m = 1$.

$$\sin\theta = \frac{\lambda}{a} = \frac{600\times10^{-9}\text{ m}}{0.300\times10^{-3}\text{ m}} = 2.00\times10^{-3}\ ;\ \theta = 2.00\times10^{-3}\text{ rad}$$

The distance y_1 from the center of the central maximum to the first minimum is given by $y_1 = x\tan\theta = (1.40\text{ m})\tan(2.00\times10^{-3}\text{ rad}) = 2.80\times10^{-3}\text{ m} = \underline{2.80\text{ mm}}$. (Note that θ is small enough for $\theta \approx \sin\theta \approx \tan\theta$, and Eq. (38-3) applies.)

b) From part (a) of Problem 38-37, $I = \frac{1}{2}I_0$ when $\beta = 2.78$ rad.

$\beta = \frac{2\pi}{\lambda}a\sin\theta$ (Eq. (38-6)), so $\sin\theta = \frac{\beta\lambda}{2\pi a}$.

$$y = x\tan\theta \approx x\sin\theta \approx \frac{\beta\lambda x}{2\pi a} = \frac{(2.78\text{ rad})(600\times10^{-9}\text{ m})(1.40\text{ m})}{2\pi(0.300\times10^{-3}\text{ m})} = 1.24\times10^{-3}\text{ m} = \underline{1.24\text{ mm}}$$

38-41

The phase difference between adjacent slits is $\phi = \frac{2\pi d}{\lambda}\sin\theta \approx \frac{2\pi d\theta}{\lambda}$ when θ is small and $\sin\theta \approx \theta$. Thus $\theta = \frac{\lambda\phi}{2\pi d}$.

A principal maximum occurs when $\phi = \phi_{max} = m2\pi$, where m is an integer, since then all the phasors add. The first minima on either side of the mth

38-41 (cont)

principal maximum occurs when $\phi = \phi^{\pm}_{min} = m2\pi \pm \frac{2\pi}{N}$ and the phasor diagram for N slits forms a closed loop and the resultant phasor is zero.

The angular position of a principal maximum is $\theta = (\frac{\lambda}{2\pi d})\phi_{max}$. The angular position of the adjacent minimum is $\theta^{\pm}_{min} = (\frac{\lambda}{2\pi d})\phi^{\pm}_{min}$.

$\theta^{+}_{min} = \frac{\lambda}{2\pi d}(\phi_{max} + \frac{2\pi}{N}) = \theta + \frac{\lambda}{2\pi d}(\frac{2\pi}{N}) = \theta + \frac{\lambda}{Nd}$

$\theta^{-}_{min} = \frac{\lambda}{2\pi d}(\phi_{max} - \frac{2\pi}{N}) = \theta - \frac{\lambda}{Nd}$

The angular width of the principal maximum is $\theta^{+}_{min} - \theta^{-}_{min} = \frac{2\lambda}{Nd}$, as was to be shown.

38-47

The condition for an intensity maximum is $d\sin\theta = m\lambda$, $m = 0, \pm1, \pm2, \ldots$

fourth order $\Rightarrow$ $m = 4$

7000 lines/cm $\Rightarrow$ 7.00×10^5 lines/m, so $d = \frac{1}{7.00\times10^5}\text{ m} = 1.429\times10^{-6}\text{ m}$

The longest observable wavelength is the one that gives $\theta = 90°$ and hence $\sin\theta = 1$.

$\lambda = \frac{d\sin\theta}{m} = \frac{(1.429\times10^{-6}\text{ m})(1)}{4} = 3.57\times10^{-7}\text{ m} = \underline{357\text{ nm}}$

38-51

Rayleigh's criterion says that the two objects are resolved if the center of one diffraction pattern coincides with the first minimum of the other.

By Eq.(38-2) the angular position of the first minimum relative to the center of the central maximum is $\sin\theta = \frac{\lambda}{a}$, where a is the slit width. Hence if the objects are resolved according to Rayleigh's criterion, the angular separation between the centers of the images of the two objects must be at least $\frac{\lambda}{a}$.

But as discussed in Example 38-8, the angular separation of the image points equals the angular separation of the object points. The angular separation of the object points is $\frac{y}{s}$, where $y = 3.00\text{ m}$ is the linear separation of the two points and s is their distance from the observer.

Thus $\frac{y}{s} = \frac{\lambda}{a} \Rightarrow s = \frac{ya}{\lambda} = \frac{(3.00\text{ m})(0.400\times10^{-3}\text{ m})}{500\times10^{-9}\text{ m}} = \underline{2.40\text{ km}}$.

38-53

Resolved by Rayleigh's criterion $\Rightarrow$ angular separation θ of the objects is given by $\theta = 1.22\frac{\lambda}{D}$. $\theta = \frac{y}{s}$ where $y = 55.0\text{ m}$ is the distance between the two objects and s is their distance from the astronaut (her altitude).

$\frac{y}{s} = 1.22\frac{\lambda}{D}$

$s = \frac{yD}{1.22\lambda} = \frac{(55.0\text{ m})(4.00\times10^{-3}\text{ m})}{1.22(550\times10^{-9}\text{ m})} = 3.28\times10^5\text{ m} = \underline{328\text{ km}}$

CHAPTER 39

Exercises 1, 3, 7, 11, 13, 19, 21, 23, 25, 27, 31, 33, 39, 41

Problems 43, 45, 47, 53, 55, 57, 63

Exercises

39-1

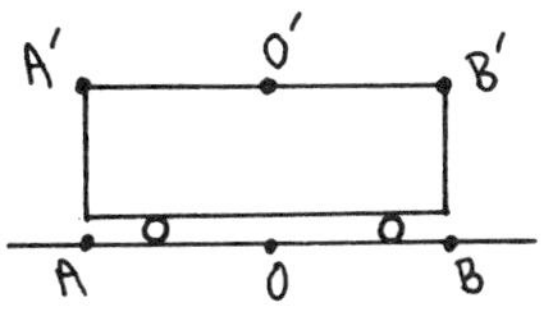

Simultaneous to observer on train ⇒ light pulses from A' and B' arrive at O' at the same time. To observer at O light from A' has a shorter distance to travel than light from B', so O will conclude that the pulse from A (A') started before the pulse at B (B'). To observer at O bolt A appears to strike first.

39-3

a) $\Delta t_0 = 2.2\times10^{-6}\,\text{s}$; $\Delta t = 1.9\times10^{-5}\,\text{s}$

$$\Delta t = \frac{\Delta t_0}{\sqrt{1-u^2/c^2}} \Rightarrow 1-\frac{u^2}{c^2} = \left(\frac{\Delta t_0}{\Delta t}\right)^2$$

$$\frac{u}{c} = \sqrt{1-\left(\frac{\Delta t_0}{\Delta t}\right)^2} = \sqrt{1-\left(\frac{2.2\times10^{-6}\,\text{s}}{1.9\times10^{-5}\,\text{s}}\right)^2} = 0.993\ ;\ u = \underline{0.993c}$$

b) The speed in the laboratory frame is $u = 0.993c$; the time measured in this frame is Δt, so the distance as measured in this frame is
$d = u\Delta t = (0.993)(2.998\times10^8\,\text{m/s})(1.9\times10^{-5}\,\text{s}) = 5.7\times10^3\,\text{m} = \underline{5.7\,\text{km}}$

39-7

$$\Delta t = 5.00\,\text{h} = 5.00\,\text{h}\left(\frac{3600\,\text{s}}{1\,\text{h}}\right) = 1.80\times10^4\,\text{s}$$

The elapsed time for the clock on the plane is Δt_0.

$$\Delta t = \frac{\Delta t_0}{\sqrt{1-u^2/c^2}} \Rightarrow \Delta t_0 = \Delta t\sqrt{1-u^2/c^2}$$

$$\frac{u}{c}\text{ small} \Rightarrow \sqrt{1-\frac{u^2}{c^2}} = \left(1-\frac{u^2}{c^2}\right)^{1/2} \simeq 1-\frac{1}{2}\frac{u^2}{c^2} \Rightarrow \Delta t_0 = \Delta t\left(1-\frac{1}{2}\frac{u^2}{c^2}\right)$$

The difference in the clock readings is $\Delta t - \Delta t_0 = \frac{1}{2}\frac{u^2}{c^2}\Delta t = \frac{1}{2}\left(\frac{400\,\text{m/s}}{2.998\times10^8\,\text{m/s}}\right)^2(1.80\times10^4\,\text{s})$
$= \underline{1.60\times10^{-8}\,\text{s}}$. The clock on the plane shows the shorter elapsed time.

39-11

$l = l_0\sqrt{1-\frac{u^2}{c^2}}$. The length measured when the spacecraft is moving is $l = 86.5\,\text{m}$; l_0 is the length measured in a frame at rest relative to the spacecraft.

$$l_0 = \frac{l}{\sqrt{1-u^2/c^2}} = \frac{86.5\,\text{m}}{\sqrt{1-(0.800c/c)^2}} = \underline{144\,\text{m}}$$

39-13

a) The distance measured in the earth's frame is the proper length $l_0 = 45.0 \times 10^3$ m.

$$l = l_0\sqrt{1-\frac{u^2}{c^2}} = (45.0\times10^3\,\text{m})\sqrt{1-\left(\frac{0.9954c}{c}\right)^2} = 4.31\times10^3\,\text{m} = \underline{4.31\,\text{km}}$$

b) Use the lifetime measured in the muon's frame as the time of travel to calculate the distance traveled as measured in that frame.

$$d = u\Delta t = (0.9954)(2.998\times10^8\,\text{m/s})(2.2\times10^{-6}\,\text{s}) = 660\,\text{m} = \underline{0.66\,\text{km}}$$

The muon's original height as measured in the muon's frame (part (a)) is 4.31 km, so the fraction is $\frac{0.66\,\text{km}}{4.31\,\text{km}} = 0.15 = \underline{15\%}$.

c) $\Delta t_0 = 2.2\times10^{-6}$ s ; $\Delta t = ?$

$$\Delta t = \frac{\Delta t_0}{\sqrt{1-u^2/c^2}} = \frac{2.2\times10^{-6}\,\text{s}}{\sqrt{1-\left(\frac{0.9954c}{c}\right)^2}} = 2.3\times10^{-5}\,\text{s} = 23\,\mu\text{s}$$

Use the lifetime in the earth's frame to find the distance traveled in that frame:

$$d = u\Delta t = (0.9954)(2.998\times10^8\,\text{m/s})(2.3\times10^{-5}\,\text{s}) = 6.9\times10^3\,\text{m} = \underline{6.9\,\text{km}}$$

The fraction is $\frac{6.9\,\text{km}}{45.0\,\text{km}} = 0.15 = 15\%$, the same fraction as in the muon's frame.

39-19

Use the Lorentz velocity transformation equation, Eq.(39-23): $v' = \frac{v-u}{1-uv/c^2}$

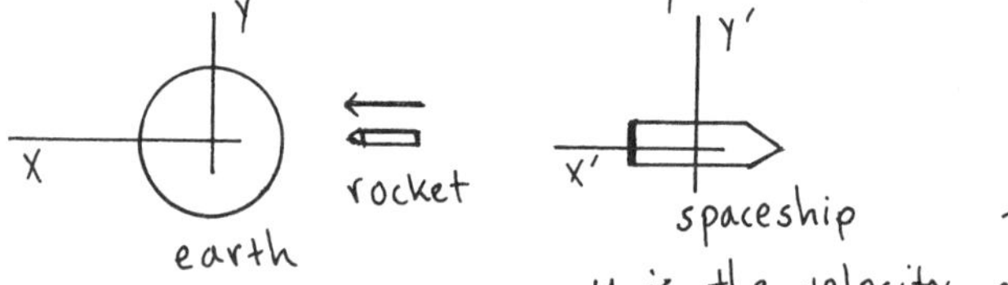

S = earth frame
S' = spaceship frame
The object is the rocket.
u is the velocity of the spaceship relative to the earth.

$v = +0.290c$
$v' = +0.840c$

(In each frame the rocket is moving in the positive coordinate direction.)

$$v' = \frac{v-u}{1-uv/c^2} \Rightarrow v' - u\frac{vv'}{c^2} = v-u \Rightarrow u\left(1-\frac{vv'}{c^2}\right) = v-v'$$

$$u = \frac{v-v'}{1-\frac{vv'}{c^2}} = \frac{0.290c-0.840c}{1-\frac{(0.290c)(0.840c)}{c^2}} = -\frac{0.550c}{0.7564} = \underline{-0.727c}$$

The speed of the spaceship relative to the earth is $0.727c = 2.18\times10^8$ m/s. The minus sign in our result for u means that the spaceship is moving in the $-x$-direction, so it is moving away from the earth.

39-21

Use the Lorentz velocity transformation equation, Eq.(39-23): $v' = \frac{v-u}{1-uv/c^2}$.

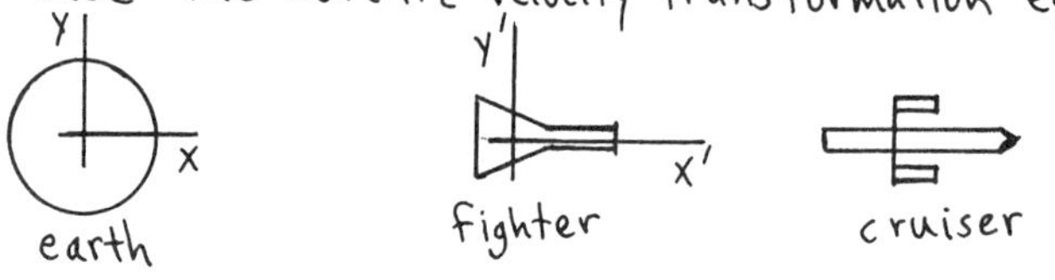

S = earth frame
S' = fighter frame
The object is the cruiser.

39-21 (cont)

With the coordinates shown, each ship is moving in the positive coordinate direction in the earth frame.

u is the velocity of the fighter relative to the earth; $u = +0.900c$

$v = +0.700c$

$v' = ?$ (velocity of the cruiser relative to the fighter)

$$v' = \frac{v - u}{1 - uv/c^2} = \frac{0.700c - 0.900c}{1 - (0.900c)(0.700c)/c^2} = \frac{-0.200c}{0.370} = -0.541c$$

The cruiser is moving toward the fighter with a speed of $\underline{0.541c} = \underline{1.62 \times 10^8 \text{ m/s}}$.

39-23

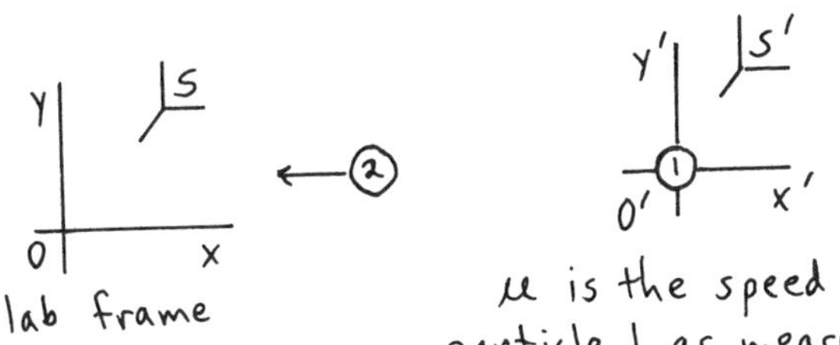

Frame S is at rest in the laboratory.
Frame S' is attached to particle 1.

u is the speed of S' relative to S; this is the speed of particle 1 as measured in the laboratory. Thus $u = +0.700c$.

The speed of particle 2 in S' is $0.900c$. Also, since the two particles move in opposite directions, 2 moves in the $-x'$ direction and $v' = -0.900c$.

We want to calculate v, the speed of particle 2 in frame S.

$$\text{Eq.(39-24)}: \; v = \frac{v' + u}{1 + uv'/c^2} = \frac{-0.900c + 0.700c}{1 + (0.900c)(-0.700c)/c^2} = \frac{-0.200c}{1 - 0.630} = -0.357c$$

The speed of the second particle, as measured in the laboratory, is $\underline{0.357c}$.

39-25

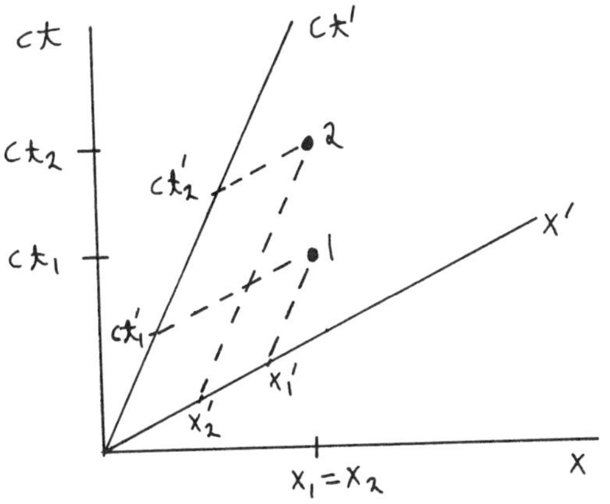

a) To find t_2' and t_1' on the spacetime diagram draw lines through points 2 and 1 parallel to the x' axis. The intercepts with the ct' axis gives ct_2' and ct_1'. We see that $t_2' > t_1'$, so Mavis agrees that event 1 occurs first.

b) To find x_2' and x_1' on the spacetime diagram draw lines through points 2 and 1 parallel to the ct' axis. The intercepts with the x' axis give x_2' and x_1'. We see that $x_1' \neq x_2'$ (in fact $x_1' > x_2'$), so according to Mavis the two events do not occur at the same position.

39-27

Source and observer approaching $\Rightarrow$ use Eq. (39-26): $f = \sqrt{\frac{c+u}{c-u}}\, f_0$.
Solve for u:

$$f^2 = \left(\frac{c+u}{c-u}\right) f_0^2$$

$$(c-u)f^2 = (c+u)f_0^2 \Rightarrow u = \frac{c(f^2 - f_0^2)}{f^2 + f_0^2} = c\left(\frac{(f/f_0)^2 - 1}{(f/f_0)^2 + 1}\right)$$

Convert to an equation in terms of λ, λ_0 rather than f, f_0:

$$f = \frac{c}{\lambda},\ f_0 = \frac{c}{\lambda_0} \Rightarrow \frac{f}{f_0} = \frac{\lambda_0}{\lambda} \Rightarrow u = c\left(\frac{(\lambda_0/\lambda)^2 - 1}{(\lambda_0/\lambda)^2 + 1}\right)$$

$\lambda_0 = 675\text{ nm}$, $\lambda = 525\text{ nm}$

$$u = \left[\frac{(675\text{nm}/525\text{nm})^2 - 1}{(675\text{nm}/525\text{nm})^2 + 1}\right] c = 0.246c = (0.246)(2.998\times10^8\text{ m/s}) = \underline{7.38\times10^7\text{ m/s}};$$ definitely speeding

39-31

$$\text{Eq. (39-30): } F = \frac{dp}{dt} = \frac{d}{dt}\left(\frac{mv}{\sqrt{1-v^2/c^2}}\right) = \frac{m}{(1-v^2/c^2)^{1/2}}\frac{dv}{dt} + \frac{mv}{(1-v^2/c^2)^{3/2}}\left(-\frac{1}{2}\right)\left(-\frac{2v}{c^2}\right)\frac{dv}{dt}$$

$$F = \frac{dv}{dt}\frac{m}{(1-v^2/c^2)^{3/2}}\left[1 - \frac{v^2}{c^2} + \frac{v^2}{c^2}\right] = \frac{dv}{dt}\frac{m}{(1-v^2/c^2)^{3/2}}$$

But $\frac{dv}{dt} = a$, so $F = \frac{ma}{(1-v^2/c^2)^{3/2}}$, which is Eq. (39-31).

39-33

$$\text{Eq. 39-38: } K = \frac{1}{2}mv^2 + \frac{3}{8}\frac{mv^4}{c^2}$$

We want $\frac{K - \frac{1}{2}mv^2}{\frac{1}{2}mv^2} = 0.010$. So $\frac{\frac{3}{8}mv^4/c^2}{\frac{1}{2}mv^2} = 0.010$.

$$\frac{3}{4}\frac{v^2}{c^2} = 0.010 \Rightarrow v = \sqrt{\frac{4}{3}(0.010)}\ c = 0.12c = \underline{3.5\times10^7\text{ m/s}}$$

39-39

a) $K = 0.380\text{ MeV} = 3.80\times10^5\text{ eV}$

b) $E = K + mc^2$ (Eq. 39-39)

For an electron $mc^2 = (9.109\times10^{-31}\text{ kg})(2.998\times10^8\text{ m/s})^2 = 8.187\times10^{-14}\text{ J}\left(\frac{1\text{ eV}}{1.602\times10^{-19}\text{ J}}\right)$

$mc^2 = 5.110\times10^5\text{ eV} = 0.511\text{ MeV}$

Then $E = K + mc^2 = 0.380\text{ MeV} + 0.511\text{ MeV} = 0.891\text{ MeV} = \underline{8.91\times10^5\text{ eV}}$

c) $E = \frac{mc^2}{\sqrt{1-v^2/c^2}} \Rightarrow 1 - \frac{v^2}{c^2} = \left(\frac{mc^2}{E}\right)^2$

$$v = c\sqrt{1 - \left(\frac{mc^2}{E}\right)^2} = 2.998\times10^8\text{ m/s}\sqrt{1 - \left(\frac{0.511\text{ MeV}}{0.891\text{ MeV}}\right)^2} = \underline{2.46\times10^8\text{ m/s}}$$

d) The classical relation between speed and kinetic energy is

$$K = \frac{1}{2}mv^2 \Rightarrow v = \sqrt{\frac{2K}{m}} = \sqrt{\frac{2(3.80\times10^5\text{ eV})(1.602\times10^{-19}\text{ J/1 eV})}{9.109\times10^{-31}\text{ kg}}} = \underline{3.66\times10^8\text{ m/s}}$$

39-39 (cont)

The classical result is too large by about 50%.

39-41

a) The total energy is given in terms of the momentum by Eq. (39-40):

$$E=\sqrt{(mc^2)^2+(pc)^2}=\sqrt{[(3.32\times10^{-27}\text{ kg})(2.998\times10^8\text{ m/s})^2]^2+[(8.25\times10^{-19}\text{ kg}\cdot\text{m/s})(2.998\times10^8\text{ m/s})]^2}$$

$E = \underline{3.88\times10^{-10}\text{ J}}$

b) $E = K + mc^2 \Rightarrow K = E - mc^2$

$mc^2 = 3.32\times10^{-27}\text{ kg}\,(2.998\times10^8\text{ m/s})^2 = 2.98\times10^{-10}\text{ J}$

$K = E - mc^2 = 3.88\times10^{-10}\text{ J} - 2.98\times10^{-10}\text{ J} = \underline{9.0\times10^{-11}\text{ J}}$

c) $\dfrac{K}{mc^2} = \dfrac{9.0\times10^{-11}\text{ J}}{2.98\times10^{-10}\text{ J}} = \underline{0.30}$

Problems

39-43

There must be a length contraction such that the length a becomes the same as b; $l_0 = a$, $l = b$.

Eq. (39-13): $\dfrac{l}{l_0} = \sqrt{1-u^2/c^2} \Rightarrow \dfrac{b}{a} = \sqrt{1-u^2/c^2}$; $a = 1.50b \Rightarrow \dfrac{b}{1.50b} = \sqrt{1-u^2/c^2}$

$1-\dfrac{u^2}{c^2} = \dfrac{1}{(1.50)^2} \Rightarrow u = \sqrt{1-\dfrac{1}{(1.50)^2}}\; c = 0.745c = \underline{2.23\times10^8\text{ m/s}}$

39-45

$\Delta t_0 = 2.6\times10^{-8}\text{ s}$

The time measured in the lab must satisfy $d = c\Delta t \Rightarrow \Delta t = \dfrac{d}{c} = \dfrac{2.20\times10^3\text{ m}}{2.998\times10^8\text{ m/s}} = 7.338\times10^{-6}\text{ s}$

$\Delta t = \dfrac{\Delta t_0}{\sqrt{1-u^2/c^2}} \Rightarrow \left(1-\dfrac{u^2}{c^2}\right)^{1/2} = \dfrac{\Delta t_0}{\Delta t} \Rightarrow \left(1-\dfrac{u^2}{c^2}\right) = \left(\dfrac{\Delta t_0}{\Delta t}\right)^2$

Write $u = (1-\Delta)c \Rightarrow \left(\dfrac{u}{c}\right)^2 = (1-\Delta)^2 = 1-2\Delta+\Delta^2 \simeq 1-2\Delta$ since Δ small.

Use this in the above $\Rightarrow 1-(1-2\Delta) = \left(\dfrac{\Delta t_0}{\Delta t}\right)^2$

$\Delta = \dfrac{1}{2}\left(\dfrac{\Delta t_0}{\Delta t}\right)^2 = \dfrac{1}{2}\left(\dfrac{2.6\times10^{-8}\text{ s}}{7.338\times10^{-6}\text{ s}}\right)^2 = \underline{6.28\times10^{-6}}$

An alternative calculation is to say that the length of the tube must contract relative to the moving pion so that the pion travels that length before decaying. The contracted length must be $l = c\,\Delta t_0 = (2.998\times10^8\text{ m/s})(2.6\times10^{-8}\text{ s}) = 7.8\text{ m}$.

$l = l_0\sqrt{1-u^2/c^2} \Rightarrow 1-u^2/c^2 = \left(\dfrac{l}{l_0}\right)^2$

Then $u = (1-\Delta)c \Rightarrow \Delta = \dfrac{1}{2}\left(\dfrac{l}{l_0}\right)^2 = \dfrac{1}{2}\left(\dfrac{7.8\text{ m}}{2.20\times10^3\text{ m}}\right)^2 = 6.28\times10^{-6}$, which checks.

b) $E = \gamma mc^2$

39-45 (cont)

$$\gamma = \frac{1}{\sqrt{1-u^2/c^2}} = \frac{1}{\sqrt{2\Delta}} = 282$$

$$E = 282\,(139.6 \text{ MeV}) = 3.94\times10^4 \text{ MeV} = \underline{39.4 \text{ GeV}}$$

39-47

$E = mc^2$; the mass increase is due to the heat flow into the ice to melt it.

$$E = Q = mL_f = (5.00 \text{ kg})(334\times10^3 \text{ J/kg}) = 1.67\times10^6 \text{ J}$$

$$m = \frac{E}{c^2} = \frac{1.67\times10^6 \text{ J}}{(2.998\times10^8 \text{ m/s})^2} = \underline{1.86\times10^{-11} \text{ kg}}$$

39-53

In flint glass the speed of light is $v = \frac{c}{n} = \frac{2.998\times10^8 \text{ m/s}}{1.62} = 1.851\times10^8 \text{ m/s}$.
Calculate the kinetic energy of an electron that has this speed:

$$K = mc^2(\gamma-1)$$

$$mc^2 = (9.109\times10^{-31} \text{ kg})(2.998\times10^8 \text{ m/s})^2 = 8.187\times10^{-14} \text{ J}\left(\frac{1 \text{ eV}}{1.602\times10^{-19} \text{ J}}\right) = 0.5111 \text{ MeV}$$

$$\gamma = \frac{1}{\sqrt{1-v^2/c^2}} = \frac{1}{\sqrt{1-\left(\frac{1.851\times10^8 \text{ m/s}}{2.998\times10^8 \text{ m/s}}\right)^2}} = 1.271$$

$$K = mc^2(\gamma-1) = 0.5111 \text{ MeV}\,(1.271-1) = \underline{0.139 \text{ MeV}}$$

39-55

An increase in wavelength corresponds to a decrease in frequency $\left(f = \frac{c}{\lambda}\right)$, so the atoms are moving away from the earth.

Receding $\Rightarrow$ use Eq. (39-27): $f = \sqrt{\frac{c-u}{c+u}}\, f_0$

Solve for u: $\left(\frac{f}{f_0}\right)^2 (c+u) = c-u \Rightarrow u = c\left(\frac{1-(f/f_0)^2}{1+(f/f_0)^2}\right)$

$$f = \frac{c}{\lambda},\quad f_0 = \frac{c}{\lambda_0} \Rightarrow \frac{f}{f_0} = \frac{\lambda_0}{\lambda}$$

$$u = c\left(\frac{1-(\lambda_0/\lambda)^2}{1+(\lambda_0/\lambda)^2}\right) = c\left(\frac{1-(121.6/346.2)^2}{1+(121.6/346.2)^2}\right) = 0.780\,c = \underline{2.34\times10^8 \text{ m/s}}$$

39-57

According to Eq. (39-31), $a = \frac{dv}{dt} = \frac{F}{m}\left(1 - v^2/c^2\right)^{3/2}$.
(One-dimensional motion is assumed, and all the F, v, and a refer to x-components.)

$$\Rightarrow \frac{dv}{(1-v^2/c^2)^{3/2}} = \left(\frac{F}{m}\right) dt$$

Integrate from $t=0$ when $v=0$ to time t when the velocity is v.

$$\int_0^v \frac{dv}{(1-v^2/c^2)^{3/2}} = \int_0^t \frac{F}{m}\, dt$$

Since F is constant, $\int_0^t \frac{F}{m}\, dt = \frac{Ft}{m}$.

39-57 (cont)

In the velocity integral make the change of variable $y = \frac{v}{c}$; $dy = \frac{dv}{c}$

$$c\int_0^{v/c} \frac{dy}{(1-y^2)^{3/2}} = c\left(\frac{y}{(1-y^2)^{1/2}}\Big|_0^{v/c}\right) = c\frac{v/c}{\sqrt{1-(v/c)^2}} = \frac{v}{\sqrt{1-v^2/c^2}}$$

Thus $\frac{v}{\sqrt{1-v^2/c^2}} = \frac{Ft}{m}$.

Solve this equation for v:

$$\frac{v^2}{1-v^2/c^2} = \left(\frac{Ft}{m}\right)^2 \Rightarrow v^2 = \left(\frac{Ft}{m}\right)^2\left(1-\frac{v^2}{c^2}\right)$$

$$v^2\left(1+\left(\frac{Ft}{mc}\right)^2\right) = \left(\frac{Ft}{m}\right)^2$$

$$v = \frac{Ft/m}{\sqrt{1+(Ft/mc)^2}} = c\frac{Ft}{\sqrt{m^2c^2+F^2t^2}}$$

As $t \to \infty$, $\frac{Ft}{\sqrt{m^2c^2+F^2t^2}} \to \frac{Ft}{\sqrt{F^2t^2}} \to 1$, so $v \to c$.

Note also that $\frac{Ft}{\sqrt{m^2c^2+F^2t^2}}$ is always less than 1, so $v<c$ always and approaches c only when $t \to \infty$.

39-63

a) Use the Lorentz coordinate transformation (Eq. 39-22) for (x_1, t_1) and (x_2, t_2)

$$x_1' = \frac{x_1 - ut_1}{\sqrt{1-u^2/c^2}}, \quad x_2' = \frac{x_2 - ut_2}{\sqrt{1-u^2/c^2}}$$

$$t_1' = \frac{t_1 - ux_1/c^2}{\sqrt{1-u^2/c^2}}, \quad t_2' = \frac{t_2 - ux_2/c^2}{\sqrt{1-u^2/c^2}}$$

Same point in $S' \Rightarrow x_1' = x_2'$. What then is $\Delta t' = t_2' - t_1'$?

$$x_1' = x_2' \Rightarrow x_1 - ut_1 = x_2 - ut_2$$

$$u(t_2 - t_1) = x_2 - x_1 \Rightarrow u = \frac{x_2 - x_1}{t_2 - t_1} = \frac{\Delta x}{\Delta t}$$

From the time transformation equations,

$$\Delta t' = t_2' - t_1' = \frac{1}{\sqrt{1-u^2/c^2}}(\Delta t - u\Delta x/c^2)$$

Use the result that $u = \frac{\Delta x}{\Delta t}$

$$\Rightarrow \Delta t' = \frac{1}{\sqrt{1-\frac{(\Delta x)^2}{(\Delta t)^2c^2}}}\left(\Delta t - \frac{(\Delta x)^2}{(\Delta t)c^2}\right) = \frac{\Delta t}{\sqrt{(\Delta t)^2 - \frac{(\Delta x)^2}{c^2}}}\left(\Delta t - \frac{(\Delta x)^2}{(\Delta t)c^2}\right)$$

$$\Delta t' = \frac{(\Delta t)^2 - (\Delta x)^2/c^2}{\sqrt{(\Delta t)^2 - (\Delta x)^2/c^2}} = \sqrt{(\Delta t)^2 - \left(\frac{\Delta x}{c}\right)^2},$$ as was to be shown.

This equation doesn't have a physical solution (because of negative square root) if $\left(\frac{\Delta x}{c}\right)^2 > (\Delta t)^2$ or $\Delta x \geq c\Delta t$.

b) Now require that $t_2' = t_1'$ (the two events are simultaneous in S') and use the Lorentz coordinate transformation equations.

$$t_1' = t_2' \Rightarrow t_1 - ux_1/c^2 = t_2 - ux_2/c^2$$

$$t_2 - t_1 = \frac{x_2 - x_1}{c^2}u \Rightarrow \Delta t = \frac{\Delta x}{c^2}u \Rightarrow u = \frac{c^2\,\Delta t}{\Delta x}$$

From the Lorentz transformation equations, $\Delta x' = x_2' - x_1' = \frac{1}{\sqrt{1-u^2/c^2}}(\Delta x - u\Delta t)$.
Use the result that $u = c^2\Delta t/\Delta x$

$$\Rightarrow \Delta x' = \frac{1}{\sqrt{1 - \frac{c^2(\Delta t)^2}{(\Delta x)^2}}}\left(\Delta x - \frac{c^2(\Delta t)^2}{\Delta x}\right) = \frac{\Delta x}{\sqrt{(\Delta x)^2 - c^2(\Delta t)^2}}\left(\Delta x - \frac{c^2(\Delta t)^2}{\Delta x}\right)$$

$$\Delta x' = \frac{(\Delta x)^2 - c^2(\Delta t)^2}{\sqrt{(\Delta x)^2 - c^2(\Delta t)^2}} = \sqrt{(\Delta x)^2 - c^2(\Delta t)^2}$$

c) $\Delta x' = \sqrt{(\Delta x)^2 - c^2(\Delta t)^2}$

Solve for Δt: $(\Delta x')^2 = (\Delta x)^2 - c^2(\Delta t)^2$

$$\Delta t = \frac{\sqrt{(\Delta x)^2 - (\Delta x')^2}}{c} = \frac{\sqrt{(2.50\text{ m})^2 - (1.25\text{ m})^2}}{2.998\times10^8\text{ m/s}} = \underline{7.22\times10^{-9}\text{ s}}$$

CHAPTER 40

Exercises 1, 5, 7, 9, 11, 13, 17, 21, 23, 25, 27, 33, 39

Problems 43, 45, 49, 51, 55, 57, 59, 61, 65

Exercises

40-1

a) $E = hf \Rightarrow f = \frac{E}{h} = \frac{(3.25\times10^{6}\ \text{eV})\left(\frac{1.602\times10^{-19}\ \text{J}}{1\ \text{eV}}\right)}{6.626\times10^{-34}\ \text{J}\cdot\text{s}} = \underline{7.86\times10^{20}\ \text{Hz}}$

b) $c = f\lambda \Rightarrow \lambda = \frac{c}{f} = \frac{2.998\times10^{8}\ \text{m/s}}{7.86\times10^{20}\ \text{Hz}} = \underline{3.81\times10^{-13}\ \text{m}}$

c) λ is comparable to a nuclear radius

40-5

a) Eq. (39-41): $E = pc = (4.25\times10^{-27}\ \text{kg}\cdot\text{m/s})(2.998\times10^{8}\ \text{m/s}) = 1.27\times10^{-18}\ \text{J}$

$E = (1.27\times10^{-18}\ \text{J})\left(\frac{1\ \text{eV}}{1.602\times10^{-19}\ \text{J}}\right) = \underline{7.93\ \text{eV}}$

b) Eq. (40-5): $p = \frac{h}{\lambda} \Rightarrow \lambda = \frac{h}{p} = \frac{6.626\times10^{-34}\ \text{J}\cdot\text{s}}{4.25\times10^{-27}\ \text{kg}\cdot\text{m/s}} = 1.56\times10^{-7}\ \text{m} = \underline{156\ \text{nm}}$

This wavelength is shorter than visible wavelengths; it is in the ultraviolet region of the electromagnetic spectrum.

40-7

Eq. (40-3): $\frac{1}{2}mv_{max}^2 = hf - \phi = (6.626\times10^{-34}\ \text{J}\cdot\text{s})(3.4\times10^{15}\ \text{Hz}) - (5.1\ \text{eV})\left(\frac{1.602\times10^{-19}\ \text{J}}{1\ \text{eV}}\right)$

$\frac{1}{2}mv_{max}^2 = 2.253\times10^{-18}\ \text{J} - 8.170\times10^{-19}\ \text{J} = 1.436\times10^{-18}\ \text{J}$

$v_{max} = \sqrt{\frac{2(1.436\times10^{-18}\ \text{J})}{m}} = \sqrt{\frac{2(1.436\times10^{-18}\ \text{J})}{9.109\times10^{-31}\ \text{kg}}} = \underline{1.78\times10^{6}\ \text{m/s}}$

40-9

a) First find the work function ϕ:

$eV_0 = hf - \phi \Rightarrow \phi = hf - eV_0 = \frac{hc}{\lambda} - eV_0$

$\phi = \frac{(6.626\times10^{-34}\ \text{J}\cdot\text{s})(2.998\times10^{8}\ \text{m/s})}{254\times10^{-9}\ \text{m}} - (1.602\times10^{-19})(0.181\ \text{V})$

$\phi = 7.821\times10^{-19}\ \text{J} - 2.900\times10^{-20}\ \text{J} = 7.531\times10^{-19}\ \text{J}\left(\frac{1\ \text{eV}}{1.602\times10^{-19}\ \text{J}}\right) = 4.70\ \text{eV}$

The threshold frequency f_{th} is the smallest frequency that still produces photoelectrons. It corresponds to $K_{max} = 0$ in Eq. (40-3), so $hf_{th} = \phi$.

$f = \frac{c}{\lambda} \Rightarrow \frac{hc}{\lambda_{th}} = \phi$

$\lambda_{th} = \frac{hc}{\phi} = \frac{(6.626\times10^{-34}\ \text{J}\cdot\text{s})(2.998\times10^{8}\ \text{m/s})}{7.531\times10^{-19}\ \text{J}} = 2.64\times10^{-7}\ \text{m} = \underline{264\ \text{nm}}$

40-9 (cont)

b) As calculated in part (a), $\phi = 4.70\text{ eV}$. This is the value given in Table 40-1.

40-11

a) Eq. (40-7): $\frac{1}{\lambda} = R\left(\frac{1}{2^2} - \frac{1}{n^2}\right)$

$n = 4 \Rightarrow \frac{1}{\lambda} = R\left(\frac{1}{2^2} - \frac{1}{4^2}\right) = R\,\frac{4-1}{16} = R\,\frac{3}{16}$

$\lambda = \frac{16}{3R} = \frac{16}{3(1.097\times10^7)}\text{ m} = 4.86\times10^{-7}\text{ m} = 486\text{ nm}$

$f = \frac{c}{\lambda} = \frac{2.998\times10^8\text{ m/s}}{4.86\times10^{-7}\text{ m}} = \underline{6.17\times10^{14}\text{ Hz}}$

b) $\lambda = \underline{486\text{ nm}}$, as calculated in part (a).

40-13

$hf = \Delta E$; $f = \frac{c}{\lambda} \Rightarrow \frac{hc}{\lambda} = \Delta E \Rightarrow \lambda = \frac{hc}{\Delta E}$

$\Delta E = 3.40\text{ eV}\left(\frac{1.602\times10^{-19}\text{ J}}{1\text{ eV}}\right) = 5.447\times10^{-19}\text{ J}$

$\lambda = \frac{hc}{\Delta E} = \frac{(6.626\times10^{-34}\text{ J}\cdot\text{s})(2.998\times10^8\text{ m/s})}{5.447\times10^{-19}\text{ J}} = 3.65\times10^{-7}\text{ m} = \underline{365\text{ nm}}$

40-17

a) If the particles are treated as point charges, $U = \frac{1}{4\pi\epsilon_0}\frac{q_1 q_2}{r}$.

$q_1 = 2e$ (alpha particle)

$q_2 = 79e$ (gold nucleus)

$U = (8.987\times10^9\text{ N}\cdot\text{m}^2/\text{C}^2)\frac{(2)(79)(1.602\times10^{-19}\text{ C})^2}{5.60\times10^{-14}\text{ m}} = \underline{6.51\times10^{-13}\text{ J}}$

$U = 6.51\times10^{-13}\text{ J}\left(\frac{1\text{ eV}}{1.602\times10^{-19}\text{ J}}\right) = 4.06\times10^6\text{ eV} = \underline{4.06\text{ MeV}}$

b) $K_1 + U_1 = K_2 + U_2$

Alpha particle is initially far from the gold nucleus $\Rightarrow r_1 \simeq \infty$ and $U_1 = 0$.

Alpha particle stops $\Rightarrow K_2 = 0$.

Thus $K_1 = U_2 = 6.51\times10^{-13}\text{ J} = 4.06\text{ MeV}$

c) $K = \frac{1}{2}mv^2 \Rightarrow v = \sqrt{\frac{2K}{m}} = \sqrt{\frac{2(6.51\times10^{-13}\text{ J})}{6.65\times10^{-27}\text{ kg}}} = \underline{1.40\times10^7\text{ m/s}}$

40-21

The force between the electron and the nucleus in Li^{2+} is $F = \frac{1}{4\pi\epsilon_0}\frac{Ze^2}{r^2}$, where $Z = 3$ is the nuclear charge. All the equations for the hydrogen atom apply to Li^{2+} if we replace e^2 by Ze^2.

a) $E_n = -\frac{1}{\epsilon_0^2}\frac{me^4}{8n^2h^2}$ (hydrogen) becomes

40-21 (cont)

$$E_n = -\frac{1}{\epsilon_0^2}\frac{m(Ze^2)^2}{8n^2h^2} = Z^2\left(-\frac{1}{\epsilon_0^2}\frac{me^4}{8n^2h^2}\right) = Z^2\left(-\frac{13.60\text{ eV}}{n^2}\right) \quad (Li^{2+}).$$

The ground-level energy of Li^{2+} is $E_1 = 9\left(\frac{-13.60\text{ eV}}{1^2}\right) = -122\text{ eV}$.
The ground-level energy of Li^{2+} is $Z^2 = 9$ times the ground-level energy of H.

b) The ionization energy is the energy difference between the $n\to\infty$ level energy and the $n=1$ level energy. The $n\to\infty$ level energy is zero, so the ionization energy of Li^{2+} is 122 eV; this is 9 times the ionization energy of hydrogen.

c) $\frac{1}{\lambda} = R\left(\frac{1}{n_1^2} - \frac{1}{n_2^2}\right)$ just as for hydrogen but now R has a different value.

$R_H = \frac{me^4}{8\epsilon_0^2 h^3 c} = 1.097\times10^7\text{ m}^{-1}$ for hydrogen becomes

$R_{Li} = Z^2\frac{me^4}{8\epsilon_0^2 h^3 c} = 9(1.097\times10^7\text{ m}^{-1}) = 9.873\times10^7\text{ m}^{-1}$ for Li^{2+}.

For $n=2$ to $n=1$, $\frac{1}{\lambda} = R_{Li}\left(\frac{1}{1^2} - \frac{1}{2^2}\right) = \frac{3R}{4}$

$\lambda = \frac{4}{3R_{Li}} = \frac{4}{3(9.873\times10^7\text{ m}^{-1})} = 1.35\times10^{-8}\text{ m} = 13.5\text{ nm}$

This wavelength is smaller by a factor of 9 compared to the wavelength for the corresponding transition in the hydrogen atom.

d) Eq. (40-12): $r_n = \epsilon_0\frac{n^2h^2}{\pi m e^2}$ (hydrogen). Thus $r_n = \epsilon_0\frac{n^2h^2}{\pi m(Ze^2)}$ (Li^{2+})

For a given n the orbit radius for Li^{2+} is smaller by a factor of $Z=3$ compared to the corresponding radius for hydrogen.

40-23

Power is energy per unit time, so in 1.00 s the energy emitted by the laser is $(2.50\times10^{-3}\text{ W})(1.00\text{ s}) = 2.50\times10^{-3}\text{ J}$.

The energy of each photon is $E = \frac{hc}{\lambda} = \frac{(6.626\times10^{-34}\text{ J}\cdot\text{s})(2.998\times10^8\text{ m/s})}{633\times10^{-9}\text{ m}} = 3.138\times10^{-19}\text{ J}$.

The number of photons emitted each second is the total energy emitted divided by the energy of one photon: $\frac{2.50\times10^{-3}\text{ J/s}}{3.138\times10^{-19}\text{ J/photon}} = \underline{7.97\times10^{15}\text{ photons/s}}$

40-25

$$\frac{n_{5s}}{n_{3p}} = e^{-(E_{5s}-E_{3p})/kT}$$

From Fig. 40-21a, $E_{5s} = 20.66\text{ eV}$ and $E_{3p} = 18.70\text{ eV}$

$E_{5s} - E_{3p} = 20.66\text{ eV} - 18.70\text{ eV} = 1.96\text{ eV}\left(\frac{1.602\times10^{-19}\text{ J}}{1\text{ eV}}\right) = 3.140\times10^{-19}\text{ J}$

$$\frac{n_{5s}}{n_{3p}} = e^{-(3.140\times10^{-19}\text{ J})/[(1.381\times10^{-23}\text{ J/K})(300\text{ K})]} = e^{-75.79} = \underline{1.2\times10^{-33}}$$

40-27

The kinetic energy K of an electron after acceleration is
$K = q\Delta V = (1.602\times10^{-19}\text{ C})(17.0\times10^{3}\text{ V}) = 2.723\times10^{-15}\text{ J}$.
The shortest wavelength x rays are those for which the kinetic energy of an electron is converted entirely to the photon energy $hf = \frac{hc}{\lambda}$.
$$\frac{hc}{\lambda} = K \Rightarrow \lambda = \frac{hc}{K} = \frac{(6.626\times10^{-34}\text{ J}\cdot\text{s})(2.998\times10^{8}\text{ m/s})}{2.723\times10^{-15}\text{ J}} = 7.30\times10^{-11}\text{ m} = \underline{0.0730\text{ nm}}$$

40-33

Eq. (40-21): $\lambda' - \lambda = \frac{h}{mc}(1-\cos\phi) = \lambda_c(1-\cos\phi)$
$\lambda' = \lambda + \lambda_c(1-\cos\phi)$
The largest λ' corresponds to $\phi = 180°$, so $\cos\phi = -1$.
Then $\lambda' = \lambda + 2\lambda_c = 0.0840\times10^{-9}\text{ m} + 2(2.426\times10^{-12}\text{ m}) = 8.885\times10^{-11}\text{ m} = \underline{0.0889\text{ nm}}$.

40-39

The wavelength λ_m where the Planck distribution peaks is given by Eq. (40-28):
$$\lambda_m = \frac{2.90\times10^{-3}\text{ m}\cdot\text{K}}{T} = \frac{2.90\times10^{-3}\text{ m}\cdot\text{K}}{2.726\text{ K}} = 1.06\times10^{-3}\text{ m} = \underline{1.06\text{ mm}}$$
(This wavelength is in the microwave portion of the electromagnetic spectrum.)

Problems

40-43

a) One photon dissociates one AgBr molecule, so we need to find the energy required to dissociate a single molecule. The problem states that it requires 1.00×10^{5} J to dissociate one mole of AgBr, and one mole contains Avogadro's number (6.02×10^{23}) of molecules, so the energy required to dissociate one AgBr is
$$\frac{1.00\times10^{5}\text{ J/mol}}{6.02\times10^{23}\text{ molecules/mol}} = 1.66\times10^{-19}\text{ J/molecule}$$
The photon is to have this energy, so $E = 1.66\times10^{-19}\text{ J}\left(\frac{1\text{ eV}}{1.602\times10^{-19}\text{ J}}\right) = \underline{1.04\text{ eV}}$

b) $E = \frac{hc}{\lambda} \Rightarrow \lambda = \frac{hc}{E} = \frac{(6.626\times10^{-34}\text{ J}\cdot\text{s})(3.00\times10^{8}\text{ m/s})}{1.66\times10^{-19}\text{ J}} = 1.20\times10^{-6}\text{ m} = \underline{1200\text{ nm}}$

c) $c = f\lambda \Rightarrow f = \frac{c}{\lambda} = \frac{3.00\times10^{8}\text{ m/s}}{1.20\times10^{-6}\text{ m}} = \underline{2.50\times10^{14}\text{ Hz}}$

d) $E = hf = (6.626\times10^{-34}\text{ J}\cdot\text{s})(99.9\times10^{6}\text{ Hz}) = 6.62\times10^{-26}\text{ J}$
$E = 6.62\times10^{-26}\text{ J}\left(\frac{1\text{ eV}}{1.602\times10^{-19}\text{ J}}\right) = \underline{4.13\times10^{-7}\text{ eV}}$

e) A photon with frequency $f = 99.9$ MHz has too little energy, by a large factor, to dissociate a AgBr molecule. The photons in the visible light from a firefly do individually have enough energy to dissociate AgBr. The huge number of 99.9 MHz TV radiation photons can't compensate for the fact that individually they have too little energy.

40-45

The stopping potential V_0 is given by Eq. (40-4): $eV_0 = hf - \phi = \frac{hc}{\lambda} - \phi$.

Call the two wavelengths λ_1 and λ_2, and the corresponding stopping potentials V_{01} and V_{02}. Thus $eV_{01} = \frac{hc}{\lambda_1} - \phi$ and $eV_{02} = \frac{hc}{\lambda_2} - \phi$. Note that the work function ϕ is a property of the material and is independent of the wavelength of the light.

Subtracting one equation from the other gives $e(V_{02} - V_{01}) = hc\left(\frac{1}{\lambda_2} - \frac{1}{\lambda_1}\right)$.
The change in stopping potential is $\Delta V_0 = V_{02} - V_{01} = \frac{hc}{e}\left(\frac{\lambda_1 - \lambda_2}{\lambda_1 \lambda_2}\right)$.

$$\Delta V_0 = \frac{(6.626\times10^{-34}\ \text{J}\cdot\text{s})(2.998\times10^{8}\ \text{m/s})}{1.602\times10^{-19}\ \text{C}}\left[\frac{380\times10^{-9}\ \text{m} - 310\times10^{-9}\ \text{m}}{(380\times10^{-9}\ \text{m})(310\times10^{-9}\ \text{m})}\right] = 0.737\ \text{V}.$$

The stopping potential will increase by 0.737 V.

40-49

$hf = E_f - E_i$, the energy given to the electron in the atom when a photon is absorbed.

The energy of one photon is $\frac{hc}{\lambda} = \frac{(6.626\times10^{-34}\ \text{J}\cdot\text{s})(2.998\times10^{8}\ \text{m/s})}{53.4\times10^{-9}\ \text{m}}$

$$\frac{hc}{\lambda} = 3.720\times10^{-18}\ \text{J}\left(\frac{1\ \text{eV}}{1.602\times10^{-19}\ \text{J}}\right) = 23.22\ \text{eV}.$$

The final energy of the electron is $E_f = E_i + hf$. In the ground state of the hydrogen atom the energy of the electron is $E_i = -13.60$ eV. Thus
$E_f = -13.60\ \text{eV} + 23.22\ \text{eV} = \underline{9.6\ \text{eV}}$.

40-51

a) Eq. (40-18): $m_r = \frac{m_1 m_2}{m_1 + m_2} = \frac{207 m_e m_p}{207 m_e + m_p} = \frac{207(9.109\times10^{-31}\ \text{kg})(1.673\times10^{-27}\ \text{kg})}{207(9.109\times10^{-31}\ \text{kg}) + (1.673\times10^{-27}\ \text{kg})} = 1.69\times10^{-28}\ \text{kg}$

(We have used m_e to clearly identify when we mean the electron mass.)

b) In Eq. (40-16) replace $m = m_e$ by m_r: $E_n = -\frac{1}{\epsilon_0^2}\frac{m_r e^4}{8n^2h^2}$.

Write as $E_n = \left(\frac{m_r}{m_e}\right)\left(-\frac{1}{\epsilon_0^2}\frac{m_e e^4}{8n^2h^2}\right)$, since we know that $\frac{1}{\epsilon_0^2}\frac{m_e e^4}{8h^2} = 13.60$ eV.

$$E_n = \left(\frac{m_r}{m_e}\right)\left(-\frac{13.60\ \text{eV}}{n^2}\right)$$

$$E_1 = \frac{1.69\times10^{-28}\ \text{kg}}{9.109\times10^{-31}\ \text{kg}}(-13.60\ \text{eV}) = 186(-13.60\ \text{eV}) = \underline{-2.53\ \text{keV}}$$

c) From part (a), $E_n = \left(\frac{m_r}{m_e}\right)\left(-\frac{R_H ch}{n^2}\right)$, where $R_H = 1.097\times10^{7}\ \text{m}^{-1}$ is the Rydberg constant for the hydrogen atom.

$\frac{hc}{\lambda} = E_i - E_f$

The initial level for the transition is the $n_i = 2$ level and the final level is the $n_f = 1$ level.

$$\frac{hc}{\lambda} = \frac{m_r}{m_e}\left(-\frac{R_H ch}{n_i^2} - \left(-\frac{R_H ch}{n_f^2}\right)\right)$$

40-51 (cont)

$$\frac{1}{\lambda} = \frac{m_r}{m_e} R_H\left(\frac{1}{n_f^2} - \frac{1}{n_i^2}\right) = \frac{1.69\times10^{-28}\text{ kg}}{9.109\times10^{-31}\text{ kg}}(1.097\times10^{7}\text{ m}^{-1})\left(\frac{1}{1^2} - \frac{1}{2^2}\right) = 1.526\times10^{9}\text{ m}^{-1}$$

$\lambda = 0.655$ nm

Note: From Example 40-7 the wavelength of the radiation emitted in this transition in hydrogen is 122nm. The wavelength for muonium is $\frac{m_e}{m_r} = 5.39\times10^{-3}$ times this.

40-55

From Chapter 16 (Eq. 16-16) the average kinetic energy of a gas atom or molecule is $\frac{3}{2}kT$.

$$\frac{3}{2}kT = 4.0\text{ eV} = 6.408\times10^{-19}\text{ J} \Rightarrow T = \frac{2(6.408\times10^{-19}\text{ J})}{3k} = \frac{2(6.408\times10^{-19}\text{ J})}{3(1.381\times10^{-23}\text{ J/K})} = 3.1\times10^{4}\text{ K}$$

If the temperature of the metal is raised to a value of this order of magnitude, the random kinetic energy of the electrons is large enough for them to escape from the metal (thermionic emission). However, this temperature is well above the melting point of any metal.

40-57

a) Bohr's angular momentum result: $L = mvr = n\frac{h}{2\pi}$

$$v = \frac{nh}{2\pi mr}$$

$$\Sigma\vec{F} = m\vec{a} \Rightarrow Dr = m\frac{v^2}{r}$$

$$r = \sqrt{\frac{m}{D}}\,v$$

Use the Bohr equation to replace v: $r = \sqrt{\frac{m}{D}}\,\frac{nh}{2\pi mr}$

$$r^2 = \frac{nh}{2\pi D^{1/2}m^{1/2}} \Rightarrow r_n = \left(\frac{n^2h^2}{4\pi^2 mD}\right)^{1/4}$$

b) The force has the same dependence on position as $F=-kx$ for a one-dimensional simple harmonic oscillator. For $F=-kx$ the potential energy is $U=\frac{1}{2}kx^2$, so for $F=-Dr$, it is $U=\frac{1}{2}Dr^2$. (This does give $U=0$ when $r=0$.)

$$E = K+U = \frac{1}{2}mv^2 + \frac{1}{2}Dr^2$$

$$v = \sqrt{\frac{D}{m}}\,r, \text{ so } E = \frac{1}{2}m\left(\frac{D}{m}r^2\right) + \frac{1}{2}Dr^2 = Dr^2$$

Use the result for r obtained in part (a):

$$E_n = D\frac{nh}{2\pi\sqrt{m}\sqrt{D}} = \sqrt{\frac{D}{m}}\,\frac{nh}{2\pi}$$

c) $\Delta E = E_i - E_f = (n_i - n_f)\sqrt{\frac{D}{m}}\,\frac{h}{2\pi}$

$n_i - n_f$ can be any positive integer, so can write the possible photon energies as $n\sqrt{\frac{D}{m}}\,\frac{h}{2\pi}$, where $n=1,2,\ldots$

40-59

Let E_{tr} be the transition energy, E_{ph} be the energy of the photon with wavelength λ', and E_r be the kinetic energy of the recoiling atom.

$$E_{ph} + E_r = E_{tr}$$

$$E_{ph} = \frac{hc}{\lambda'} \Rightarrow \frac{hc}{\lambda'} = E_{tr} - E_r \Rightarrow \lambda' = \frac{hc}{E_{tr} - E_r}$$

If the recoil energy is neglected then the photon wavelength is $\lambda = \frac{hc}{E_{tr}}$.

$$\Delta\lambda = \lambda' - \lambda = hc\left(\frac{1}{E_{tr} - E_r} - \frac{1}{E_{tr}}\right) = \frac{hc}{E_{tr}}\left(\frac{1}{1 - E_r/E_{tr}} - 1\right)$$

$$\frac{1}{1 - E_r/E_{tr}} = \left(1 - \frac{E_r}{E_{tr}}\right)^{-1} \simeq 1 + \frac{E_r}{E_{th}} \text{ since } \frac{E_r}{E_{th}} \ll 1$$ (We have used the binomial theorem, Appendix B.)

Thus $\Delta\lambda = \frac{hc}{E_{tr}}\left(\frac{E_r}{E_{tr}}\right)$, or since $E_{tr} = \frac{hc}{\lambda}$, $\Delta\lambda = \left(\frac{E_r}{hc}\right)\lambda^2$.

Use conservation of linear momentum to find E_r:
Assuming that the atom is initially at rest, the momentum p_r of the recoiling atom must be equal in magnitude and opposite in direction to the momentum $p_{ph} = \frac{h}{\lambda}$ of the emitted photon: $\frac{h}{\lambda} = p_r$.

$E_r = \frac{p_r^2}{2M}$, where M is the mass of the atom $\Rightarrow E_r = \frac{h^2}{2M\lambda^2}$.

Use this result in the above equation:

$\Delta\lambda = \left(\frac{E_r}{hc}\right)\lambda^2 = \left(\frac{h^2}{2M\lambda^2}\right)\left(\frac{\lambda^2}{hc}\right) = \frac{h}{2Mc}$; note that this result for $\Delta\lambda$ is independent of the atomic transition energy.

For a hydrogen atom $M = m_p$ and

$$\Delta\lambda = \frac{h}{2m_pc} = \frac{6.626\times10^{-34}\text{ J}\cdot\text{s}}{2(1.673\times10^{-27}\text{ kg})(2.998\times10^8\text{ m/s})} = \underline{6.6\times10^{-16}\text{ m}}$$

Note: For the $n=5$ to $n=1$ transition the wavelength of the emitted photon is

$\frac{1}{\lambda} = R\left(\frac{1}{1^2} - \frac{1}{5^2}\right) = \frac{24R}{25} \Rightarrow \lambda = \frac{25}{24R} = 9.50\times10^{-8}\text{ m}$, so $\frac{\Delta\lambda}{\lambda} = \frac{6.60\times10^{-16}\text{ m}}{9.50\times10^{-8}\text{ m}} = 6.9\times10^{-9}$;
the correction is extremely small.

40-61

a) $\Delta\lambda = \frac{h}{mc}(1-\cos\phi) = \lambda_c(1-\cos\phi)$

Largest $\Delta\lambda$ is for $\phi = 180°$: $\Delta\lambda = 2\lambda_c = 2(2.426\text{ pm}) = \underline{4.85\text{ pm}}$

b) $\lambda' - \lambda = \lambda_c(1-\cos\phi)$

Wavelength doubles $\Rightarrow \lambda' = 2\lambda$ so $\lambda' - \lambda = \lambda$. Thus $\lambda = \lambda_c(1-\cos\phi)$.

$E = \frac{hc}{\lambda}$, so smallest-energy photon means largest wavelength photon $\Rightarrow \phi = 180°$ and $\lambda = 2\lambda_c = 4.85\text{ pm}$.

Then $E = \frac{hc}{\lambda} = \frac{(6.626\times10^{-34}\text{ J}\cdot\text{s})(2.998\times10^8\text{ m/s})}{4.85\times10^{-12}\text{ m}} = 4.096\times10^{-14}\text{ J}\left(\frac{1\text{ eV}}{1.602\times10^{-19}\text{ J}}\right) = \underline{0.256\text{ MeV}}$

40-65

Find the wavelengths of the incident and scattered photons:

$$\lambda = \frac{hc}{E} = \frac{(6.626\times10^{-34}\ \text{J}\cdot\text{s})(2.998\times10^{8}\ \text{m/s})}{(2.00\times10^{6}\ \text{eV})\left(\frac{1.602\times10^{-19}\ \text{J}}{1\ \text{eV}}\right)} = 6.200\times10^{-13}\ \text{m}$$

$$\lambda' = \frac{hc}{E'} = \frac{(6.626\times10^{-34}\ \text{J}\cdot\text{s})(2.998\times10^{8}\ \text{m/s})}{(0.750\times10^{6}\ \text{eV})\left(\frac{1.602\times10^{-19}\ \text{J}}{1\ \text{eV}}\right)} = 1.653\times10^{-12}\ \text{m}$$

$$\lambda' - \lambda = \lambda_c(1-\cos\phi)$$

$$1-\cos\phi = \frac{\lambda'-\lambda}{\lambda_c}$$

$$\cos\phi = 1 - \frac{\lambda'-\lambda}{\lambda_c} = 1 - \frac{1.653\times10^{-12}\ \text{m} - 6.200\times10^{-13}\ \text{m}}{2.426\times10^{-12}\ \text{m}} = 0.4258$$

$$\phi = \underline{64.8^\circ}$$

CHAPTER 41

Exercises 3, 5, 7, 9, 15, 17, 21

Problems 27, 29, 31, 35, 37, 39, 41

Exercises

41-3

a) The visible spectrum is approximately from 400 nm to 700 nm.

$\lambda = 700\text{ nm} \Rightarrow E = \dfrac{hc}{\lambda} = \dfrac{(6.626\times10^{-34}\text{ J}\cdot\text{s})(2.998\times10^{8}\text{ m/s})}{400\times10^{-9}\text{ m}} = 4.966\times10^{-19}\text{ J}\left(\dfrac{1\text{ eV}}{1.602\times10^{-19}\text{ J}}\right) = 3.10\text{ eV}$

$\lambda = 400\text{ nm} \Rightarrow E = \dfrac{hc}{\lambda} = \dfrac{(6.626\times10^{-34}\text{ J}\cdot\text{s})(2.998\times10^{8}\text{ m/s})}{700\times10^{-9}\text{ m}} = 2.838\times10^{-19}\text{ J}\left(\dfrac{1\text{ eV}}{1.602\times10^{-19}\text{ J}}\right) = 1.77\text{ eV}$

The visible spectrum corresponds to photon energies between 1.8 eV to 3.1 eV.

b) $\lambda = \dfrac{h}{p}$

$E = \frac{1}{2}mv^2 = \dfrac{p^2}{2m} \Rightarrow p = \sqrt{2mE}$. Thus $\lambda = \dfrac{h}{\sqrt{2mE}}$.

$E = 1.77\text{ eV} = 2.838\times10^{-19}\text{ J} \Rightarrow \lambda = \dfrac{6.626\times10^{-34}\text{ J}\cdot\text{s}}{\sqrt{2(9.109\times10^{-31}\text{ kg})(2.838\times10^{-19}\text{ J})}} = 9.21\times10^{-10}\text{ m} = 0.921\text{ nm}$

$E = 3.10\text{ eV} = 4.966\times10^{-19}\text{ J} \Rightarrow \lambda = \dfrac{6.626\times10^{-34}\text{ J}\cdot\text{s}}{\sqrt{2(9.109\times10^{-31}\text{ kg})(4.966\times10^{-19}\text{ J})}} = 6.97\times10^{-10}\text{ m} = 0.697\text{ nm}$

The range of electron wavelengths is 0.70 nm to 0.92 nm.

41-5

$\lambda = \dfrac{h}{p} = \dfrac{h}{mv} = \dfrac{6.626\times10^{-34}\text{ J}\cdot\text{s}}{(2000\text{ kg})(24.0\text{ m/s})} = \underline{1.38\times10^{-38}\text{ m}}$

This wavelength is extremely short; the car will not exhibit wavelike properties.

41-7

a) The de Broglie wavelength is $\lambda = \dfrac{h}{p} = \dfrac{h}{mv}$.
In the Bohr model, $mvr_n = n\dfrac{h}{2\pi}$, so $mv = \dfrac{nh}{2\pi r_n}$.

Then $\lambda = h\left(\dfrac{2\pi r_n}{nh}\right) = \dfrac{2\pi r_n}{n}$.

$n = 1 \Rightarrow \lambda = 2\pi r_1$; $r_1 = a_0 = 0.5292\times10^{-10}\text{ m}$

$\lambda = 2\pi(0.5292\times10^{-10}\text{ m}) = \underline{3.32\times10^{-10}\text{ m}}$

$\lambda = 2\pi r_1$; the de Broglie wavelength equals the circumference of the orbit.

b) $n = 4 \Rightarrow \lambda = \dfrac{2\pi r_4}{4}$

$r_n = n^2 a_0$ so $r_4 = 16a_0$

41-7 (cont) $\lambda = \frac{2\pi(16a_0)}{4} = 4(2\pi a_0) = 4(3.32\times10^{-10}\,m) = \underline{1.33\times10^{-9}\,m}$

$\lambda = \frac{2\pi r_4}{4}$; the de Broglie wavelength is $\frac{1}{n} = \frac{1}{4}$ times the circumference of the orbit.

41-9

Since the alpha particles are scattered from the surface plane of the crystal Eq. (41-3) applies: $d\sin\theta = m\lambda$.

$$m = 1 \Rightarrow \sin\theta = \frac{\lambda}{d}$$

Use the de Broglie relation to calculate the wavelength of the particles: $\lambda = \frac{h}{p}$.

$E = \frac{1}{2}mv^2 = \frac{p^2}{2m} \Rightarrow p = \sqrt{2mE}$ so $\lambda = \frac{h}{\sqrt{2mE}} = \frac{6.626\times10^{-34}\,J.s}{\sqrt{2(6.64\times10^{-27}\,kg)(1.20\times10^3\,eV)\left(\frac{1.602\times10^{-19}\,J}{1\,eV}\right)}}$

$$\lambda = 4.147\times10^{-13}\,m$$

$\sin\theta = \frac{\lambda}{d} = \frac{4.147\times10^{-13}\,m}{0.0634\times10^{-9}\,m} = 6.54\times10^{-3} \Rightarrow \theta = \underline{0.375°}$

41-15

Find the uncertainty ΔE in the particle's energy:

$E = mc^2 \Rightarrow \Delta E = (\Delta m)c^2$

$m = 5m_p$; $\Delta m = (2.8\%)\,m = 0.028(5m_p) = (0.14)m_p = 0.14(1.673\times10^{-27}\,kg)$

$\Delta m = 2.342\times10^{-28}\,kg$

Then $\Delta E = (\Delta m)c^2 = (2.342\times10^{-28}\,kg)(2.998\times10^8\,m/s)^2 = 2.105\times10^{-11}\,J$.

Then use the energy uncertainty principle $\Delta E\Delta t \geq \frac{h}{2\pi}$ to estimate the lifetime:

$$\Delta t \simeq \frac{h}{2\pi\,\Delta E} = \frac{6.626\times10^{-34}\,J.s}{2\pi(2.105\times10^{-11}\,J)} = \underline{5.0\times10^{-24}\,s}$$

41-17

a) The relation between the kinetic energy K of the particle of charge q and the potential difference through which it has been accelerated is $K = |q|\,\Delta V$.

And $K = \frac{p^2}{2m}$, so $p = \sqrt{2mK} = \sqrt{2m|q|\,\Delta V}$.

The de Broglie wavelength is $\lambda = \frac{h}{p} = \frac{h}{\sqrt{2m|q|\Delta V}} = \frac{6.626\times10^{-34}\,J.s}{\sqrt{2(9.109\times10^{-31}\,kg)(1.602\times10^{-19}\,C)(700\,V)}}$

$$\lambda = 4.64\times10^{-11}\,m = \underline{46.4\,pm}$$

b) The only change is the mass m of the particle:

$\lambda = \frac{h}{\sqrt{2m|q|\Delta V}} = \frac{6.626\times10^{-34}\,J.s}{\sqrt{2(1.673\times10^{-27}\,kg)(1.602\times10^{-19}\,C)(700\,V)}} = 1.08\times10^{-12}\,m = \underline{1.08\,pm}$

(The proton wavelength is smaller than the electron wavelength by a factor of $\sqrt{\frac{m_e}{m_p}} = \frac{1}{\sqrt{1836}}$.)

41-21

a) $\psi(x) = A \sin kx$

The position probability density is given by $|\psi|^2 = A^2 \sin^2 kx$.

The probability is highest where $\sin kx = 1 \Rightarrow kx = \frac{2\pi x}{\lambda} = \frac{n\pi}{2}$, $n = 1, 3, 5, \ldots$

$x = n\frac{\lambda}{4}$, $n = 1, 3, 5, \ldots$ so $x = \frac{\lambda}{4}, \frac{3\lambda}{4}, \frac{5\lambda}{4}, \ldots$

b) The probability of finding the particle is zero where $|\psi|^2 = 0$, which occurs where $\sin kx = 0 \Rightarrow kx = \frac{2\pi x}{\lambda} = n\pi$, $n = 0, 1, 2, \ldots$

$x = n\frac{\lambda}{2}$, $n = 0, 1, 2, \ldots$ so $x = 0, \frac{\lambda}{2}, \lambda, \frac{3\lambda}{2}, \ldots$

Problems

41-27

$$E^2 = (pc)^2 + (mc^2)^2 \Rightarrow p = \frac{\sqrt{E^2 - (mc^2)^2}}{c}$$

$$\lambda = \frac{h}{p} = \frac{hc}{\sqrt{E^2 - (mc^2)^2}}$$

$E = K + mc^2$

For an electron $mc^2 = 8.187 \times 10^{-14}\ \text{J} = 0.511\ \text{MeV}$ (Example 39-13), so

$$E = 1.02\ \text{MeV} + 0.511\ \text{MeV} = 1.53\ \text{MeV} = 1.53 \times 10^6\ \text{eV}\left(\frac{1.602 \times 10^{-19}\ \text{J}}{1\ \text{eV}}\right) = 2.45 \times 10^{-13}\ \text{J}$$

Then $$\lambda = \frac{(6.626 \times 10^{-34}\ \text{J.s})(2.998 \times 10^8\ \text{m/s})}{\sqrt{(2.45 \times 10^{-13}\ \text{J})^2 - (8.187 \times 10^{-14}\ \text{J})^2}} = 8.60 \times 10^{-13}\ \text{m} = 860\ \text{fm}$$

41-29

$$\Delta x = 0.40\lambda = 0.40\frac{h}{p}$$

$\Delta x \Delta p_x \geq \frac{h}{2\pi}$. So $(\Delta p_x)_{min} = \frac{h}{2\pi \Delta x}$ (as in Example 41-3).

$$(\Delta p_x)_{min} = \frac{h}{2\pi(0.40\, h/p)} = \frac{p}{2\pi(0.40)} = 0.40p$$

41-31

a) $\Delta x \Delta p_x \geq \frac{h}{2\pi}$

Estimate Δx as $\Delta x \simeq 5.0 \times 10^{-15}\ \text{m}$. Then the minimum allowed Δp_x is

$$\Delta p_x \simeq \frac{h}{2\pi\Delta x} = \frac{6.626 \times 10^{-34}\ \text{J.s}}{2\pi(5.0 \times 10^{-15}\ \text{m})} = 2.1 \times 10^{-20}\ \text{kg.m/s}$$

b) Assume $p \simeq 2.1 \times 10^{-20}\ \text{kg.m/s}$

Eq. (39-40): $E = \sqrt{(mc^2)^2 + (pc)^2}$

$mc^2 = (9.109 \times 10^{-31}\ \text{kg})(2.998 \times 10^8\ \text{m/s})^2 = 8.187 \times 10^{-14}\ \text{J}$

$pc = (2.1 \times 10^{-20}\ \text{kg.m/s})(2.998 \times 10^8\ \text{m/s}) = 6.296 \times 10^{-12}\ \text{J}$

$E = \sqrt{(8.187 \times 10^{-14}\ \text{J})^2 + (6.296 \times 10^{-12}\ \text{J})^2} = 6.297 \times 10^{-12}\ \text{J}$

$$K = E - mc^2 = 6.297 \times 10^{-12}\ \text{J} - 8.187 \times 10^{-14}\ \text{J} = 6.215 \times 10^{-12}\ \text{J}\left(\frac{1\ \text{eV}}{1.602 \times 10^{-19}\ \text{J}}\right) = 39\ \text{MeV}$$

41-31 (cont)

c) $|U| = \frac{1}{4\pi\epsilon_0}\frac{e^2}{r} = 8.987\times10^9\ N\cdot m^2/C^2 \frac{(1.602\times10^{-19}\ C)^2}{5.0\times10^{-15}\ m} = 4.613\times10^{-14}\ J\left(\frac{1\ eV}{1.602\times10^{-19}\ J}\right) = \underline{0.29\ MeV}$

The estimate from the uncertainty principle of the kinetic energy of an electron confined within the nucleus is much larger than the binding energy from the attractive Coulomb interaction with a proton, so there cannot be electrons within the nucleus.

41-35

a) $\lambda = \frac{h}{mv} \Rightarrow v = \frac{h}{m\lambda} = \frac{6.626\times10^{-34}\ J\cdot s}{(60.0\ kg)(1.0\ m)} = \underline{1.1\times10^{-35}\ m/s}$

b) $t = \frac{\text{distance}}{\text{velocity}} = \frac{0.80\ m}{1.1\times10^{-35}\ m/s} = 7.3\times10^{34}\ s\left(\frac{1\ y}{3.156\times10^7\ s}\right) = \underline{2.3\times10^{27}\ y}$

Since you walk through doorways much more quickly than this, you will not experience diffraction effects.

41-37

$h = 6.63\times10^{-22}\ J\cdot s$

$\Delta E\,\Delta t \geq \frac{h}{2\pi} \Rightarrow \Delta E \simeq \frac{h}{2\pi\Delta t} = \frac{6.63\times10^{-22}\ J\cdot s}{2\pi(4.50\times10^{-3}\ s)} = 2.345\times10^{-20}\ J\left(\frac{1\ eV}{1.602\times10^{-19}\ J}\right) = \underline{0.146\ eV}$

41-39

a) $U = A|x|$

$F = -\frac{dU}{dx}$ (Eq. 7-17)

For $x > 0$, $|x| = x$ so $U = Ax$ and $F = -\frac{d(Ax)}{dx} = -A$

For $x < 0$, $|x| = -x$ so $U = -Ax$ and $F = -\frac{d(-Ax)}{dx} = +A$

We can write this result as $F = A\frac{|x|}{x}$, valid for all x except for $x = 0$.

b) $E = K + U = \frac{p^2}{2m} + A|x|$

$px \simeq h$, so $p \simeq \frac{h}{x}$

Then $E \simeq \frac{h^2}{2mx^2} + A|x|$.

For $x > 0$, $E = \frac{h^2}{2mx^2} + Ax$.

To find the value of x that gives minimum E set $\frac{dE}{dx} = 0$

$\Rightarrow\ 0 = -\frac{2h^2}{2mx^3} + A$

$x^3 = \frac{h^2}{mA}$ and $x = \left(\frac{h^2}{mA}\right)^{1/3}$

$E = \frac{h^2}{2m}\left(\frac{mA}{h^2}\right)^{2/3} + A\left(\frac{h^2}{mA}\right)^{1/3} = \frac{1}{2}h^{2/3}m^{-1/3}A^{2/3} + h^{2/3}m^{-1/3}A^{2/3} = \frac{3}{2}\left(\frac{h^2A^2}{m}\right)^{1/3}$

41-41

a) Let the y-direction be from the thrower to the catcher, and let the x-direction be horizontal and perpendicular to the y-direction.

A cube with volume $V = 1000\ cm^3 = 1.00 \times 10^{-3}\ m^3$ has side length $l = V^{1/3} = (1.00 \times 10^{-3}\ m^3)^{1/3} = 0.100\ m$. Thus estimate Δx as $\Delta x \simeq 0.10\ m$.

$$\Delta x \Delta p_x \geq \frac{h}{2\pi} \Rightarrow \Delta p_x \simeq \frac{h}{2\pi \Delta x} = \frac{0.0663\ J \cdot s}{2\pi (0.10\ m)} = \underline{0.11\ kg \cdot m/s}$$

(The value of h in this other universe has been used.)

The uncertainty in the ball's horizontal velocity is $\Delta v_x = \frac{\Delta p_x}{m} = \frac{0.11\ kg \cdot m/s}{0.30\ kg} = 0.37\ m/s$

b) The time it takes the ball to travel to the second student is $t = \frac{20\ m}{5.0\ m/s} = 4.0\ s$.

The uncertainty in the x-coordinate of the ball when it reaches the second student that is introduced by Δv_x is $\Delta x = (\Delta v_x) t = (0.37\ m/s)(4.0\ s) = 1.5\ m$.

The ball could miss the second student by about 1.5 m.

CHAPTER 42

Exercises 3, 7, 9, 17, 19, 21, 23, 25

Problems 27, 31, 33, 37, 41, 43

Exercises

42-3

Eq. (42-5): $E_n = \frac{n^2h^2}{8mL^2}$

Ground state energy $E_1 = \frac{h^2}{8mL^2}$; first excited state energy $E_2 = \frac{4h^2}{8mL^2}$.

$\Delta E = E_2 - E_1 = \frac{3h^2}{8mL^2}$

$L = h\sqrt{\frac{3}{8m\,\Delta E}} = 6.626\times 10^{-34}\ \text{J}\cdot\text{s}\sqrt{\frac{3}{8(9.109\times10^{-31}\ \text{kg})(4.0\ \text{eV})\left(\frac{1.602\times10^{-19}\ \text{J}}{1\ \text{eV}}\right)}} = 5.3\times10^{-10}\ \text{m} = \underline{0.53\ \text{nm}}$

42-7

For the $n=2$ first excited state the normalized wave function is given by Eq. (42-9): $\psi_2(x) = \sqrt{\frac{2}{L}}\sin\left(\frac{2\pi x}{L}\right)$

$|\psi_2(x)|^2 dx = \frac{2}{L}\sin^2\left(\frac{2\pi x}{L}\right)dx$

a) $|\psi_2|^2 dx = 0 \Rightarrow \sin\left(\frac{2\pi x}{L}\right) = 0 \Rightarrow \frac{2\pi x}{L} = m\pi,\ m = 0, 1, 2, \ldots\ ;\ x = m\left(\frac{L}{2}\right)$

$m = 0 \Rightarrow x = 0$

$m = 1 \Rightarrow x = \frac{L}{2}$

$m = 2 \Rightarrow x = L$

The probability of finding the particle is zero at $x = 0, \frac{L}{2}$, and L.

b) $|\psi_2|^2 dx$ is maximum when $\sin\left(\frac{2\pi x}{L}\right) = \pm 1 \Rightarrow \frac{2\pi x}{L} = m\left(\frac{\pi}{2}\right),\ m = 1, 3, 5, \ldots$

$x = m\left(\frac{L}{4}\right)$

$m = 1 \Rightarrow x = \frac{L}{4}$

$m = 3 \Rightarrow x = \frac{3L}{4}$

The probability of finding the particle is largest at $x = \frac{L}{4}$ and $\frac{3L}{4}$.

c) The answers to part (a) correspond to the zeros of $|\psi|^2$ shown in Fig. 42-4 and the answers to part (b) correspond to the two values of x where $|\psi|^2$ in the figure is maximum.

42-9

a) $\psi = A\cos kx$

Eq. (42-13): $-\frac{h^2}{8\pi^2 m}\frac{d^2\psi}{dx^2} = E\psi$

$\frac{d\psi}{dx} = A(-k\sin kx) = -Ak\sin kx$

$\frac{d^2\psi}{dx^2} = -Ak(k\cos kx) = -Ak^2\cos kx$

42-9 (cont)

Thus Eq. (42-13) requires $-\frac{h^2}{8\pi^2 m}(-AK^2\cos kx) = E(A\cos kx)$

$$\frac{h^2k^2}{8\pi^2 m} = E \;;\; k = \frac{\sqrt{2mE}}{(h/2\pi)} = \frac{\sqrt{2mE}}{\hbar}$$

$\psi = A\cos kx$ is a solution to Eq. (42-13) if $k = \frac{\sqrt{2mE}}{\hbar}$.

b) The wave function for a particle in a box with rigid walls at $x=0$ and $x=L$ must satisfy the boundary conditions $\psi=0$ at $x=0$ and $\psi=0$ at $x=L$.

$\psi(0) = A\cos 0 = A$, since $\cos 0 = 1$.
Thus $\psi \neq 0$ at $x=0$ and this wave function isn't acceptable because it doesn't satisfy the required boundary condition.

42-17

$U_0 = 6E_\infty$, as in Fig. 42-8, so $E_1 = 0.625\,E_\infty$ and $E_3 = 5.09\,E_\infty$ with $E_\infty = \frac{\pi^2\hbar^2}{2mL^2}$. In this problem the particle bound in the well is a proton, so $m = 1.673\times10^{-27}$ kg. Then

$$E_\infty = \frac{\pi^2\hbar^2}{2mL^2} = \frac{\pi^2(1.054\times10^{-34}\text{ J}\cdot\text{s})^2}{2(1.673\times10^{-27}\text{ kg})(8.0\times10^{-15}\text{ m})^2} = 5.12\times10^{-13}\text{ J}.$$

The transition energy is $\Delta E = E_3 - E_1 = (5.09 - 0.625)E_\infty = 4.465\,E_\infty$.
$\Delta E = 4.465(5.12\times10^{-13}\text{ J}) = 2.286\times10^{-12}\text{ J}$
The wavelength of the photon that is absorbed is related to the transition energy by $\Delta E = \frac{hc}{\lambda}$, so

$$\lambda = \frac{hc}{\Delta E} = \frac{(6.626\times10^{-34}\text{ J}\cdot\text{s})(2.998\times10^8\text{ m/s})}{2.286\times10^{-12}\text{ J}} = 8.7\times10^{-14}\text{ m} = \underline{87\text{ fm}}$$

42-19

The probability is $T = Ae^{-2\kappa L}$, with $A = 16\frac{E}{U_0}\left(1-\frac{E}{U_0}\right)$ and $\kappa = \frac{\sqrt{2m(U_0-E)}}{\hbar}$.
$E = 46$ eV, $U_0 = 55$ eV, $L = 0.20\times10^{-9}$ m

a) $A = 16\frac{E}{U_0}\left(1-\frac{E}{U_0}\right) = 16\frac{46}{55}\left(1-\frac{46}{55}\right) = 2.190$

$$\kappa = \frac{\sqrt{2m(U_0-E)}}{\hbar} = \frac{\sqrt{2(9.109\times10^{-31}\text{ kg})(55\text{ eV}-46\text{ eV})(1.602\times10^{-19}\text{ J/eV})}}{1.054\times10^{-34}\text{ J}\cdot\text{s}} = 1.538\times10^{10}\text{ m}^{-1}$$

$$T = Ae^{-2\kappa L} = (2.190)e^{-2(1.538\times10^{10}\text{ m}^{-1})(0.20\times10^{-9}\text{ m})} = 2.190e^{-6.152} = \underline{0.0047}$$

b) The only change is the mass m, which appears in κ.

$$\kappa = \frac{\sqrt{2m(U_0-E)}}{\hbar} = \frac{\sqrt{2(1.673\times10^{-27}\text{ kg})(55\text{ eV}-46\text{ eV})(1.602\times10^{-19}\text{ J/eV})}}{1.054\times10^{-34}\text{ J}\cdot\text{s}} = 6.590\times10^{11}\text{ m}^{-1}$$

Then $T = Ae^{-2\kappa L} = 2.190\,e^{-2(6.590\times10^{11})(0.20\times10^{-9}\text{ m})} = 2.190\,e^{-263.6} \simeq \underline{10^{-114}}$

42-21

$$\omega = \sqrt{\frac{k'}{m}} = \sqrt{\frac{130\ \text{N/m}}{0.200\ \text{kg}}} = 25.5\ \text{rad/s}$$

The ground state energy is given by Eq. (42-29):

$$E = \tfrac{1}{2}\hbar\omega = \tfrac{1}{2}(1.054\times10^{-34}\ \text{J}\cdot\text{s})(25.5\ \text{rad/s}) = 1.344\times10^{-33}\ \text{J}\left(\frac{1\ \text{eV}}{1.602\times10^{-19}\ \text{J}}\right) = \underline{8.39\times10^{-15}\ \text{eV}}$$

$$E_n = (n+\tfrac{1}{2})\hbar\omega$$
$$E_{n+1} = (n+1+\tfrac{1}{2})\hbar\omega$$

The energy separation between these adjacent levels is

$$\Delta E = E_{n+1} - E_n = \hbar\omega = 2E_0 = 2(8.39\times10^{-15}\ \text{eV}) = \underline{1.68\times10^{-14}\ \text{eV}}$$

These energies are extremely small; quantum effects are not important for this oscillator.

42-23

The photon wavelength λ is related to the transition energy by $\Delta E = \frac{hc}{\lambda}$. Ground state $E_0 = \frac{1}{2}\hbar\omega$; first excited state $E_1 = \frac{3}{2}\hbar\omega$, so the transition energy is $\Delta E = (\frac{3}{2} - \frac{1}{2})\hbar\omega = \hbar\omega = \frac{h}{2\pi}\sqrt{\frac{k'}{m}}$.

Equate these two expressions for $\Delta E \Rightarrow \frac{hc}{\lambda} = \frac{h}{2\pi}\sqrt{\frac{k'}{m}}$

$$\Rightarrow k' = m\left(\frac{2\pi c}{\lambda}\right)^2 = (4.7\times10^{-26}\ \text{kg})\left(\frac{2\pi\ (2.998\times10^{8}\ \text{m/s})}{340\times10^{-6}\ \text{m}}\right)^2 = \underline{1.4\ \text{N/m}}$$

42-25

The total energy of a Newtonian oscillator is given by $E = \frac{1}{2}k'A^2$ where k' is the force constant and A is the amplitude of the oscillator. Set this equal to the energy $E = (n+\frac{1}{2})\hbar\omega$ of an excited level that has quantum number n, where $\omega = \sqrt{\frac{k'}{m}}$, and solve for A: $\frac{1}{2}k'A^2 = (n+\frac{1}{2})\hbar\omega$

$$A = \sqrt{\frac{(2n+1)\hbar\omega}{k'}}$$

The total energy of the Newtonian oscillator can also be written as $E = \frac{1}{2}mv_{max}^2$. Set this equal to $E = (n+\frac{1}{2})\hbar\omega$ and solve for v_{max}: $\frac{1}{2}mv_{max}^2 = (n+\frac{1}{2})\hbar\omega$

$$v_{max} = \sqrt{\frac{(2n+1)\hbar\omega}{m}}$$

Thus the maximum linear momentum of the oscillator is $p_{max} = mv_{max} = \sqrt{(2n+1)\hbar m\omega}$

Assume that A represents the uncertainty Δx in position and that p_{max} is the corresponding uncertainty Δp_x in momentum. Then the uncertainty product is

$$\Delta x\,\Delta p_x = \sqrt{\frac{(2n+1)\hbar\omega}{k'}}\sqrt{(2n+1)\hbar m\omega} = (2n+1)\hbar\omega\sqrt{\frac{m}{k'}} = (2n+1)\hbar\omega\left(\frac{1}{\omega}\right) = (2n+1)\hbar.$$

For $n=1$ this gives $\Delta x\,\Delta p_x = \hbar$, in agreement with the result derived in Section 42-6. The uncertainty product $\Delta x\,\Delta p_x$ increases with n.

Problems

42-27

The energy levels are given by Eq (42-5): $E_n = \frac{n^2 h^2}{8mL^2}$.

Ground state, $n=1 \Rightarrow E_1 = \frac{h^2}{8mL^2}$

First excited state, $n=2 \Rightarrow E_2 = \frac{4h^2}{8mL^2}$

The transition energy is $\Delta E = E_2 - E_1 = \frac{3h^2}{8mL^2}$.

The transition energy is related to the wavelength of the emitted photon by $\Delta E = \frac{hc}{\lambda}$. Combine these two equations for $\Delta E \Rightarrow \frac{hc}{\lambda} = \frac{3h^2}{8mL^2}$.

$$\lambda = \frac{8mcL^2}{3h} = \frac{8(9.109\times10^{-31}\text{ kg})(2.998\times10^{8}\text{ m/s})(3.52\times10^{-9}\text{ m})^2}{3\ (6.626\times10^{-34}\text{ J}\cdot\text{s})} = 1.36\times10^{-5}\text{ m} = \underline{13.6\ \mu\text{m}}$$

42-31

a) The normalized wave function for the ground state is $\psi_1 = \sqrt{\frac{2}{L}}\sin\left(\frac{\pi x}{L}\right)$.

The probability P of the particle being between $x = \frac{L}{4}$ and $x = \frac{3L}{4}$ is

$$P = \int_{L/4}^{3L/4} |\psi_1|^2 dx = \frac{2}{L}\int_{L/4}^{3L/4} \sin^2\left(\frac{\pi x}{L}\right) dx$$

Let $y = \frac{\pi x}{L}$; $dx = \frac{L}{\pi}dy$ and the integration limits become $\frac{\pi}{4}$ and $\frac{3\pi}{4}$.

$$P = \frac{2}{L}\left(\frac{L}{\pi}\right)\int_{\pi/4}^{3\pi/4} \sin^2 y\, dy = \frac{2}{\pi}\left[\frac{1}{2}y - \frac{1}{4}\sin 2y\right]\Big|_{\pi/4}^{3\pi/4} = \frac{2}{\pi}\left[\frac{3\pi}{8} - \frac{\pi}{8} - \frac{1}{4}\sin\frac{3\pi}{2} + \frac{1}{4}\sin\frac{\pi}{2}\right]$$

$$P = \frac{2}{\pi}\left[\frac{\pi}{4} - \frac{1}{4}(-1) + \frac{1}{4}(1)\right] = \frac{1}{2} + \frac{1}{\pi} = \underline{0.818}$$

(Note: The integral formula $\int \sin^2 y\, dy = \frac{1}{2}y - \frac{1}{4}\sin 2y$ was used.)

b) The normalized wave function for the first excited state is $\psi_2 = \sqrt{\frac{2}{L}}\sin\left(\frac{2\pi x}{L}\right)$.

$$P = \int_{L/4}^{3L/4} |\psi_2|^2 dx = \frac{2}{L}\int_{L/4}^{3L/4} \sin^2\left(\frac{2\pi x}{L}\right) dx$$

Let $y = \frac{2\pi x}{L}$; $dx = \frac{L}{2\pi}dy$ and the integration limits become $\frac{\pi}{2}$ and $\frac{3\pi}{2}$.

$$P = \frac{2}{L}\left(\frac{L}{2\pi}\right)\int_{\pi/2}^{3\pi/2} \sin^2 y\, dy = \frac{1}{\pi}\left[\frac{1}{2}y - \frac{1}{4}\sin 2y\right]\Big|_{\pi/2}^{3\pi/2} = \frac{1}{\pi}\left[\frac{3\pi}{4} - \frac{\pi}{4} - \frac{1}{4}\sin 3\pi + \frac{1}{4}\sin\pi\right] = \frac{1}{2}$$

(with $\sin 3\pi = 0$ and $\sin\pi = 0$)

$P = \underline{0.500}$

42-33

Eq. (42-9): $\psi_n = \sqrt{\frac{2}{L}}\sin\left(\frac{n\pi x}{L}\right)$ for particle in a box with walls at $x=0$ and $x=L$. (This is the wave function for $0 \le x \le L$. For $x<0$ or $x>L$, $\psi_n = 0$.)

a) $n=1 \Rightarrow \psi_1 = \sqrt{\frac{2}{L}}\sin\left(\frac{\pi x}{L}\right)$

$$\frac{d\psi_1}{dx} = \sqrt{\frac{2}{L}}\,\frac{\pi}{L}\cos\left(\frac{\pi x}{L}\right)$$

42-33 (cont)

$$\cos\theta = 1 - \frac{\theta^2}{2} + \ldots$$

So for x slightly greater than 0, $\cos\frac{\pi x}{L} = 1$ and $\frac{d\psi_1}{dx} = \sqrt{\frac{2}{L}}\,\frac{\pi}{L} = \sqrt{\frac{2\pi^2}{L^3}}$

b) $n = 2 \Rightarrow \psi_2 = \sqrt{\frac{2}{L}}\sin\left(\frac{2\pi x}{L}\right)$

$$\frac{d\psi_2}{dx} = \sqrt{\frac{2}{L}}\left(\frac{2\pi}{L}\right)\cos\left(\frac{2\pi x}{L}\right)$$

For x slightly greater than 0, $\cos\frac{\pi x}{L} = 1$ and $\frac{d\psi_2}{dx} = \sqrt{\frac{2}{L}}\,\frac{2\pi}{L} = \sqrt{\frac{8\pi^2}{L^3}}$.

c) For x very close to L, $\cos\left(\frac{\pi x}{L}\right) \simeq \cos\pi = -1$ and, from part (a),

$$\frac{d\psi_1}{dx} = -\sqrt{\frac{2\pi^2}{L^3}}.$$

d) For x very close to L, $\cos\left(\frac{2\pi x}{L}\right) \simeq \cos(2\pi) = 1$ and, from part (b),

$$\frac{d\psi_2}{dx} = \sqrt{\frac{8\pi^2}{L^3}}.$$

e) The magnitude of the slope of the wave function near the walls is greater for the $n=2$ level.

These results all agree with Fig. 42-4. For $n=1$ the slope near both walls is positive, for $n=2$ the slope is positive near $x=0$ and negative near $x=L$, and the magnitude of the slope near the walls is larger for $n=2$ than for $n=1$.

42-37

a) $$T = \left[1 + \frac{(U_0 \sinh \kappa L)^2}{4E(U_0 - E)}\right]^{-1}$$

$$\sinh \kappa L = \frac{e^{\kappa L} - e^{-\kappa L}}{2}$$

For $\kappa L \gg 1$, $\sinh \kappa L \rightarrow \frac{e^{\kappa L}}{2}$

$$T \rightarrow \left[1 + \frac{U_0^2 e^{2\kappa L}}{16E(U_0 - E)}\right]^{-1} = \frac{16E(U_0 - E)}{16E(U_0 - E) + U_0^2 e^{2\kappa L}}$$

For $\kappa L \gg 1$, $16E(U_0 - E) + U_0^2 e^{2\kappa L} \rightarrow U_0^2 e^{2\kappa L}$

$$T \rightarrow \frac{16E(U_0 - E)}{U_0^2 e^{2\kappa L}} = 16\frac{E}{U_0}\left(1 - \frac{E}{U_0}\right)e^{-2\kappa L},$$ which is Eq. (42-24).

b) $\kappa = \frac{\sqrt{2m(U_0 - E)}}{\hbar}$; κ becomes small as E approaches U_0.

For κ small, $\sin \kappa L \rightarrow \kappa L$ and

$$T \rightarrow \left[1 + \frac{U_0^2 \kappa^2 L^2}{4E(U_0 - E)}\right]^{-1} = \left[1 + \frac{U_0^2\, 2m(U_0 - E)L^2}{\hbar^2 4E(U_0 - E)}\right]^{-1}$$ (where we have used the definition of κ)

$$T \rightarrow \left[1 + \frac{2U_0^2 L^2 m}{4E\hbar^2}\right]^{-1}$$

42-37 (cont)

$U_0 \to E$ so $\frac{U_0^2}{E} \to E$ and $T \to \left[1 + \frac{2EL^2 m}{4\hbar^2}\right]^{-1}$

But $k^2 = \frac{2mE}{\hbar^2}$, so $T \to \left[1 + \left(\frac{kL}{2}\right)^2\right]^{-1}$, as was to be shown.

42-41

Calculate the angular frequency ω of the pendulum:

$$\omega = \frac{2\pi}{T} = \frac{2\pi}{0.500\,s} = 4\pi\ s^{-1}$$

The ground-state energy is $E_0 = \frac{1}{2}\hbar\omega = \frac{1}{2}(1.054\times10^{-34}\,J\cdot s)(4\pi\,s^{-1}) = \underline{6.62\times10^{-34}\,J}$

$$E_0 = 6.62\times10^{-34}\,J\left(\frac{1\,eV}{1.602\times10^{-19}\,J}\right) = \underline{4.13\times10^{-15}\,eV}$$

$E_n = (n+\frac{1}{2})\hbar\omega$

$E_{n+1} = (n+1+\frac{1}{2})\hbar\omega$

The energy difference between the adjacent energy levels is

$\Delta E = E_{n+1} - E_n = \hbar\omega = 2E_0 = \underline{1.32\times10^{-33}\,J} = \underline{8.26\times10^{-15}\,eV}$

These energies are much too small to detect.

42-43

a) Eq (42-32): $-\frac{\hbar^2}{2m}\left(\frac{\partial^2\psi}{\partial x^2} + \frac{\partial^2\psi}{\partial y^2} + \frac{\partial^2\psi}{\partial z^2}\right) + U\psi = E\psi$

$\psi_{n_x}, \psi_{n_y}, \psi_{n_z}$ are each solutions of Eq. (42-25)

$$\Rightarrow -\frac{\hbar^2}{2m}\frac{d^2\psi_{n_x}}{dx^2} + \frac{1}{2}k'x^2\psi_{n_x} = E_{n_x}\psi_{n_x}$$

$$-\frac{\hbar^2}{2m}\frac{d^2\psi_{n_y}}{dy^2} + \frac{1}{2}k'y^2\psi_{n_y} = E_{n_y}\psi_{n_y}$$

$$-\frac{\hbar^2}{2m}\frac{d^2\psi_{n_z}}{dz^2} + \frac{1}{2}k'z^2\psi_{n_z} = E_{n_z}\psi_{n_z}$$

$\psi = \psi_{n_x}(x)\,\psi_{n_y}(y)\,\psi_{n_z}(z)$, $U = \frac{1}{2}k'x^2 + \frac{1}{2}k'y^2 + \frac{1}{2}k'z^2$

$\frac{\partial^2\psi}{\partial x^2} = \left(\frac{d^2\psi_{n_x}}{dx^2}\right)\psi_{n_y}\psi_{n_z}$, $\frac{\partial^2\psi}{\partial y^2} = \left(\frac{d^2\psi_{n_y}}{dy^2}\right)\psi_{n_x}\psi_{n_z}$, $\frac{\partial^2\psi}{\partial z^2} = \left(\frac{d^2\psi_{n_z}}{dz^2}\right)\psi_{n_x}\psi_{n_y}$

So $-\frac{\hbar^2}{2m}\left(\frac{\partial^2\psi}{\partial x^2} + \frac{\partial^2\psi}{\partial y^2} + \frac{\partial^2\psi}{\partial z^2}\right) + U\psi = \left(-\frac{\hbar^2}{2m}\frac{d^2\psi_{n_x}}{dx^2} + \frac{1}{2}k'x^2\psi_{n_x}\right)\psi_{n_y}\psi_{n_z}$

$+\left(-\frac{\hbar^2}{2m}\frac{d^2\psi_{n_y}}{dy^2} + \frac{1}{2}k'y^2\psi_{n_y}\right)\psi_{n_x}\psi_{n_z} + \left(-\frac{\hbar^2}{2m}\frac{d^2\psi_{n_z}}{dz^2} + \frac{1}{2}k'z^2\psi_{n_z}\right)\psi_{n_x}\psi_{n_y}$

$$-\frac{\hbar^2}{2m}\left(\frac{\partial^2\psi}{\partial x^2} + \frac{\partial^2\psi}{\partial y^2} + \frac{\partial^2\psi}{\partial z^2}\right) + U\psi = (E_{n_x} + E_{n_y} + E_{n_z})\psi$$

Therefore, we have shown that this ψ is a solution to Eq. (42-32), with energy $E_{n_x n_y n_z} = E_{n_x} + E_{n_y} + E_{n_z} = (n_x + n_y + n_z + \frac{3}{2})\hbar\omega$

b) and c) Ground state $n_x = n_y = n_z = 0$; $E_{000} = \frac{3}{2}\hbar\omega$

There is only one set of $n_x, n_y,$ and n_z that give this energy.

First-excited state: $n_x = 1, n_y = n_z = 0$ or $n_y = 1, n_x = n_z = 0$ or $n_z = 1, n_x = n_y = 0$

$$E_{100} = E_{010} = E_{001} = \frac{5}{2}\hbar\omega$$

There are three different sets of n_x, n_y, n_z quantum numbers that give this energy, so there are three different quantum states that have this same energy.

CHAPTER 43

Exercises 3, 5, 11, 15, 17, 19, 21, 25, 27, 29

Problems 33, 35, 37, 39, 43, 47

Exercises

43-3

Eq. (43-4): $L = \sqrt{l(l+1)}\,\hbar$, $l = 0, 1, 2, \ldots$

$l(l+1) = \left(\frac{L}{\hbar}\right)^2 = \left(\frac{3.653 \times 10^{-34}\ \text{kg} \cdot \text{m/s}}{1.054 \times 10^{-34}\ \text{J} \cdot \text{s}}\right)^2 = 12.0$

$l(l+1) = 12 \Rightarrow \underline{l = 3}$

43-5

$L = \sqrt{l(l+1)}\,\hbar$

The maximum l, l_{max}, for a given n is $l_{max} = n - 1$.

$n = 1 \Rightarrow l_{max} = 0 \Rightarrow L = 0$

$n = 10 \Rightarrow l_{max} = 9 \Rightarrow L = \sqrt{9(10)}\,\hbar = 9.49\,\hbar$

$n = 100 \Rightarrow l_{max} = 99 \Rightarrow L = \sqrt{99(100)}\,\hbar = 99.5\,\hbar$

As n increases the maximum L gets closer to the value $n\hbar$ postulated in the Bohr model.

43-11

Eq. (43-8): $a = \frac{4\pi\epsilon_0 \hbar^2}{m_r e^2} = \frac{\epsilon_0 h^2}{\pi m_r e^2}$

a) $m_r = m$

$a = \frac{\epsilon_0 h^2}{\pi m e^2} = \frac{(8.854 \times 10^{-12}\ \text{C}^2/\text{N} \cdot \text{m}^2)(6.626 \times 10^{-34}\ \text{J} \cdot \text{s})^2}{\pi (9.109 \times 10^{-31}\ \text{kg})(1.602 \times 10^{-19}\ \text{C})^2} = \underline{0.5292 \times 10^{-10}\ \text{m}}$

b) $m_r = m/2$

$a = 2\left(\frac{\epsilon_0 h^2}{\pi m e^2}\right) = \underline{1.058 \times 10^{-11}\ \text{m}}$

c) $m_r = 186 m$

$a = \frac{1}{186}\left(\frac{\epsilon_0 h^2}{\pi m e^2}\right) = \underline{2.85 \times 10^{-13}\ \text{m}}$

43-15

a) 4f level $\Rightarrow l = 3 \Rightarrow 2l + 1 = 7$ different m_l states.
The 4f level is split into 7 levels by the magnetic field.

43-15 (cont)

b) Each m_l level is shifted in energy an amount given by Eq. (43-18): $U = m_l \mu_B B$.
Adjacent levels differ in m_l by one, so $\Delta U = \mu_B B$.

$$\mu_B = \frac{e\hbar}{2m} = \frac{(1.602\times10^{-19}\,\text{C})(1.054\times10^{-34}\,\text{J}\cdot\text{s})}{2(9.109\times10^{-31}\,\text{kg})} = 9.274\times10^{-24}\,\text{A}\cdot\text{m}^2$$

$$\Delta U = \mu_B B = (9.274\times10^{-24}\,\text{A}\cdot\text{m}^2)(1.25\,\text{T}) = 1.159\times10^{-23}\,\text{J}\left(\frac{1\,\text{eV}}{1.602\times10^{-19}\,\text{J}}\right) = \underline{7.24\times10^{-5}\,\text{eV}}$$

c) The level of highest energy is for the largest m_l, which is $m_l = l = 3$;
$U_3 = 3\mu_B B$.
The level of lowest energy is for the smallest m_l, which is $m_l = -l = -3$;
$U_{-3} = -3\mu_B B$.

The energy separation between these two levels is $U_3 - U_{-3} = 6\mu_B B$
$= 6(7.24\times10^{-5}\,\text{eV}) = \underline{4.34\times10^{-4}\,\text{eV}}$.

43-17

a) For a classical particle $L = I\omega$.
For a uniform sphere with mass m and radius R, $I = \frac{2}{5}mR^2$, so $L = (\frac{2}{5}mR^2)\omega$.

$$L = \sqrt{\tfrac{3}{4}}\,\hbar \Rightarrow \tfrac{2}{5}mR^2\omega = \sqrt{\tfrac{3}{4}}\,\hbar$$

$$\omega = \frac{5\sqrt{\frac{3}{4}}\,\hbar}{2mR^2} = \frac{5\sqrt{\frac{3}{4}}(1.054\times10^{-34}\,\text{J}\cdot\text{s})}{2(9.109\times10^{-31}\,\text{kg})(1.0\times10^{-17}\,\text{m})^2} = \underline{2.5\times10^{30}\,\text{rad/s}}$$

b) $v = r\omega = (1.0\times10^{-17}\,\text{m})(2.5\times10^{30}\,\text{rad/s}) = 2.5\times10^{13}\,\text{m/s}$. This is much greater than the speed of light c, so the model cannot be valid.

43-19

$$U = -\mu_z B$$

$$\mu_z = -(2.00232)\left(\frac{e}{2m}\right)S_z, \text{ so } U = +(2.00232)\left(\frac{e}{2m}\right)S_z B$$

$$S_z = m_s\hbar, \text{ so } U = +2.00232\left(\frac{e\hbar}{2m}\right)m_s B$$

$$\frac{e\hbar}{2m} = \mu_B = 5.788\times10^{-5}\,\text{eV/T}$$

$$U = 2.00232\,\mu_B m_s B$$

The $m_s = -\frac{1}{2}$ level has lower energy (as shown in Fig. 43-12).

$$\Delta U = U(m_s = +\tfrac{1}{2}) - U(m_s = -\tfrac{1}{2}) = 2.00232\mu_B B\left(\tfrac{1}{2} - (-\tfrac{1}{2})\right) = 2.00232\mu_B B$$

$$\Delta U = 2.00232(5.788\times10^{-5}\,\text{eV/T})(0.730\,\text{T}) = \underline{8.46\times10^{-5}\,\text{eV}}$$

43-21

j can have the values $l + \frac{1}{2}$ and $l = -\frac{1}{2}$. Thus if j takes the values $\frac{5}{2}$ and $\frac{7}{2}$ it must be that $l - \frac{1}{2} = \frac{5}{2} \Rightarrow l = \frac{6}{2} = 3$. The letter that labels this l is f.

43-25

$$\text{Eq. (43-27)}: E_n = -\frac{Z_{eff}^2}{n^2}(13.6\text{ eV})$$

$n=5$ and $Z_{eff} = 2.771 \Rightarrow E_5 = -\frac{(2.771)^2}{5^2}(13.6\text{ eV}) = -4.18\text{ eV}$

The ionization energy is $\underline{4.18\text{ eV}}$.

43-27

$$E_n = -\frac{Z_{eff}^2}{n^2}(13.6\text{ eV})$$

a) The element Be has nuclear charge $Z=4$. The ion Be^+ has 3 electrons. The outermost electron sees the nuclear charge screened by the other two electrons so $Z_{eff} = 4-2 = 2$.

$$E_2 = -\frac{2^2}{2^2}(13.6\text{ eV}) = \underline{-13.6\text{ eV}}$$

b) The outermost electron in Ca^+ sees a $Z_{eff} = 2$.

$$E_4 = -\frac{2^2}{4^2}(13.6\text{ eV}) = \underline{-3.4\text{ eV}}$$

43-29

a) $Z=7$ for nitrogen so nitrogen atom has 7 electrons.
N^{2+} has 5 electrons: $1s^2 2s^2 2p$.

b) $Z_{eff} = 7-4 = 3$ for the $2p$ level.

$$E = -\frac{Z_{eff}^2}{n^2}(13.6\text{ eV}) = -\frac{3^2}{2^2}(13.6\text{ eV}) = \underline{-30.6\text{ eV}}$$

c) $Z=15$ for phosphorus so phosphorus atom has 15 electrons.
P^{2+} has 13 electrons: $1s^2 2s^2 2p^6 3s^2 3p$

d) $Z_{eff} = 15-12 = 3$ for the $3p$ level.

$$E = -\frac{Z_{eff}^2}{n^2}(13.6\text{ eV}) = -\frac{3^2}{3^2}(13.6\text{ eV}) = \underline{-13.6\text{ eV}}$$

Problems

43-33

a) $\cos\theta_L = \frac{L_z}{L}$ so $\theta_L = \arccos\left(\frac{L_z}{L}\right)$

The smallest angle $(\theta_L)_{min}$ is for the state with the largest L and the largest L_z. This is the state with $l=n-1$ and $m_l = l = n-1$.

$$L_z = m_l\hbar = (n-1)\hbar$$

$$L = \sqrt{l(l+1)}\,\hbar = \sqrt{(n-1)n}\,\hbar$$

43-33 (cont)

$$(\theta_L)_{min} = \arccos\left(\frac{(n-1)\hbar}{\sqrt{(n-1)n}\ \hbar}\right) = \arccos\left(\frac{n-1}{\sqrt{n(n-1)}}\right) = \arccos\left(\sqrt{\frac{n-1}{n}}\right) = \arccos\left(\sqrt{1-\frac{1}{n}}\right)$$

Note that $(\theta_L)_{min}$ approaches $0°$ as $n \to \infty$.

b) The largest angle $(\theta_L)_{max}$ is for $l = n-1$ and $m_l = -l = -(n-1)$

A similar calculation to part (a) yields $(\theta_L)_{max} = \arccos\left(-\sqrt{1-\frac{1}{n}}\right)$

Note that $(\theta_L)_{max}$ approaches $180°$ as $n \to \infty$.

43-35

a) $E_{1s} = -\frac{1}{(4\pi\epsilon_0)^2}\frac{me^4}{2\hbar^2}$; $U(r) = -\frac{1}{4\pi\epsilon_0}\frac{e^2}{r}$

$$E_{1s} = U(r) \Rightarrow -\frac{1}{(4\pi\epsilon_0)^2}\frac{me^4}{2\hbar^2} = -\frac{1}{4\pi\epsilon_0}\frac{e^2}{r}$$

$$r = \frac{(4\pi\epsilon_0)2\hbar^2}{me^2} = \underline{2a}$$

b) For the 1s state the probability that the electron is in the classically forbidden region is $P(r>2a) = \int_{2a}^{\infty} |\psi_{1s}|^2\,dV = 4\pi\int_{2a}^{\infty}|\psi_{1s}|^2 r^2\,dr$

The normalized wave function of the 1s state of hydrogen is given in Example 43-3:

$$\psi_{1s}(r) = \frac{1}{\sqrt{\pi a^3}}e^{-r/a}$$

$$P(r>2a) = 4\pi\left(\frac{1}{\pi a^3}\right)\int_{2a}^{\infty} r^2 e^{-2r/a}\,dr$$

Use the integral formula $\int r^2 e^{-\alpha r}dr = -e^{-\alpha r}\left(\frac{r^2}{\alpha}+\frac{2r}{\alpha^2}+\frac{2}{\alpha^3}\right)$, with $\alpha = \frac{2}{a}$.

$$P(r>2a) = -\frac{4}{a^3}\left[e^{-2r/a}\left(\frac{ar^2}{2}+\frac{a^2r}{2}+\frac{a^3}{4}\right)\right]\Big|_{2a}^{\infty} = +\frac{4}{a^3}e^{-4}\left(2a^3+a^3+\frac{a^3}{4}\right)$$

$$P(r>2a) = 4e^{-4}\left(\frac{13}{4}\right) = 13e^{-4} = \underline{0.238}.$$

43-37

$$\psi_{2s} = \frac{1}{\sqrt{32\pi a^3}}\left(2-\frac{r}{a}\right)e^{-r/2a}$$

a) Let $I = \int_0^{\infty}|\psi_{2s}|^2\,dV = 4\pi\int_0^{\infty}|\psi_{2s}|^2 r^2\,dr$. If ψ_{2s} is normalized then we will find that $I = 1$.

$$I = 4\pi\left(\frac{1}{32\pi a^3}\right)\int_0^{\infty}\left(2-\frac{r}{a}\right)^2 e^{-r/a} r^2\,dr = \frac{1}{8a^3}\int_0^{\infty}\left(4r^2-4\frac{r^3}{a}+\frac{r^4}{a^2}\right)e^{-r/a}\,dr$$

Use the integral formula $\int_0^{\infty} x^n e^{-\alpha x}dx = \frac{n!}{\alpha^{n+1}}$, with $\alpha = \frac{1}{a}$

$$I = \frac{1}{8a^3}\left(4(2!)(a)^3 - \frac{4}{a}(3!)(a)^4 + \frac{1}{a^2}(4!)(a)^5\right) = \frac{1}{8}(8-24+24) = 1;$$

ψ_{2s} is normalized.

43-37 (cont)

b) $P(r<4a) = \int_0^{4a} |\psi_{2s}|^2 dV = 4\pi \int_0^{4a} |\psi_{2s}|^2 r^2 dr = \frac{1}{8a^3}\int_0^{4a}\left(4r^2 - 4\frac{r^3}{a} + \frac{r^4}{a^2}\right)e^{-r/a}dr$

Let $P(r<4a) = \frac{1}{8a^3}(I_1 + I_2 + I_3)$.

$I_1 = 4\int_0^{4a} r^2 e^{-r/a} dr$

Use the integral formula $\int r^2 e^{-\alpha r} dr = -e^{-\alpha r}\left(\frac{r^2}{\alpha} + \frac{2r}{\alpha^2} + \frac{2}{\alpha^3}\right)$ with $\alpha = \frac{1}{a}$.

$I_1 = -4\left[e^{-r/a}(r^2 a + 2ra^2 + 2a^3)\right]\Big|_0^{4a} = (-104e^{-4} + 8)a^3$

$I_2 = -\frac{4}{a}\int_0^{4a} r^3 e^{-r/a} dr$

Use the integral formula $\int r^3 e^{-\alpha r} dr = -e^{-\alpha r}\left(\frac{r^3}{\alpha} + 3\frac{r^2}{\alpha^2} + 6\frac{r}{\alpha^3} + 6\frac{1}{\alpha^4}\right)$ with $\alpha = \frac{1}{a}$.

$I_2 = \frac{4}{a}\left[e^{-r/a}(r^3 a + 3r^2a^2 + 6ra^3 + 6a^4)\right]\Big|_0^{4a} = (568e^{-4} - 24)a^3$

$I_3 = \frac{1}{a^2}\int_0^{4a} r^4 e^{-r/a} dr$

Use the integral formula $\int r^4 e^{-\alpha r} dr = -e^{-\alpha r}\left(\frac{r^4}{\alpha} + 4\frac{r^3}{\alpha^2} + 12\frac{r^2}{\alpha^3} + 24\frac{r}{\alpha^4} + 24\frac{1}{\alpha^5}\right)$

$I_3 = -\frac{1}{a^2}\left[e^{-r/a}(r^4 a + 4r^3a^2 + 12r^2a^3 + 24ra^4 + 24a^5)\right]\Big|_0^{4a} = a^3(-824e^{-4} + 24)$

thus $P(r<4a) = \frac{1}{8a^3}(I_1 + I_2 + I_3) = \frac{1}{8a^3}a^3\left([8-24+24] + e^{-4}[-104+568-824]\right)$

$P(r<4a) = \frac{1}{8}(8 - 360e^{-4}) = 1 - 45e^{-4} = \underline{0.176}$

43-39

a) The energy of the photon equals the transition energy of the atom: $\Delta E = \frac{hc}{\lambda}$.

$E_n = -\frac{13.60\text{ eV}}{n^2} \Rightarrow E_2 = -\frac{13.60\text{ eV}}{4}, \quad E_1 = -\frac{13.60\text{ eV}}{1}$

$\Delta E = E_2 - E_1 = 13.60\text{ eV}\left(-\frac{1}{4}+1\right) = \frac{3}{4}(13.60\text{ eV}) = 10.20\text{ eV}(1.602\times10^{-19}\text{ J/eV}) = 1.634\times10^{-18}\text{ J}$

$\lambda = \frac{hc}{\Delta E} = \frac{(6.626\times10^{-34}\text{ J}\cdot\text{s})(2.998\times10^{8}\text{ m/s})}{1.634\times10^{-18}\text{ J}} = 1.22\times10^{-7}\text{ m} = \underline{122\text{ nm}}$

b) The shift of a level due to the energy of interaction with the magnetic field in the z-direction is $U = m_l \mu_B B$.
The ground state has $m_l = 0$ so is unaffected by the magnetic field.
The n=2 initial state has $m_l = -1$ so its energy is shifted downward an amount
$U = m_l \mu_B B = (-1)(9.274\times10^{-24}\text{ A}\cdot\text{m}^2)(1.75\text{ T}) = -1.623\times10^{-23}\text{ J}\left(1\text{ eV}/1.602\times10^{-19}\text{ J}\right) = 1.013\times10^{-4}\text{ eV}$

Note that the shift in energy due to the magnetic field is a very small fraction of the 10.2 eV transition energy. Problem 41-34 (c) shows that in this situation $\left|\frac{\Delta\lambda}{\lambda}\right| = \left|\frac{\Delta E}{E}\right|$.

this gives $|\Delta\lambda| = \lambda\left|\frac{\Delta E}{E}\right| = 122\text{ nm}\left(\frac{1.013\times10^{-4}\text{ eV}}{10.2\text{ eV}}\right) = 1.21\times10^{-3}\text{ nm} = \underline{1.21\text{ pm}}$.

The upper level in the transition is lowered in energy so the transition energy is decreased. A smaller ΔE means a larger λ; the magnetic field increases the wavelength.

43-43

Eq. (40-19): $\frac{n_1}{n_0} = e^{-(E_1 - E_0)/kT}$, where 1 is the higher energy state and 0 is the lower energy state.

The interaction energy with the magnetic field is

$U = -\mu_z B = 2.00232 \frac{e\hbar}{2m} m_s B$ (Example 43-6).

The energy of the $m_s = +\frac{1}{2}$ level is increased and the energy of the $m_s = -\frac{1}{2}$ level is decreased.

$$\frac{n_{1/2}}{n_{-1/2}} = e^{-(U_{1/2} - U_{-1/2})/kT}$$

$$U_{\frac{1}{2}} - U_{-1/2} = (2.00232)\frac{e\hbar}{2m} B\left(\tfrac{1}{2} - \left(-\tfrac{1}{2}\right)\right) = (2.00232)\frac{e\hbar}{2m} B = 2.00232 \frac{e\hbar}{2m} B$$

$$= 2.00232\, \mu_B B$$

$$\frac{n_{1/2}}{n_{-1/2}} = e^{-(2.00232)\mu_B B/kT}$$

a) $B = 5.00 \times 10^{-5}\ \text{T} \Rightarrow \frac{n_{1/2}}{n_{-1/2}} = e^{-2.00232(9.274\times 10^{-24}\ \text{A}\cdot\text{m}^2)(5.00\times 10^{-5}\ \text{T})/[(1.381\times 10^{-23}\ \text{J/K})(300\ \text{K})]}$

$$\frac{n_{1/2}}{n_{-1/2}} = e^{-2.238\times 10^{-7}} = 0.999\,999\,78 = \underline{1 - 2.2\times 10^{-7}}$$

b) $B = 0.500\ \text{T} \Rightarrow \frac{n_{1/2}}{n_{-1/2}} = e^{-2.238\times 10^{-3}} = \underline{0.9978}$

c) $B = 5.00\ \text{T} \Rightarrow \frac{n_{1/2}}{n_{-1/2}} = e^{-2.238\times 10^{-2}} = \underline{0.978}$

43-47

a) calcium, $Z = 20$

minimum wavelength ⇒ largest transition energy

The highest occupied shell is the N shell ($n = 4$). The highest energy transition is N → K, with transition energy $\Delta E = E_N - E_K$. Since the shell energies scale like $\frac{1}{n^2}$ neglect E_N relative to E_K, so $\Delta E = E_K = (Z-1)^2(13.6\ \text{eV}) = (20-1)^2(13.6\ \text{eV}) = 4.910\times 10^3\ \text{eV} = 7.866\times 10^{-16}\ \text{J}$.

The energy of the emitted photon equals this transition energy, so the photon's wavelength is given by $\Delta E = \frac{hc}{\lambda} \Rightarrow \lambda = \frac{hc}{\Delta E}$.

$$\lambda = \frac{(6.626\times 10^{-34}\ \text{J}\cdot\text{s})(2.998\times 10^{8}\ \text{m/s})}{7.866\times 10^{-16}\ \text{J}} = 2.5\times 10^{-10}\ \text{m} = \underline{0.25\ \text{nm}}$$

maximum wavelength ⇒ smallest transition energy ⇒ K_α transition

The frequency of the photon emitted in this transition is given by Moseley's law (Eq. 43-29):

$$f = (2.48\times 10^{15}\ \text{Hz})(Z-1)^2 = (2.48\times 10^{15}\ \text{Hz})(20-1)^2 = 8.953\times 10^{17}\ \text{Hz}$$

$$\lambda = \frac{c}{f} = \frac{2.998\times 10^{8}\ \text{m/s}}{8.953\times 10^{17}\ \text{Hz}} = 3.3\times 10^{-10}\ \text{m} = \underline{0.33\ \text{nm}}$$

43-47 (cont)

b) niobium, $Z = 41$

Apply the analysis of part (a), just with this different value of Z.

<u>minimum wavelength</u>

$$\Delta E = E_K = (Z-1)^2 (13.6\text{ eV}) = (41-1)^2 (13.6\text{ eV}) = 2.176\times10^{4}\text{ eV} = 3.486\times10^{-15}\text{ J}$$

$$\lambda = \frac{hc}{\Delta E} = \frac{(6.626\times10^{-34}\text{ J}\cdot\text{s})(2.998\times10^{8}\text{ m/s})}{3.486\times10^{-15}\text{ J}} = 5.7\times10^{-11}\text{ m} = \underline{57\text{ pm}}$$

<u>maximum wavelength</u>

$$f = (2.48\times10^{15}\text{ Hz})(Z-1)^2 = (2.48\times10^{15}\text{ Hz})(41-1)^2 = 3.968\times10^{18}\text{ Hz}$$

$$\lambda = \frac{c}{f} = \frac{2.998\times10^{8}\text{ m/s}}{3.968\times10^{18}\text{ Hz}} = 7.6\times10^{-11}\text{ m} = \underline{76\text{ pm}}$$

CHAPTER 44

Exercises 5, 7, 9, 11, 13, 19, 23, 25, 27

Problems 29, 31, 35, 37, 39

Exercises

44-5

a) The energy of a rotational level with quantum number l is $E_l = l(l+1)\frac{\hbar^2}{2I}$ (Eq. 44-3).

$I = m_r r^2$

$m_r = \frac{m_1 m_2}{m_1+m_2} = \frac{m_{Li} m_H}{m_{Li}+m_H} = \frac{(1.17\times10^{-26}\text{ kg})(1.67\times10^{-27}\text{ kg})}{1.17\times10^{-26}\text{ kg}+1.67\times10^{-27}\text{ kg}} = 1.461\times10^{-27}\text{ kg}$

$I = m_r r^2 = (1.461\times10^{-27}\text{ kg})(0.159\times10^{-9}\text{ m})^2 = 3.694\times10^{-47}\text{ kg}\cdot\text{m}^2$

$l=2$: $E_2 = 2(3)\frac{\hbar^2}{2I} = 3\left(\frac{\hbar^2}{I}\right)$

$l=3$: $E_3 = 3(4)\frac{\hbar^2}{2I} = 6\left(\frac{\hbar^2}{I}\right)$

$\Delta E = E_3 - E_2 = 3\left(\frac{\hbar^2}{I}\right) = 3\frac{(1.054\times10^{-34}\text{ J}\cdot\text{s})^2}{3.694\times10^{-47}\text{ kg}\cdot\text{m}^2} = 9.02\times10^{-22}\text{ J} = \underline{5.63\times10^{-3}\text{ eV}}$

b) $\Delta E = \frac{hc}{\lambda} \Rightarrow \lambda = \frac{hc}{\Delta E} = \frac{(4.136\times10^{-15}\text{ eV})(2.998\times10^{8}\text{ m/s})}{5.63\times10^{-3}\text{ eV}} = \underline{220\,\mu\text{m}}$

44-7

a) $f = \frac{\omega}{2\pi} = \frac{1}{2\pi}\sqrt{\frac{k'}{m_r}} \Rightarrow k' = m_r(2\pi f)^2$

$m_r = \frac{m_1 m_2}{m_1+m_2} = \frac{m_H m_F}{m_H+m_F} = \frac{(1.67\times10^{-27}\text{ kg})(3.15\times10^{-26}\text{ kg})}{1.67\times10^{-27}\text{ kg}+3.15\times10^{-26}\text{ kg}} = 1.586\times10^{-27}\text{ kg}$

$k' = m_r(2\pi f)^2 = 1.586\times10^{-27}\text{ kg}\,(2\pi\,[1.24\times10^{14}\text{ Hz}])^2 = \underline{963\text{ N/m}}$

b) $E_n = (n+\frac{1}{2})\hbar\omega = (n+\frac{1}{2})hf$, since $\hbar\omega = \frac{h}{2\pi}\omega$ and $\frac{\omega}{2\pi} = f$.
The energy spacing between adjacent levels is $\Delta E = E_{n+1} - E_n = (n+1+\frac{1}{2}-n-\frac{1}{2})hf = hf$, independent of n.
$\Delta E = hf = (6.626\times10^{-34}\text{ J}\cdot\text{s})(1.24\times10^{14}\text{ Hz}) = 8.22\times10^{-20}\text{ J} = \underline{0.513\text{ eV}}$

c) The photon energy equals the transition energy $\Rightarrow \Delta E = \frac{hc}{\lambda}$.
$hf = \frac{hc}{\lambda} \Rightarrow \lambda = \frac{c}{f} = \frac{2.998\times10^{8}\text{ m/s}}{1.24\times10^{14}\text{ Hz}} = 2.42\times10^{-6}\text{ m} = \underline{2.42\,\mu\text{m}}$
This photon is in the infrared.

44-9

From Example 44-2, $E_l = l(l+1)\frac{\hbar^2}{2I}$, with $\frac{\hbar^2}{2I} = 0.2395\times10^{-3}\text{ eV}$.
From Example 44-3, $\Delta E = 0.2690$ eV is the spacing between vibrational levels.
Thus $E_n = (n+\frac{1}{2})\hbar\omega$, with $\hbar\omega = 0.2690$ eV.

44-9 (cont)

By Eq. (44-9), $E = E_n + E_\ell = (n+\frac{1}{2})\hbar\omega + \ell(\ell+1)\frac{\hbar^2}{2I}$.

a) $n=0 \to n=1$ and $\ell=1 \to \ell=2$

For $n=0, \ell=1$, $E_i = \frac{1}{2}\hbar\omega + 2\left(\frac{\hbar^2}{2I}\right)$

For $n=1, \ell=2$, $E_f = \frac{3}{2}\hbar\omega + 6\left(\frac{\hbar^2}{2I}\right)$

$$\Delta E = E_f - E_i = \hbar\omega + 4\left(\frac{\hbar^2}{2I}\right) = 0.2690 \text{ eV} + 4(0.2395\times10^{-3}\text{ eV}) = 0.2700 \text{ eV}$$

$$\frac{hc}{\lambda} = \Delta E \Rightarrow \lambda = \frac{hc}{\Delta E} = \frac{(4.136\times10^{-15}\text{ eV}\cdot\text{s})(2.998\times10^{8}\text{ m/s})}{0.2700\text{ eV}} = 4.592\times10^{-6}\text{ m} = \underline{4.592\ \mu\text{m}}$$

b) $n=0 \to n=1$ and $\ell=2 \to \ell=1$

For $n=0, \ell=2$, $E_i = \frac{1}{2}\hbar\omega + 6\left(\frac{\hbar^2}{2I}\right)$

For $n=1, \ell=1$, $E_f = \frac{3}{2}\hbar\omega + 2\left(\frac{\hbar^2}{2I}\right)$

$$\Delta E = E_f - E_i = \hbar\omega - 4\left(\frac{\hbar^2}{2I}\right) = 0.2690 \text{ eV} - 4(0.2395\times10^{-3}\text{ eV}) = 0.2680 \text{ eV}$$

$$\lambda = \frac{hc}{\Delta E} = \frac{(4.136\times10^{-15}\text{ eV}\cdot\text{s})(2.998\times10^{8}\text{ m/s})}{0.2680\text{ eV}} = 4.627\times10^{-6}\text{ m} = \underline{4.627\ \mu\text{m}}$$

c) $n=0 \to n=1$ and $\ell=3 \to \ell=2$

For $n=0, \ell=3$, $E_i = \frac{1}{2}\hbar\omega + 12\left(\frac{\hbar^2}{2I}\right)$

For $n=1, \ell=2$, $E_f = \frac{3}{2}\hbar\omega + 6\left(\frac{\hbar^2}{2I}\right)$

$$\Delta E = E_f - E_i = \hbar\omega - 6\left(\frac{\hbar^2}{2I}\right) = 0.2690 \text{ eV} - 6(0.2395\times10^{-3}\text{ eV}) = 0.2676 \text{ eV}$$

$$\lambda = \frac{hc}{\Delta E} = \frac{(4.136\times10^{-15}\text{ eV}\cdot\text{s})(2.998\times10^{8}\text{ m/s})}{0.2676\text{ eV}} = 4.634\times10^{-6}\text{ m} = \underline{4.634\ \mu\text{m}}$$

<u>44-11</u>

Each atom occupies a cube with side length 0.282 nm. Therefore, the volume occupied by each atom is $V = (0.282\times10^{-9}\text{ m})^3 = 2.24\times10^{-29}\text{ m}^3$.

In NaCl there are equal numbers of Na and Cl atoms, so the average mass of the atoms in the crystal is

$$m = \tfrac{1}{2}(m_{Na} + m_{Cl}) = \tfrac{1}{2}(3.82\times10^{-26}\text{ kg} + 5.89\times10^{-26}\text{ kg}) = 4.855\times10^{-26}\text{ kg}$$

The density then is $\rho = \frac{m}{V} = \frac{4.855\times10^{-26}\text{ kg}}{2.24\times10^{-29}\text{ m}^3} = \underline{2.16\times10^{3}\text{ kg/m}^3}$.

<u>44-13</u>

The photon energy must be at least $E = 1.12\text{ eV} = 1.794\times10^{-19}$ J.

The wavelength is given by $E = \frac{hc}{\lambda}$, so

$$\lambda = \frac{hc}{E} = \frac{(6.626\times10^{-34}\text{ J}\cdot\text{s})(2.998\times10^{8}\text{ m/s})}{1.794\times10^{-19}\text{ J}} = 1.11\times10^{-6}\text{ m} = \underline{1.11\ \mu\text{m}}$$

44-19

a) The three-dimensional Schrodinger equation is $-\frac{\hbar^2}{2m}\left(\frac{\partial^2\psi}{\partial x^2}+\frac{\partial^2\psi}{\partial y^2}+\frac{\partial^2\psi}{\partial z^2}\right)+U\psi=E\psi$ (Eq. 42-32)

(Eq. 44-10): $\psi = A\sin\frac{n_x\pi x}{L}\sin\frac{n_y\pi y}{L}\sin\frac{n_z\pi z}{L}$

Free electrons $\Rightarrow U=0.$

$$\frac{\partial\psi}{\partial x} = \frac{n_x\pi}{L}A\cos\left(\frac{n_x\pi x}{L}\right)\sin\left(\frac{n_y\pi y}{L}\right)\sin\left(\frac{n_z\pi z}{L}\right)$$

$$\frac{\partial^2\psi}{\partial x^2} = -\left(\frac{n_x\pi}{L}\right)^2 A\sin\left(\frac{n_x\pi x}{L}\right)\sin\left(\frac{n_y\pi y}{L}\right)\sin\left(\frac{n_z\pi z}{L}\right) = -\left(\frac{n_x\pi}{L}\right)^2\psi$$

Similarly $\frac{\partial^2\psi}{\partial y^2} = -\left(\frac{n_y\pi}{L}\right)^2\psi$ and $\frac{\partial^2\psi}{\partial z^2} = -\left(\frac{n_z\pi}{L}\right)^2\psi$

Therefore, $-\frac{\hbar^2}{2m}\left(\frac{\partial^2\psi}{\partial x^2}+\frac{\partial^2\psi}{\partial y^2}+\frac{\partial^2\psi}{\partial z^2}\right)+\cancel{U\psi}^{\,0} = -\frac{\hbar^2}{2m}\left(-\frac{\pi^2}{L^2}\right)(n_x^2+n_y^2+n_z^2)\psi$

$$= \frac{(n_x^2+n_y^2+n_z^2)\pi^2\hbar^2}{2mL^2}\psi$$

This equals $E\psi$, with $E = \frac{(n_x^2+n_y^2+n_z^2)\pi^2\hbar^2}{2mL^2}$, which is Eq. (44-11).

b) Ground level $\Rightarrow$ lowest $E \Rightarrow n_x = n_y = n_z = 1$ and $E = \frac{3\pi^2\hbar^2}{2mL^2}$.

No other combination of n_x, n_y, and n_z gives this same E so the only degeneracy is the degeneracy of two due to spin.

First excited level $\Rightarrow$ next lowest $E \Rightarrow$ one n equals 2 and the others equal 1.

$$E = (2^2+1^2+1^2)\frac{\pi^2\hbar^2}{2mL^2} = \frac{6\pi^2\hbar^2}{2mL^2}$$

There are three different sets of n_x, n_y, n_z values that give this E:
$n_x=2, n_y=1, n_z=1$; $n_y=2, n_x=1, n_z=1$; and $n_z=2, n_x=1, n_y=1$.
This gives a degeneracy of 3 so the total degeneracy, with the factor of 2 from spin, is 6.

Second excited level $\Rightarrow$ next lowest $E \Rightarrow$ two of n_x, n_y, n_z equal 2 and the other equals 1.

$$E = (2^2+2^2+1^2)\frac{\pi^2\hbar^2}{2mL^2} = \frac{9\pi^2\hbar^2}{2mL^2}$$

There are three different sets of n_x, n_y, n_z values that give this E:
$n_x=2, n_y=2, n_z=1$; $n_x=2, n_z=2, n_y=1$; $n_y=2, n_z=2, n_x=1$.
Thus, as in part (b), the total degeneracy, including spin, is 6.

44-23

The electron contribution to the heat capacity of a metal is $C=\left(\frac{\pi^2 kT}{2E_F}\right)R$.

$$C = \frac{\pi^2(1.381\times10^{-23}\ \text{J/K})(300\ \text{K})}{2(3.23\ \text{eV})(1.602\times10^{-19}\ \text{J/eV})}R = \underline{0.0395\ R}.$$

44-25

Eq. (44-17): $f(E) = \dfrac{1}{e^{(E-E_F)/kT} + 1}$

$e^{(E-E_F)/kT} = \dfrac{1}{f(E)} - 1$

The problem states that $f(E) = 4.4\times10^{-4}$ for E at the bottom of the conduction band.

$e^{(E-E_F)/kT} = \dfrac{1}{4.4\times10^{-4}} - 1 = 2.272\times10^{3}$

$E - E_F = kT\ln(2.272\times10^3) = (1.3807\times10^{-23}\ J/T)(300\ K)\ln(2.272\times10^3) = 3.201\times10^{-20}\ J = 0.20\ eV$

$E_F = E - 0.20\ eV$; the Fermi level is 0.20 eV below the bottom of the conduction band.

44-27

The voltage-current relation is given by Eq. (44-23): $I = I_s(e^{eV/kT} - 1)$

a) Find I_s: $V = +15.0\times10^{-3}\ V$ gives $I = 9.25\times10^{-3}\ A$

$\dfrac{eV}{kT} = \dfrac{(1.602\times10^{-19}\ C)(15.0\times10^{-3}\ V)}{(1.381\times10^{-23}\ J/K)(300\ K)} = 0.5800$

$I_s = \dfrac{I}{e^{eV/kT} - 1} = \dfrac{9.25\times10^{-3}\ A}{e^{0.5800} - 1} = 1.177\times10^{-2}\ A = 11.77\ mA$

Then can calculate I for V = 10.0 mV:

$\dfrac{eV}{kT} = \dfrac{(1.602\times10^{-19}\ C)(10.0\times10^{-3}\ V)}{(1.381\times10^{-23}\ J/K)(300\ K)} = 0.3867$

$I = I_s(e^{eV/kT} - 1) = (11.77\ mA)(e^{0.3867} - 1) = \underline{5.56\ mA}$

b) $\frac{eV}{kT}$ has the same magnitudes as in part (a) but now V is negative so $\frac{eV}{kT}$ is negative.

$V = -15.0\ mV \Rightarrow \dfrac{eV}{kT} = -0.5800$

$I = I_s(e^{eV/kT} - 1) = (11.77\ mA)(e^{-0.5800} - 1) = \underline{-5.18\ mA}$

$V = -10.0\ mV \Rightarrow \dfrac{eV}{kT} = -0.3867$

$I = I_s(e^{eV/kT} - 1) = (11.77\ mA)(e^{-0.3867} - 1) = \underline{-3.77\ mA}$

Problems

44-29

a) U must make up for the difference between the ionization energy of Na (5.1 eV) and the electron affinity of Cl (3.6 eV).

$\Rightarrow U = -(5.1\ eV - 3.6\ eV) = -1.5\ eV = -2.40\times10^{-19}\ J$

$-\dfrac{1}{4\pi\epsilon_0}\dfrac{e^2}{r} = -2.40\times10^{-19}\ J$

$r = \dfrac{1}{4\pi\epsilon_0}\dfrac{e^2}{2.40\times10^{-19}\ J} = 8.987\times10^9\ N\cdot m^2/C^2\,\dfrac{(1.602\times10^{-19}\ C)^2}{2.40\times10^{-19}\ J} = 9.60\times10^{-10}\ m = \underline{0.96\ nm}$

44-29 (cont)

b) ionization energy of K = 4.3 eV
electron affinity of Br = 3.5 eV
$\Rightarrow U = -(4.3\text{ eV} - 3.5\text{ eV}) = -0.8\text{ eV} = -1.28\times10^{-19}\text{ J}$

$$-\frac{1}{4\pi\epsilon_0}\frac{e^2}{r} = -1.28\times10^{-19}\text{ J}$$

$$r = \left(\frac{1}{4\pi\epsilon_0}\right)\frac{e^2}{1.28\times10^{-19}\text{ J}} = 8.987\times10^9\text{ N}\cdot\text{m}^2/\text{C}^2\ \frac{(1.602\times10^{-19}\text{ C})^2}{1.28\times10^{-19}\text{ J}} = 1.8\times10^{-9}\text{ m} = \underline{1.8\text{ nm}}$$

44-31

a) Eq. (22-14): electric dipole moment $p = qd$
Point charges +e and −e separated by distance d
$\Rightarrow p = ed = (1.602\times10^{-19}\text{ C})(0.24\times10^{-9}\text{ m}) = \underline{3.8\times10^{-29}\text{ C}\cdot\text{m}}$

b) $p = qd \Rightarrow q = \frac{p}{d} = \frac{3.0\times10^{-29}\text{ C}\cdot\text{m}}{0.24\times10^{-9}\text{ m}} = \underline{1.3\times10^{-19}\text{ C}}$

c) $\frac{q}{e} = \frac{1.3\times10^{-19}\text{ C}}{1.602\times10^{-19}\text{ C}} = \underline{0.78}$

d) $q = \frac{p}{d} = \frac{1.5\times10^{-30}\text{ C}\cdot\text{m}}{0.16\times10^{-9}\text{ m}} = 9.37\times10^{-21}\text{ C}$

$\frac{q}{e} = \frac{9.37\times10^{-21}\text{ C}}{1.602\times10^{-19}\text{ C}} = \underline{0.058}$

The fractional ionic character for the bond in HI is much less than the fractional ionic character for the bond in NaCl. The bond in HI is mostly covalent and not very ionic.

44-35

a) $E_l = l(l+1)\frac{\hbar^2}{2I}$

calculate I for $Na^{35}Cl$:

$$m_r = \frac{m_1 m_2}{m_1+m_2} = \frac{m_{Na}m_{Cl}}{m_{Na}+m_{Cl}} = \frac{(3.8176\times10^{-26}\text{ kg})(5.8068\times10^{-26}\text{ kg})}{3.8176\times10^{-26}\text{ kg} + 5.8068\times10^{-26}\text{ kg}} = 2.303\times10^{-26}\text{ kg}$$

$$I = m_r r^2 = 2.303\times10^{-26}\text{ kg}\,(0.2361\times10^{-9}\text{ m})^2 = 1.284\times10^{-45}\text{ kg}\cdot\text{m}^2$$

$l=2 \to l=1$ transition

$$\Delta E = E_2 - E_1 = (6-2)\frac{\hbar^2}{2I} = \frac{2\hbar^2}{I} = \frac{2(1.054\times10^{-34}\text{ J}\cdot\text{s})^2}{1.284\times10^{-45}\text{ kg}\cdot\text{m}^2} = 1.730\times10^{-23}\text{ J}$$

$$\Delta E = \frac{hc}{\lambda} \Rightarrow \lambda = \frac{hc}{\Delta E} = \frac{(6.626\times10^{-34}\text{ J}\cdot\text{s})(2.998\times10^8\text{ m/s})}{1.730\times10^{-23}\text{ J}} = 1.148\times10^{-2}\text{ m} = \underline{1.148\text{ cm}}$$

$l=1 \to l=0$ transition

$$\Delta E = E_1 - E_0 = (2-0)\frac{\hbar^2}{2I} = \frac{\hbar^2}{I} = \frac{1}{2}(1.730\times10^{-23}\text{ J}) = 8.650\times10^{-24}\text{ J}$$

$$\lambda = \frac{hc}{\Delta E} = \frac{(6.626\times10^{-34}\text{ J}\cdot\text{s})(2.998\times10^8\text{ m/s})}{8.650\times10^{-24}\text{ J}} = 2.297\times10^{-2}\text{ m} = \underline{2.297\text{ cm}}$$

44-35 (cont)

b) Calculate I for $Na^{37}Cl$:

$$m_r = \frac{m_{Na} m_{Cl}}{m_{Na} + m_{Cl}} = \frac{(3.8176\times10^{-26}\ kg)(6.1384\times10^{-26}\ kg)}{3.8176\times10^{-26}\ kg + 6.1384\times10^{-26}\ kg} = 2.354\times10^{-26}\ kg$$

$$I = m_r r^2 = 2.354\times10^{-26}\ kg\,(0.2361\times10^{-9}\ m)^2 = 1.312\times10^{-45}\ kg\cdot m^2$$

$l = 2 \to l = 1$ transition

$$\Delta E = \frac{2\hbar^2}{I} = \frac{2(1.054\times10^{-34}\ J\cdot s)^2}{1.312\times10^{-45}\ kg\cdot m^2} = 1.693\times10^{-23}\ J$$

$$\lambda = \frac{hc}{\Delta E} = \frac{(6.626\times10^{-34}\ J\cdot s)(2.998\times10^8\ m/s)}{1.693\times10^{-23}\ J} = \underline{1.173\ cm}$$

$l = 1 \to l = 0$ transition

$$\Delta E = \frac{\hbar^2}{I} = \frac{1}{2}(1.693\times10^{-23}\ J) = 8.465\times10^{-24}\ J$$

$$\lambda = \frac{hc}{\Delta E} = \frac{(6.626\times10^{-34}\ J\cdot s)(2.998\times10^8\ m/s)}{8.465\times10^{-24}\ J} = \underline{2.347\ cm}$$

The differences in the wavelengths for the two isotopes are:

$l = 2 \to l = 1$ transition: $1.173\ cm - 1.148\ cm = \underline{0.025\ cm}$

$l = 1 \to l = 0$ transition: $2.347\ cm - 2.297\ cm = \underline{0.050\ cm}$

44-37

a) Problem 44-36 gives $I = 2.71\times10^{-47}\ kg\cdot m^2$.

$I = m_r r^2$

$$m_r = \frac{m_H m_{Cl}}{m_H + m_{Cl}} = \frac{(1.67\times10^{-27}\ kg)(5.81\times10^{-26}\ kg)}{1.67\times10^{-27}\ kg + 5.81\times10^{-26}\ kg} = 1.623\times10^{-27}\ kg$$

$$r = \sqrt{\frac{I}{m_r}} = \sqrt{\frac{2.71\times10^{-47}\ kg\cdot m^2}{1.623\times10^{-27}\ kg}} = 1.29\times10^{-10}\ m = \underline{0.129\ nm}$$

b) Each transition is between the level l to the level $l-1$.

$E_l = l(l+1)\frac{\hbar^2}{2I}$, so $\Delta E = E_l - E_{l-1} = [l(l+1) - l(l-1)]\frac{\hbar^2}{2I} = l\frac{\hbar^2}{I}$.

The transition energy is related to the photon wavelength by $\Delta E = \frac{hc}{\lambda}$.

Combine these two equations for $\Delta E \Rightarrow l\frac{\hbar^2}{I} = \frac{hc}{\lambda}$

$$l = \frac{2\pi c I}{\hbar\lambda} = \frac{2\pi(2.998\times10^8\ m/s)(2.71\times10^{-47}\ kg\cdot m^2)}{(1.054\times10^{-34}\ J\cdot s)\ \lambda} = \frac{4.843\times10^{-4}\ m}{\lambda}$$

$$\lambda = 60.4\ \mu m \Rightarrow l = \frac{4.843\times10^{-4}\ m}{60.4\times10^{-6}\ m} = 8$$

$$\lambda = 69.0\ \mu m \Rightarrow l = \frac{4.843\times10^{-4}\ m}{69.0\times10^{-6}\ m} = 7$$

$$\lambda = 80.4\ \mu m \Rightarrow l = \frac{4.843\times10^{-4}\ m}{80.4\times10^{-6}\ m} = 6$$

$$\lambda = 96.4\ \mu m \Rightarrow l = \frac{4.843\times10^{-4}\ m}{96.4\times10^{-6}\ m} = 5$$

44-37 (cont)

$\lambda = 120.4\ \mu m \Rightarrow l = \dfrac{4.843\times10^{-4}\ m}{120.4\times10^{-6}\ m} = 4$

c) longest $\lambda \Rightarrow$ smallest $\Delta E \Rightarrow$ transition from $l=1$ to $l=0$.

$\Delta E = l\dfrac{\hbar^2}{I} = (1)\dfrac{(1.054\times10^{-34}\ J\cdot s)^2}{2.71\times10^{-47}\ kg\cdot m^2} = 4.099\times10^{-22}\ J$

$\lambda = \dfrac{hc}{\Delta E} = \dfrac{(6.626\times10^{-34}\ J\cdot s)(2.998\times10^{8}\ m/s)}{4.099\times10^{-22}\ J} = 4.85\times10^{-4}\ m = \underline{485\ \mu m}$

d) What changes is m_r, the reduced mass of the molecule. The transition energy is $\Delta E = l\dfrac{\hbar^2}{I}$ and $\Delta E = \dfrac{hc}{\lambda} \Rightarrow \lambda = \dfrac{2\pi c I}{l\hbar}$ (part (b)).

$I = m_r r^2$, so λ is directly proportional to m_r.

$\dfrac{\lambda(HCl)}{m_r(HCl)} = \dfrac{\lambda(DCl)}{m_r(DCl)} \Rightarrow \lambda(DCl) = \lambda(HCl)\dfrac{m_r(DCl)}{m_r(HCl)}$

The mass of a deuterium atom is approximately twice the mass of a hydrogen atom $\Rightarrow m_D = 3.34\times10^{-27}$ kg.

$m_r(DCl) = \dfrac{m_D m_{Cl}}{m_D + m_{Cl}} = \dfrac{(3.34\times10^{-27}\ kg)(5.81\times10^{-26}\ kg)}{3.34\times10^{-27}\ kg + 5.81\times10^{-26}\ kg} = 3.158\times10^{-27}\ kg$

$\lambda(DCl) = \lambda(HCl)\dfrac{3.158\times10^{-27}\ kg}{1.623\times10^{-27}\ kg} = (1.946)\ \lambda(HCl)$

$l=8 \rightarrow l=7$: $\lambda = (60.4\ \mu m)(1.946) = 118\ \mu m$

$l=7 \rightarrow l=6$: $\lambda = (69.0\ \mu m)(1.946) = 134\ \mu m$

$l=6 \rightarrow l=5$: $\lambda = (80.4\ \mu m)(1.946) = 156\ \mu m$

$l=5 \rightarrow l=4$: $\lambda = (96.4\ \mu m)(1.946) = 188\ \mu m$

$l=4 \rightarrow l=3$: $\lambda = (120.4\ \mu m)(1.946) = 234\ \mu m$

44-39

a) $I = m_r r^2$

$m_r = \dfrac{m_H m_I}{m_H + m_I} = \dfrac{(1.67\times10^{-27}\ kg)(2.11\times10^{-25}\ kg)}{1.67\times10^{-27}\ kg + 2.11\times10^{-25}\ kg} = 1.657\times10^{-27}\ kg$

$I = m_r r^2 = (1.657\times10^{-27}\ kg)(0.160\times10^{-9}\ m)^2 = \underline{4.24\times10^{-47}\ kg\cdot m^2}$

b) The energy levels are $E_{nl} = l(l+1)\dfrac{\hbar^2}{2I} + (n+\tfrac{1}{2})\hbar\sqrt{\dfrac{k'}{m_r}}$ (Eq. 44-9)

$\sqrt{\dfrac{k'}{m}} = \omega = 2\pi f \Rightarrow E_{nl} = l(l+1)\dfrac{\hbar^2}{2I} + (n+\tfrac{1}{2})hf$

(i) transition $n=1 \rightarrow n=0$, $l=1 \rightarrow l=0$

$\Delta E = (2-0)\dfrac{\hbar^2}{2I} + (1+\tfrac{1}{2}-\tfrac{1}{2})hf = \dfrac{\hbar^2}{I} + hf$

$\Delta E = \dfrac{hc}{\lambda} \Rightarrow \lambda = \dfrac{hc}{\Delta E} = \dfrac{hc}{\frac{\hbar^2}{I} + hf} = \dfrac{c}{\frac{\hbar}{2\pi I} + f}$

44-39 (cont) $\frac{\hbar}{2\pi I} = \frac{1.054\times10^{-34}\ \text{J}\cdot\text{s}}{2\pi(4.24\times10^{-47}\ \text{kg}\cdot\text{m}^2)} = 3.956\times10^{11}\ \text{Hz}$

$$\lambda = \frac{c}{\frac{\hbar}{2\pi I} + f} = \frac{2.998\times10^{8}\ \text{m/s}}{3.956\times10^{11}\ \text{Hz} + 6.93\times10^{13}\ \text{Hz}} = \underline{4.30\ \mu\text{m}}$$

(ii) transition $n=1 \rightarrow n=0$, $l=2 \rightarrow l=1$

$$\Delta E = (6-2)\frac{\hbar^2}{2I} + hf = 2\frac{\hbar^2}{I} + hf$$

$$\lambda = \frac{c}{2\left(\frac{\hbar}{2\pi I}\right) + f} = \frac{2.998\times10^{8}\ \text{m/s}}{2(3.956\times10^{11}\ \text{Hz}) + 6.93\times10^{13}\ \text{Hz}} = \underline{4.28\ \mu\text{m}}$$

(iii) transition $n=2 \rightarrow n=1$, $l=2 \rightarrow l=3$

$$\Delta E = (6-12)\frac{\hbar^2}{2I} + hf = -\frac{3\hbar^2}{I} + hf$$

$$\lambda = \frac{c}{-3\left(\frac{\hbar}{2\pi I}\right) + f} = \frac{2.998\times10^{8}\ \text{m/s}}{-3(3.956\times10^{11}\ \text{Hz}) + 6.93\times10^{13}\ \text{Hz}} = \underline{4.40\ \mu\text{m}}$$

CHAPTER 45

Exercises 3, 5, 9, 11, 15, 17, 21, 25, 27, 31, 33, 35

Problems 41, 45, 47, 49, 51, 53, 55, 57

Exercises

45-3

When the spin component is parallel to the field the interaction energy is $U = -\mu_z B$. When the spin component is antiparallel to the field the interaction energy is $U = +\mu_z B$. The transition energy for a transition between these two states is $\Delta E = 2\mu_z B$, where $\mu_z = 2.7928\mu_n$.

The transition energy is related to the photon frequency by $\Delta E = hf$, so $2\mu_z B = hf$.

$$B = \frac{hf}{2\mu_z} = \frac{(6.626\times10^{-34}\ \text{J}\cdot\text{s})(36.9\times10^{6}\ \text{Hz})}{2(2.7928)(5.051\times10^{-27}\ \text{J/T})} = \underline{0.867\ \text{T}}$$

45-5

a) From Example 45-2, when the z-component of $\vec{S}$ (and $\vec{\mu}$) is parallel to $\vec{B}$, $U = -|\mu_z|B = -2.7928\,\mu_n B$.
When the z-component of $\vec{S}$ (and $\vec{\mu}$) is antiparallel to $\vec{B}$, $U = +|\mu_z|B = +2.7928\mu_n B$.
The state with the proton spin component parallel to the field lies lower in energy. The energy difference between these two states is $\Delta E = 2(2.7928\mu_n B)$.

$$\Delta E = hf \Rightarrow f = \frac{\Delta E}{h} = \frac{2(2.7928\mu_n B)}{h} = \frac{2(2.7928)(5.05079\times10^{-27}\ \text{J/T})(1.80\ \text{T})}{6.626\times10^{-34}\ \text{J}}$$

$$f = 7.66\times10^{7}\ \text{Hz} = \underline{76.6\ \text{MHz}}$$

$$\lambda = \frac{c}{f} = \frac{2.998\times10^{8}\ \text{m/s}}{7.66\times10^{7}\ \text{Hz}} = \underline{3.91\ \text{m}}$$

From Fig. 33-15, these are radio waves.

b) From Eqs. (28-27), (43-22), and Fig. 43-12, the state with the z-component of $\vec{\mu}$ parallel to $\vec{B}$ has lower energy. But, since the charge of the electron is negative, this is the state with the electron spin component antiparallel to $\vec{B}$. That is, the $m_s = -\frac{1}{2}$ state lies lower in energy.

For the $m_s = +\frac{1}{2}$ state, $U = +(2.00232)\left(\frac{e}{2m}\right)\left(+\frac{\hbar}{2}\right)B = +\frac{1}{2}(2.00232)\left(\frac{e\hbar}{2m}\right)B$

$$U = +\tfrac{1}{2}(2.00232)\mu_B B$$

For the $m_s = -\frac{1}{2}$ state, $U = -\frac{1}{2}(2.00232)\mu_B B$

The energy difference between these two states is $\Delta E = (2.00232)\mu_B B$

$$\Delta E = hf \Rightarrow f = \frac{\Delta E}{h} = \frac{2.00232\mu_B B}{h} = \frac{(2.00232)(9.274\times10^{-24}\ \text{J/T})(1.80\ \text{T})}{6.626\times10^{-34}\ \text{J}} = \underline{5.05\times10^{10}\ \text{Hz}}$$

$$\lambda = \frac{c}{f} = \frac{2.998\times10^{8}\ \text{m/s}}{5.05\times10^{10}\ \text{Hz}} = 5.94\times10^{-3}\ \text{m} = \underline{5.94\ \text{mm}}$$

From Fig. 33-15, these are microwaves.

45-9

The text calculates that the binding energy of the deuteron is 2.224 MeV. A photon that breaks the deuteron up into a proton and a neutron must have at least this much energy.

$$E = \frac{hc}{\lambda} \Rightarrow \lambda = \frac{hc}{E} = \frac{(4.136\times10^{-15}\,\text{eV}\cdot\text{s})(2.998\times10^{8}\,\text{m/s})}{2.224\times10^{6}\,\text{eV}} = 5.575\times10^{-13}\,\text{m} = \underline{0.5575\,\text{pm}}$$

45-11

a) A $^{11}_{5}B$ atom has 5 protons, $11-5=6$ neutrons, and 5 electrons. The mass defect therefore is $\Delta M = 5m_p + 6m_n + 5m_e - M(^{11}_{5}B)$

$\Delta M = 5(1.0072765\,u) + 6(1.0086649\,u) + 5(0.0005485799\,u) - 11.009305\,u = 0.08181\,u$

The energy equivalent is $E_B = (0.08181\,u)(931.5\,\text{MeV}/u) = \underline{76.21\,\text{MeV}}$.

b) Eq. (45-9): $E_B = C_1 A - C_2 A^{2/3} - C_3 \frac{Z(Z-1)}{A^{1/3}} - C_4 \frac{(A-2Z)^2}{A}$

The fifth term is zero since Z is odd but N is even.

$$E_B = (15.75\,\text{MeV})(11) - (17.80\,\text{MeV})(11)^{2/3} - (0.7100\,\text{MeV})\frac{5(4)}{11^{1/3}} - (23.69\,\text{MeV})\frac{(11-10)^2}{11}$$

$$E_B = +173.25\,\text{MeV} - 88.04\,\text{MeV} - 6.38\,\text{MeV} - 2.15\,\text{MeV} = 76.68\,\text{MeV}$$

The percentage difference between the calculated and measured E_B is

$$\frac{76.68\,\text{MeV} - 76.21\,\text{MeV}}{76.21\,\text{MeV}} = 0.6\%.$$

Eq. (45-9) has a greater percentage accuracy for ^{62}Ni.

45-15

a) α decay: Z decreases by 2, $A = N+Z$ decreases by 4 (α particle is a $^{4}_{2}He$ nucleus)

$$^{239}_{94}Pu \rightarrow\ ^{4}_{2}He +\ ^{235}_{92}U$$

b) β^- decay: Z increases by 1, $A = N+Z$ remains the same (β^- particle is an electron $^{0}_{-1}e$)

$$^{24}_{11}Na \rightarrow\ ^{0}_{-1}e +\ ^{24}_{12}Mg$$

c) β^+ decay: Z decreases by 1, $A = N+Z$ remains the same (β^+ particle is a positron $^{0}_{+1}e$)

$$^{15}_{8}O \rightarrow\ ^{0}_{+1}e +\ ^{15}_{7}N$$

Note: In each case the total charge and total number of nucleons for the decay products equals the charge and number of nucleons for the parent nucleus; these two quantities are conserved in the decay.

45-17

a) The β^- decay reaction is ${}^3_1H \rightarrow {}^{\ 0}_{-1}e + {}^3_2He$. ${}^3_1H$ is unstable with respect to β^- decay since the mass of the ${}^3_2He$ nucleus plus the emitted electron is less than the mass of the ${}^3_1H$ nucleus:

The masses in Table 45-2 are for the neutral atoms, so the mass of the ${}^3_1H$ nucleus is $3.016049u - 0.0005486u = 3.0155004\ u$ and the mass of the ${}^3_2He$ nucleus is $3.016029u - 2(0.0005486\ u) = 3.0149318\ u$.
The mass decrease in the decay is $3.0155004\ u - 3.0149318\ u - 0.0005486u = 0.000020\ u$.

b) Find the energy equivalent of the mass decrease found in part (a):
$0.000020\ u\ (931.5\ MeV/u) = 0.019\ MeV = \underline{19\ keV}$

45-21

a) Eq. (45-17): $\lambda = \frac{0.693}{T_{1/2}} = \frac{0.693}{(4.47\times10^9 y)\left(\frac{3.156\times10^7 s}{1y}\right)} = \underline{4.91\times10^{-18}\ s^{-1}}$

b) Activity of 6.00×10^{-6} Ci $\Rightarrow \left|\frac{dN}{dt}\right| = 2.22\times10^5$ decays/s

Eq. (45-14): $\left|\frac{dN}{dt}\right| = \lambda N \Rightarrow N = \frac{|dN/dt|}{\lambda} = \frac{2.22\times10^5\ \text{decays/s}}{4.91\times10^{-18}\ s^{-1}} = 4.52\times10^{22}$ nuclei
The sample must contain 4.52×10^{22} uranium nuclei.

The mass of one uranium ${}^{238}U$ atom is approximately 238u, so the mass of the sample is $m = N(238\ u) = (4.52\times10^{22})(238)(1.66054\times10^{-27} kg) = 1.79\times10^{-2} kg = \underline{17.9 g}$.

c) One α particle is emitted in each decay, so the problem is asking for the activity, in decays/s, of this sample.

Compute N, the number of nuclei in the sample. The mass of one ${}^{238}U$ atom is approximately 238u, so $N = \frac{30.0\times10^{-3} kg}{238\ u} = \frac{30.0\times10^{-3} kg}{238(1.66054\times10^{-27} kg)} = 7.591\times10^{22}$ nuclei.
$\left|\frac{dN}{dt}\right| = \lambda N = (4.91\times10^{-18} s^{-1})(7.591\times10^{22}) = \underline{3.73\times10^5\ \text{decays/s}}$

45-25

a) $\left|\frac{dN}{dt}\right| = \lambda N$

$\lambda = \frac{0.693}{T_{1/2}} = \frac{0.693}{(1.28\times10^9\ y)\left(\frac{3.156\times10^7 s}{1y}\right)} = 1.715\times10^{-17}\ s^{-1}$

The mass of one ${}^{40}K$ atom is approximately 40 u, so the number of ${}^{40}K$ nuclei in the sample is $N = \frac{3.26\times10^{-9} kg}{40\ u} = \frac{3.26\times10^{-9} kg}{40(1.66054\times10^{-27} kg)} = 4.908\times10^{16}$.
$\left|\frac{dN}{dt}\right| = \lambda N = (1.715\times10^{-17} s^{-1})(4.908\times10^{16}) = \underline{0.842\ \text{decays/s}}$

b) $\left|\frac{dN}{dt}\right| = (0.842\ \text{decays/s})\left(\frac{1 Ci}{3.70\times10^{10}\ \text{decays/s}}\right) = \underline{2.28\times10^{-11}\ Ci}$

45-27

a) The energy E of each photon is $E = \frac{hc}{\lambda} = \frac{(6.626\times10^{-34}\,J\cdot s)(2.998\times10^{8}\,m/s)}{0.0200\times10^{-9}\,m}$

$E = 9.932\times10^{-15}\,J$

The total energy absorbed is the number of photons absorbed times the energy of each photon: $(6.50\times10^{10}\,\text{photons})(9.932\times10^{-15}\,J/\text{photon}) = \underline{6.46\times10^{-4}\,J}$

b) The absorbed dose is the energy absorbed divided by the mass of the absorbing tissue:

absorbed dose $= \frac{6.46\times10^{-4}\,J}{0.600\,kg} = 1.08\times10^{-3}\,J/kg\left(\frac{1\,\text{rad}}{0.01\,J/kg}\right) = 0.108\,\text{rad}$

The equivalent dose in rem equals the RBE times the absorbed dose in rad. For x rays RBE = 1, so the equivalent dose is $\underline{0.108\,\text{rem}}$.

45-31

$^{235}_{92}U + ^{1}_{0}n \rightarrow ^{236}_{92}U$

The mass defect is $\Delta M = M(^{235}_{92}U) + M(^{1}_{0}n) - M(^{236}_{92}U)$

$\Delta M = 235.043924\,u + 1.008665\,u - 236.045563\,u = +0.007026\,u$

$Q = (\Delta M)c^2 = (0.007026\,u)\left(\frac{931.5\,MeV}{1\,u}\right) = 6.545\,MeV$

This much energy is liberated in the reaction. By conservation of linear momentum since there is essentially zero momentum before the reaction, the product nucleus must be essentially at rest and therefore have zero kinetic energy. All the energy released must go into the internal excitation of the $^{236}_{92}U$.

45-33

a) X must have $A = +2+9-4 = 7$ and $Z = +1+4-2 = 3$. Thus X is $^{7}_{3}Li$ and the reaction is

$^{2}_{1}H + ^{9}_{4}Be = ^{7}_{3}Li + ^{4}_{2}He$

b) If we use the neutral atom masses then there are the same number of electrons (five) in the reactants as in the products. Their masses cancel, so we get the same mass defect whether we use nuclear masses or neutral atom masses.

The neutral atom masses are given in Table 45-2:

$^{2}_{1}H + ^{9}_{4}Be$ has mass $2.014102\,u + 9.012182\,u = 11.026284\,u$

$^{7}_{3}Li + ^{4}_{2}He$ has mass $7.016003\,u + 4.002603\,u = 11.018606\,u$

The mass decrease is $11.026284\,u - 11.018606\,u = 0.007678\,u$.

This corresponds to an energy release of $0.007678\,u\left(\frac{931.5\,MeV}{1\,u}\right) = \underline{7.152\,MeV}$.

c) Estimate the threshold energy by calculating the Coulomb potential energy when the $^{2}_{1}H$ and $^{9}_{4}Be$ nuclei just touch.

45-33 (cont)

The radius R_{Be} of the $^{9}_{4}Be$ nucleus is $R_{Be} = 1.2\times10^{-15}\,m\,(9)^{1/3} = 2.5\times10^{-15}\,m$.
The radius R_H of the $^{2}_{1}H$ nucleus is $R_H = 1.2\times10^{-15}\,m\,(2)^{1/3} = 1.5\times10^{-15}\,m$.
The nuclei touch when their center-to-center separation is $r = R_{Be} + R_H = 4.0\times10^{-15}\,m$.

As in Example 45-12 estimate the threshold energy to be the Coulomb potential energy of the two reactant nuclei at this separation:

$$U = \frac{1}{4\pi\epsilon_0}\frac{q_1 q_2}{r} = \frac{1}{4\pi\epsilon_0}\frac{e(4e)}{r} = (8.987\times10^{9}\,N\cdot m^2/C^2)\frac{4(1.602\times10^{-19}\,C)^2}{(4.0\times10^{-15}\,m)(1.602\times10^{-19}\,C)} = \underline{1.4\,MeV}$$

45-35

a) X must have $A = 2+14-10 = 6$ and $Z = 1+7-5 = 3$. Thus X is $^{6}_{3}Li$ and the reaction is $^{2}_{1}H + ^{14}_{7}N \rightarrow ^{6}_{3}Li + ^{10}_{5}B$

b) The neutral atoms on each side of the reaction equation have a total of 8 electrons, so the electron masses cancel when neutral atom masses are used. The neutral atom masses are found in Table 45-2.

mass of $^{2}_{1}H + ^{14}_{7}N$ is $2.014102\,u + 14.003074\,u = 16.017176\,u$
mass of $^{6}_{3}Li + ^{10}_{5}B$ is $6.015121\,u + 10.012937\,u = 16.028058\,u$
The mass increases, so energy is absorbed by the reaction.
The Q value is $(16.017176\,u - 16.028058\,u)(931.5\,MeV/u) = \underline{-10.14\,MeV}$

c) The kinetic energy that must be available to cause the reaction is 10.14 MeV. Thus in Eq. (45-22) $K_{cm} = 10.14\,MeV$. The mass M of the stationary target ($^{14}_{7}N$) is $M = 14\,u$. The mass m of the colliding particle ($^{2}_{1}H$) is 2u. Then by Eq. (45-22) the minimum kinetic energy K that the $^{2}_{1}H$ must have is

$$K = \frac{M+m}{M}K_{cm} = \left(\frac{14\,u + 2\,u}{14\,u}\right)(10.14\,MeV) = \underline{11.59\,MeV}$$

Problems

45-41

a) The heavier nucleus will decay into the lighter one; $^{25}_{13}Al$ will decay into $^{25}_{12}Mg$.

b) Balance A and Z in the decay reaction: $^{25}_{13}Al \rightarrow ^{25}_{12}Mg + ^{0}_{+1}e$.
The emitted particle must have charge +e and its nucleon number must be zero. Therefore, it is a β^+ particle, or positron.

c) Calculate the energy defect ΔM for the reaction. Use the nuclear masses for $^{25}_{13}Al$ and $^{25}_{12}Mg$, to avoid confusion in including the correct number of electrons if neutral atom masses are used.
The nuclear mass for $^{25}_{13}Al$ is $M_{nuc}(^{25}_{13}Al) = 24.990429\,u - 13(0.000548580\,u) = 24.983297\,u$.
The nuclear mass for $^{25}_{12}Mg$ is $M_{nuc}(^{25}_{12}Mg) = 24.985837\,u - 12(0.000548580\,u) = 24.979254\,u$.

45-41 (cont)

The mass defect for the reaction is

$$\Delta M = M_{nuc}({}^{25}_{13}Al) - M_{nuc}({}^{25}_{12}Mg) - M({}^{0}_{+1}e)$$

$$\Delta M = 24.983297\,u - 24.979254\,u - 0.00054858\,u = 0.003494\,u$$

$$Q = (\Delta M)c^2 = 0.003494\,u\left(\frac{931.5\text{ MeV}}{1\,u}\right) = \underline{3.255\text{ MeV}}$$

The mass decreases in the decay and energy is released.

Note: ${}^{25}_{13}Al$ can also decay into ${}^{25}_{12}Mg$ by electron capture.

$${}^{25}_{13}Al + {}^{0}_{-1}e \rightarrow {}^{25}_{12}Mg$$

The ${}^{0}_{-1}e$ electron in the reaction is an orbital electron in the neutral ${}^{25}_{13}Al$ atom. The mass defect can be calculated using the nuclear masses:

$$\Delta M = M_{nuc}({}^{25}_{13}Al) + M({}^{0}_{-1}e) - M_{nuc}({}^{25}_{12}Mg) = 24.983297\,u + 0.00054858\,u - 24.979254\,u$$

$$\Delta M = 0.004592\,u$$

$$Q = (\Delta M)c^2 = (0.004592\,u)\left(\frac{931.5\text{ MeV}}{1\,u}\right) = \underline{4.277\text{ MeV}}$$

The mass decreases in the decay and energy is released.

45-45

$${}^{198}_{79}Au \rightarrow {}^{198}_{80}Hg + {}^{0}_{-1}e$$

The mass change is $197.968217\,u - 197.966743\,u = 1.474\times10^{-3}\,u$. (The neutral atom masses include 79 electrons before the decay and 80 electrons after the decay. This one additional electron in the products accounts correctly for the electron emitted by the nucleus.)

The total energy released in the decay is $(1.474\times10^{-3}\,u)(931.5\text{ MeV}/u) = 1.373\text{ MeV}$. This energy is divided between the energy of the emitted photon and the kinetic energy of the β^- particle. Thus the β^- has kinetic energy $1.373\text{ MeV} - 0.412\text{ MeV} = \underline{0.961\text{ MeV}}$.

45-47

$${}^{13}_{7}N \rightarrow {}^{0}_{+1}e + {}^{13}_{6}C$$

To avoid confusion in including the correct number of electrons with neutral atom masses use nuclear masses, obtained by subtracting the mass of the atomic electrons from the neutral atom masses.

The nuclear mass for ${}^{13}_{7}N$ is $M_{nuc}({}^{13}_{7}N) = 13.005739\,u - 7(0.00054858\,u) = 13.001899\,u$.

The nuclear mass for ${}^{13}_{6}C$ is $M_{nuc}({}^{13}_{6}C) = 13.003355\,u - 6(0.00054858\,u) = 13.000064\,u$.

The mass defect for the reaction is $\Delta M = M_{nuc}({}^{13}_{7}N) - M_{nuc}({}^{13}_{6}C) - M({}^{0}_{+1}e)$

$$\Delta M = 13.001899\,u - 13.000064\,u - 0.00054858\,u = \underline{0.001286\,u}$$

The mass decreases in the decay, so energy is released. This decay is energetically possible.

45-49

We have to be careful; after ^{87}Rb has undergone radioactive decay it is no longer a rubidium atom.
Let N_{85} be the number of ^{85}Rb atoms; this number doesn't change.
Let N_0 be the number of ^{87}Rb atoms on earth when the solar system was formed.
Let N be the present number of ^{87}Rb atoms.

The present measurements say that $0.2783 = \frac{N}{N+N_{85}}$
$(N+N_{85})(0.2783) = N \Rightarrow N = 0.3856\, N_{85}$

The percentage we are asked to calculate is $\frac{N_0}{N_0+N_{85}}$.
N and N_0 are related by $N = N_0 e^{-\lambda t} \Rightarrow N_0 = e^{+\lambda t} N$

Thus $\frac{N_0}{N_0+N_{85}} = \frac{Ne^{\lambda t}}{Ne^{\lambda t}+N_{85}} = \frac{(0.3856 e^{\lambda t}) N_{85}}{0.3856 e^{\lambda t} N_{85} + N_{85}} = \frac{0.3856 e^{\lambda t}}{0.3856 e^{\lambda t}+1}$

$t = 4.6\times10^9\ y$; $\lambda = \frac{0.693}{T_{1/2}} = \frac{0.693}{4.89\times10^{10}\ y} = 1.417\times10^{-11}\ y^{-1}$

$e^{\lambda t} = e^{(1.417\times10^{-11}\ y^{-1})(4.6\times10^9 y)} = e^{0.06518} = 1.0674$

Thus $\frac{N_0}{N_0+N_{85}} = \frac{(0.3856)(1.0674)}{(0.3856)(1.0674)+1} = \underline{29.2\%}$

45-51

a) First find the number of decays each second:

$26.0\times10^{-6}\ Ci\left(\frac{3.70\times10^{10}\ decays/s}{1\ Ci}\right) = 9.62\times10^5\ decays/s$

The average energy per decay is 1.25 MeV, and one-half of this energy is deposited in the tumor. The energy delivered to the tumor per second then is
$\frac{1}{2}(9.62\times10^5\ decays/s)(1.25\times10^6\ eV/decay)(1.602\times10^{-19}\ J/eV) = \underline{9.63\times10^{-8}\ J/s}$

b) The absorbed dose is the energy absorbed divided by the mass of the tissue:
$\frac{9.63\times10^{-8}\ J/s}{0.500\ kg} = 1.93\times10^{-7}\ J/kg\cdot s\left(\frac{1\ rad}{0.01\ J/kg}\right) = \underline{1.93\times10^{-5}\ rad/s}$

c) equivalent dose (rem) = RBE $\times$ absorbed dose (rad)
In one second the equivalent dose is $0.70(1.93\times10^{-5}\ rad) = \underline{1.35\times10^{-5}\ rem}$.

d) $\frac{200\ rem}{1.35\times10^{-5}\ rem/s} = 1.48\times10^7 s\left(\frac{1h}{3600\ s}\right) = 4110\ h = \underline{171\ days}$

45-53

$N = N_0 e^{-\lambda t}$
The problem says $\frac{N}{N_0} = 0.21$; solve for t.
$0.21 = e^{-\lambda t} \Rightarrow \ln(0.21) = -\lambda t \Rightarrow t = \frac{-\ln(0.21)}{\lambda}$

45-53 (cont)

Example 45-9 gives $\lambda = 1.209\times10^{-4}\,y^{-1}$ for ^{14}C.

Thus $t = \frac{-\ln 0.21}{1.209\times10^{-4}\,y} = \underline{1.3\times10^4\,y}$.

45-55

a) The radius of ^2_1H is $R = (1.2\times10^{-15}\,m)(2)^{1/3} = 1.51\times10^{-15}\,m$.
The barrier energy is the Coulomb potential energy of two ^2_1H nuclei with their centers separated by twice this distance:

$$U = \frac{1}{4\pi\epsilon_0}\frac{e^2}{r} = (8.987\times10^9\,N\cdot m^2/C^2)\frac{(1.602\times10^{-19}C)^2}{2(1.51\times10^{-15}\,m)} = 7.635\times10^{-14}\,J = \underline{0.48\ MeV}$$

b) $^2_1H + ^2_1H \rightarrow ^3_2He + ^1_0n$

If we use neutral atom masses there are two electrons on each side of the reaction equation, so their masses cancel. The neutral atom masses are given in Table 45-2.

$^2_1H + ^2_1H$ has mass $2(2.014102\,u) = 4.028204\,u$

$^3_2He + ^1_0n$ has mass $3.016029\,u + 1.008665\,u = 4.024694\,u$

The mass decrease is $4.028204\,u - 4.024694\,u = 3.510\times10^{-3}\,u$. This corresponds to a liberated energy of $(3.510\times10^{-3}\,u)(931.5\,MeV/u) = \underline{3.270\ MeV}$, or $(3.270\times10^6\,eV)(1.602\times10^{-19}\,J/eV) = \underline{5.239\times10^{-13}\,J}$.

c) Each reaction takes two ^2_1H nuclei. Each mole of D_2 has 6.022×10^{23} molecules so 6.022×10^{23} pairs of atoms. The energy liberated when one mole of deuterium undergoes fusion is $(6.022\times10^{23})(5.239\times10^{-13}\,J) = \underline{3.155\times10^{11}\,J/mol}$

45-57

Consider 1.00 kg of body tissue. The mass of ^{40}K in 1.00 kg of tissue is $(0.21\times10^{-2})(0.012\times10^{-2})(1.00\,kg) = 2.52\times10^{-7}\,kg$.
The mass of a ^{40}K atom is approximately 40u, so the number of ^{40}K nuclei in 1.00 kg of tissue is $\frac{2.52\times10^{-7}\,kg}{40u} = \frac{2.52\times10^{-7}\,kg}{40u(1.66054\times10^{-27}\,kg/u)} = 3.794\times10^{18}$

The activity is $\left|\frac{dN}{dt}\right| = \lambda N$. $\lambda = \frac{0.693}{T_{1/2}} = \frac{0.693}{1.25\times10^9\,y} = 5.544\times10^{-10}\,y^{-1}$.
$\left|\frac{dN}{dt}\right| = (5.544\times10^{-10}\,y^{-1})(3.794\times10^{18}) = 2.103\times10^9$ decays/y.
In 50y there are $(50\,y)(2.103\times10^9\,\text{decays}/y) = 1.052\times10^{11}$ decays.
For each decay an average of 0.50 MeV of energy is absorbed. The energy absorbed by 1.00 kg of tissue is $(1.052\times10^{11}\,\text{decays})(0.50\,MeV/\text{decay}) = 5.26\times10^{10}\,MeV = 5.26\times10^{16}\,eV(1.602\times10^{-19}\,J/eV) = 0.0084\,J$.

The absorbed dose is $0.0084\,J/kg\left(\frac{1\,rad}{0.01\,J/kg}\right) = \underline{0.84\,rad}$
RBE = 1.0, so the equivalent dose is $\underline{0.84\,rem}$.

CHAPTER 46

Exercises 3, 7, 9, 11, 15, 19, 23, 25, 27, 29, 31

Problems 37, 39, 41, 45, 49

Exercises

46-3

By momentum conservation the two photons must have equal and opposite momenta. Then $E = pc$ says the photons must have equal energies.

The mass of the pion is $270 m_e$, so the rest energy of the pion is $270(0.511 \text{ MeV}) = 138 \text{ MeV}$. Each photon has half this energy, or <u>69 MeV</u>.

$$E = hf \Rightarrow f = \frac{E}{h} = \frac{(69\times10^{6} \text{ eV})(1.602\times10^{-19} \text{ J/eV})}{6.626\times10^{-34} \text{ J}\cdot\text{s}} = \underline{1.7\times10^{22} \text{ Hz}}$$

$$\lambda = \frac{c}{f} = \frac{2.998\times10^{8} \text{ m/s}}{1.7\times10^{22} \text{ Hz}} = 1.8\times10^{-14} \text{ m} = \underline{18 \text{ fm}}$$

46-7

a) Eq. (46-7): $\omega = \frac{|q|B}{m} \Rightarrow B = \frac{m\omega}{|q|}$

$\omega = 2\pi f \Rightarrow B = \frac{2\pi m f}{|q|}$

A deuteron is a deuterium nucleus $({}^{2}_{1}H)$. Its charge is $q = +e$. Its mass is the mass of the neutral ${}^{2}_{1}H$ atom (Table 45-2) minus the mass of the one atomic electron: $m = 2.014102 u - 0.0005486 u = 2.013553 u\left(\frac{1.66054\times10^{-27} \text{ kg}}{1 u}\right) = 3.344\times10^{-27} \text{ kg}$

$$B = \frac{2\pi m f}{|q|} = \frac{2\pi(3.344\times10^{-27} \text{ kg})(11.0\times10^{6} \text{ Hz})}{1.602\times10^{-19} \text{ C}} = \underline{1.44 \text{ T}}$$

b) Eq. (46-8): $K = \frac{q^2B^2R^2}{2m} = \frac{[(1.602\times10^{-19} \text{ C})(1.44 \text{ T})(0.320 \text{ m})]^2}{2(3.344\times10^{-27} \text{ kg})} = 8.148\times10^{-13} \text{ J}$

$$K = 8.148\times10^{-13} \text{ J}\left(\frac{1 \text{ eV}}{1.602\times10^{-19} \text{ J}}\right) = \underline{5.09 \text{ MeV}}$$

$$K = \tfrac{1}{2}mv^2 \Rightarrow v = \sqrt{\frac{2K}{m}} = \sqrt{\frac{2(8.148\times10^{-13} \text{ J})}{3.344\times10^{-27} \text{ kg}}} = \underline{2.21\times10^{7} \text{ m/s}}$$

Note: $\frac{v}{c} = 0.07$, so it is ok to use the nonrelativistic expression for kinetic energy.

46-9

a) For a proton beam on a stationary proton target and since E_a is much larger than the proton rest energy we can use Eq. (46-11): $E_a^2 = 2mc^2 E_m$.

$$E_m = \frac{E_a^2}{2mc^2} = \frac{(77.4 \text{ GeV})^2}{2(0.938 \text{ GeV})} = 3190 \text{ GeV}$$

b) For colliding beams the total momentum is zero and the available energy E_a is

46-9 (cont)

the total energy of the two colliding particles. For proton-proton collisions the colliding beams each have the same energy, so the total energy of each beam is $\frac{1}{2}E_a = 38.7$ GeV.

46-11

a) The masses of the target and projectile particles are equal, so Eq. (46-10) can be used: $E_a^2 = 2mc^2(E_m + mc^2)$

$$E_m = \frac{E_a^2}{2mc^2} - mc^2$$

The mass of the alpha particle can be calculated by subtracting two electron masses from the ^4_2He atomic mass:

$m = m_\alpha = (4.002603\ u - 2(0.0005486\ u)) = 4.001506\ u$

Then $mc^2 = (4.001506\ u)(931.5\ MeV/u) = 3.727$ GeV.

$$E_m = \frac{E_a^2}{2mc^2} - mc^2 = \frac{(16.0\ GeV)^2}{2(3.727\ GeV)} - 3.727\ GeV = \underline{30.6\ GeV}$$

b) Each beam must have $\frac{1}{2}E_a = \underline{8.0\ GeV}$.

46-15

The mass decrease is $m(\Sigma^+) - m(p) - m(\pi^0)$ and the energy released is $mc^2(\Sigma^+) - mc^2(p) - mc^2(\pi^0) = 1189\ MeV - 938.3\ MeV - 135.0\ MeV = \underline{116\ MeV}$. (The mc^2 values for each particle were taken from Table 46-3.)

46-19

a) $K^+ \to \mu^+ + \nu_\mu$; $S_{K^+} = +1$, $S_{\mu^+} = 0$, $S_{\nu_\mu} = 0$

$S = 1$ initially ; $S = 0$ for the products $\Rightarrow$ S is <u>not conserved</u>

b) $n + K^+ \to p + \pi^0$; $S_n = 0$, $S_{K^+} = +1$, $S_p = 0$, $S_{\pi^0} = 0$

$S = 1$ initially ; $S = 0$ for the products $\Rightarrow$ S is <u>not conserved</u>

c) $K^+ + K^- \to \pi^0$; $S_{K^+} = +1$, $S_{K^-} = -1$, $S_{\pi^0} = 0$

$S = +1 - 1 = 0$ initially ; $S = 0$ for the product $\Rightarrow$ S <u>is conserved</u>

d) $p + K^- \to \Lambda^0 + \pi^0$; $S_p = 0$, $S_{K^-} = -1$, $S_{\Lambda^0} = -1$, $S_{\pi^0} = 0$.

$S = -1$ initially ; $S = -1$ for the products $\Rightarrow$ S <u>is conserved</u>.

46-23

Each value for the combination is the sum of the values for each quark.

a) uds

$Q = \frac{2}{3}e - \frac{1}{3}e - \frac{1}{3}e = 0$

46-23 (cont)

$B = \frac{1}{3} + \frac{1}{3} + \frac{1}{3} = 1$

$S = 0 + 0 - 1 = -1$

$C = 0 + 0 + 0 = 0$

b) $c\bar{u}$

The values for $\bar{u}$ are the negative of those for u.

$Q = \frac{2}{3}e - \frac{2}{3}e = 0$

$B = \frac{1}{3} - \frac{1}{3} = 0$

$S = 0 + 0 = 0$

$C = +1 + 0 = +1$

c) ddd

$Q = -\frac{1}{3}e - \frac{1}{3}e - \frac{1}{3}e = -e$

$B = \frac{1}{3} + \frac{1}{3} + \frac{1}{3} = +1$

$S = 0 + 0 + 0 = 0$

$C = 0 + 0 + 0 = 0$

d) $d\bar{c}$

$Q = -\frac{1}{3}e - \frac{2}{3}e = -e$

$B = \frac{1}{3} - \frac{1}{3} = 0$

$S = 0 + 0 = 0$

$C = 0 - 1 = -1$

46-25

a) Eq. (46-14): $v = \left[\frac{(\lambda_0/\lambda_s)^2 - 1}{(\lambda_0/\lambda_s)^2 + 1}\right]c = \left[\frac{(629.2\text{ nm}/590\text{ nm})^2 - 1}{(629.2\text{ nm}/590\text{ nm})^2 + 1}\right]c = 0.06424c$

$v = (0.06424)(2.998\times10^8\text{ m/s}) = \underline{1.93\times10^7\text{ m/s}}$

b) Eq. (46-15): $v = H_0 r$ (Hubble's law)

$r = \frac{v}{H_0} = \frac{1.93\times10^7\text{ m/s}}{23\times10^3\text{ (m/s)/Mly}} = \underline{839\text{ Mly}}$

46-27

a) $r = 1440$ Mly

Eq. (46-15): $v = H_0 r = \left(23\,\frac{\text{km/s}}{\text{Mly}}\right)(1440\text{ Mly}) = 3.3\times10^7\text{ m/s} = \underline{3.3\times10^4\text{ km/s}}$

b) $\frac{\lambda_0}{\lambda_s} = \sqrt{\frac{c+v}{c-v}} = \sqrt{\frac{1+v/c}{1-v/c}}$

$\frac{v}{c} = \frac{3.3\times10^7\text{ m/s}}{3.00\times10^8\text{ m/s}} = 0.11$

$\frac{\lambda_0}{\lambda_s} = \sqrt{\frac{1+0.11}{1-0.11}} = \underline{1.12}$

46-29

The Wein displacement law (Eq. 40-28) says $\lambda_m T$ equals a constant.

$\lambda_{m,1} T_1 = \lambda_{m,2} T_2$

$$\lambda_{m,1} = \lambda_{m,2}\left(\frac{T_2}{T_1}\right) = 1.063\times10^{-3}\,\text{m}\left(\frac{2.726\,\text{K}}{3000\,\text{K}}\right) = \underline{966\,\text{nm}}$$

46-31

a) $p + {}^2_1H \rightarrow {}^3_2He$, or ${}^1_1H + {}^2_1H \rightarrow {}^3_2He$

If neutral atom masses are used then the masses of the two atomic electrons on each side of the reaction equation will cancel.

Taking the atomic masses from Table 45-2 the mass decrease is $m({}^1_1H) + m({}^2_1H) - m({}^3_2He) = 1.007825\,u + 2.014102\,u - 3.016029\,u = 0.005898\,u$.
The energy released is the energy equivalent of this mass decrease:

$(0.005898\,u)(931.5\,\text{MeV}/u) = \underline{5.494\,\text{MeV}}$

b) $n + {}^3_2He \rightarrow {}^4_2He$

If neutral helium masses are used then the masses of the two atomic electrons on each side of the reaction equation will cancel. The mass decrease is $m(n) + m({}^3_2He) - m({}^4_2He) = 1.0086649\,u + 3.016029\,u - 4.002603\,u = 0.022091\,u$.
The energy released is the energy equivalent of this mass decrease:

$(0.022091\,u)(931.5\,\text{MeV}/u) = \underline{20.58\,\text{MeV}}$

Problems

46-37

a) $E_a = 2(20\,\text{TeV}) = 40\,\text{TeV}$

b) Need $E_a = 40\,\text{TeV} = 40\times10^6\,\text{MeV}$.
Since the target and projectile particles are both protons Eq. (46-10) can be used:

$E_a^2 = 2mc^2(E_m + mc^2)$

$$E_m = \frac{E_a^2}{2mc^2} - mc^2 = \frac{(40\times10^6\,\text{MeV})^2}{2(938.3\,\text{MeV})} - 938.3\,\text{MeV} = 8.5\times10^{11}\,\text{MeV} = \underline{8.5\times10^5\,\text{TeV}}$$

This shows the great advantage of colliding beams at relativistic energies.

46-39

$e^- + e^- \rightarrow e^- + e^- + \pi^0$

$Q = m_{\pi^0}c^2 = 135.0\,\text{MeV}$

The available energy E_a in the collision must be $E_a = Q + 2m_ec^2 = 135.0\,\text{MeV} + 2(0.511\,\text{MeV})$

$E_a = 136.0\,\text{MeV}$

46-39 (cont)

Since the beam and target particles are the same can use Eq. (46-10):

$E_a^2 = 2mc^2(E_m + mc^2)$, where m is the electron mass m_e.

$$E_m = \frac{E_a^2}{2mc^2} - mc^2 = \frac{(136.0\text{ MeV})^2}{2(0.511\text{ MeV})} - 0.511\text{ MeV} = 1.81\times10^4\text{ MeV}$$

$E_m = K + m_e c^2$, so $K = E_m - m_e c^2 = 1.81\times10^4\text{ MeV} - 0.511\text{ MeV} = 1.81\times10^4\text{ MeV} = \underline{18.1\text{ GeV}}$

46-41

The total available energy must be at least the total rest energy of the product particles: $E_a = mc^2(\Lambda^0) + mc^2(K^+) + mc^2(K^-) = 1116\text{ MeV} + 2(493.7\text{ MeV}) = 2103.4\text{ MeV}$

Since the target and projectile particles are different we must use Eq. (46-9):

$E_a^2 = 2Mc^2E_m + (Mc^2)^2 + (mc^2)^2$

$M = m_p, \quad m = m_{K^-}$

$E_a^2 = 2m_pc^2E_{K^-} + (m_pc^2)^2 + (m_{K^-}c^2)^2$

$$E_{K^-} = \frac{E_a^2 - (m_pc^2)^2 - (m_{K^-}c^2)^2}{2m_pc^2} = \frac{(2103.4\text{ MeV})^2 - (938.3\text{ MeV})^2 - (493.7\text{ MeV})^2}{2(938.3\text{ MeV})} = 1758.6\text{ MeV}$$

This is the total energy of the K^- particle. Its kinetic energy is

$K = E - m_{K^-}c^2 = 1758.6\text{ MeV} - 493.7\text{ MeV} = \underline{1265\text{ MeV}}$

46-45

$\phi \rightarrow K^+ + K^-$

a) The mass decrease is $m(\phi) - m(K^+) - m(K^-)$. The energy equivalent of the mass decrease is $mc^2(\phi) - mc^2(K^+) - mc^2(K^-)$. The rest mass energy mc^2 for the ϕ meson is given in Problem 46-34, and the values for K^+ and K^- are given in Table 46-3. The energy released then is $1019.4\text{ MeV} - 493.7\text{ MeV} - 493.7\text{ MeV} = 32.0\text{ MeV}$. The K^+ gets half this, $\underline{16.0\text{ MeV}}$.

b) $\phi \overset{?}{\longrightarrow} K^+ + K^- + \pi^0$

The energy equivalent of the $K^+ + K^- + \pi^0$ mass is $493.7\text{ MeV} + 493.7\text{ MeV} + 135.0\text{ MeV} = 1121\text{ MeV}$. This is greater than the energy equivalent of the ϕ mass. The mass of the decay products would be greater than the mass of the parent particle; the decay is energetically forbidden.

c) $\phi \overset{?}{\longrightarrow} K^+ + \pi^-$

The reaction $\phi \rightarrow K^+ + K^-$ is observed. K^+ has strangeness +1 and K^- has strangeness −1, so the total strangeness of the decay products is zero. But strangeness must be conserved, so we deduce that the ϕ particle has strangeness zero.

π^- has strangeness 0, so the products $K^+ + \pi^-$ has strangeness −1. The decay

46-45 (cont)

$\phi \to K^+ + \pi^-$ violates conservation of strangeness.

$\phi \overset{?}{\to} K^+ + \mu^-$

μ^- has strangeness 0, so this decay would also violate conservation of strangeness.

46-49

a) The energy equivalent of the mass decrease is

$mc^2(\Xi^-) - mc^2(\Lambda^0) - mc^2(\pi^-) = 1321\text{ MeV} - 1116\text{ MeV} - 139.6\text{ MeV} = \underline{65\text{ MeV}}$

b) The Ξ^- is at rest means the linear momentum is zero. Conservation of linear momentum then says that the Λ^0 and π^- must have equal and opposite momenta $\Rightarrow m_{\Lambda^0} v_{\Lambda^0} = m_{\pi^-} v_{\pi^-}$

$$\boxed{v_{\pi^-} = \frac{m_{\Lambda^0}}{m_{\pi^-}} v_{\Lambda^0}}$$

Also, the sum of the kinetic energies of the Λ^0 and π^- must equal the total kinetic energy $K_{tot} = 65\text{MeV}$ calculated in part (a):

$$K_{tot} = K_{\Lambda^0} + K_{\pi^-}$$

$$K_{\Lambda^0} + \tfrac{1}{2} m_{\pi^-} v_{\pi^-}^2 = K_{tot}$$

Use the momentum conservation result:

$$K_{\Lambda^0} + \tfrac{1}{2} m_{\pi^-} \left(\frac{m_{\Lambda^0}}{m_{\pi^-}}\right)^2 v_{\Lambda^0}^2 = K_{tot}$$

$$K_{\Lambda^0} + \left(\frac{m_{\Lambda^0}}{m_{\pi^-}}\right)\left(\tfrac{1}{2} m_{\Lambda^0} v_{\Lambda^0}^2\right) = K_{tot}$$

$$K_{\Lambda^0}\left(1 + \frac{m_{\Lambda^0}}{m_{\pi^-}}\right) = K_{tot}$$

$$K_{\Lambda^0} = \frac{K_{tot}}{1 + m_{\Lambda^0}/m_{\pi^-}} = \frac{65\text{MeV}}{1 + \frac{1116\text{MeV}}{139.6\text{ MeV}}} = 7.2\text{ MeV}$$

$$K_{\Lambda^0} + K_{\pi^-} = K_{tot} \Rightarrow K_{\pi^-} = K_{tot} - K_{\Lambda^0} = 65\text{MeV} - 7.2\text{ MeV} = 57.8\text{ MeV}$$

The fraction for the Λ^0 is $\frac{7.2\text{MeV}}{65\text{ MeV}} = \underline{11\%}$.

The fraction for the π^- is $\frac{57.8\text{ MeV}}{65\text{ MeV}} = \underline{89\%}$.

the lighter particle carries off more of the kinetic energy that is released in the decay.